Political Geographies of the Post-Soviet Union

This comprehensive volume observes how, after 25 years of transition and uncertainty in the countries that constituted the former Soviet Union, their political geographies remain in a state of flux. The authors explore the fluid relationship between Russia, by far the dominant economic and military power in the region, and the other former republics. They also examine new developments towards economic blocs, such as membership in the European Union or the competing Eurasian Economic Union, as well as new security arrangements in the form of military cooperation and alliance structures.

This book reflects the broad range of changes across this important world region by engaging in insightful analysis of current developments in Central Asia, Ukraine, Russia, the Caucasus, and separatist regions. The authors explore new state alliances and the evolving cultural and geopolitical orientations of former Soviet citizens. Some chapters also examine the dynamics of wars that have occurred in the post-Soviet space, as well as how local political developments are reflected in electoral preferences and struggles over control of public spaces.

The chapters in this book were originally published in the journal *Eurasian Geography and Economics*.

John O'Loughlin is College Professor of Distinction in Geography at the University of Colorado-Boulder, USA. He has conducted field work for over 25 years in the states of the former Soviet Union on state building, national conflicts, and emerging geopolitical orientations.

Ralph S. Clem is Emeritus Professor of Geography at Florida International University-Miami, USA. His research focuses on the interface between national security and geopolitics in the post-Soviet space.

Political Geographies of the Post-Soviet Union

Edited by
John O'Loughlin and Ralph S. Clem

LONDON AND NEW YORK

First published 2019
by Routledge
2 Park Square, Milton Park, Abingdon, Oxon, OX14 4RN

and by Routledge
605 Third Avenue, New York, NY 10017

First issued in paperback 2020

Routledge is an imprint of the Taylor & Francis Group, an informa business

© 2019 Taylor & Francis

All rights reserved. No part of this book may be reprinted or reproduced or
utilised in any form or by any electronic, mechanical, or other means,
now known or hereafter invented, including photocopying and recording,
or in any information storage or retrieval system, without permission in
writing from the publishers.

Trademark notice: Product or corporate names may be trademarks
or registered trademarks, and are used only for identification and
explanation without intent to infringe.

British Library Cataloguing in Publication Data
A catalogue record for this book is available from the British Library

ISBN 13: 978-0-367-72887-8 (pbk)
ISBN 13: 978-0-367-23681-6 (hbk)

Typeset in Myriad Pro
by RefineCatch Limited, Bungay, Suffolk

Publisher's Note
The publisher accepts responsibility for any inconsistencies that may have
arisen during the conversion of this book from journal articles to book chapters,
namely the inclusion of journal terminology.

Disclaimer
Every effort has been made to contact copyright holders for their permission to
reprint material in this book. The publishers would be grateful to hear from any
copyright holder who is not here acknowledged and will undertake to rectify
any errors or omissions in future editions of this book.

Contents

Citation Information	vii
Notes on Contributors	ix

1. Introduction: political geographies of the post-Soviet Union — 1
   John O'Loughlin and Ralph S. Clem

2. Who identifies with the "Russian World"? Geopolitical attitudes in southeastern Ukraine, Crimea, Abkhazia, South Ossetia, and Transnistria — 6
   John O'Loughlin, Gerard Toal and Vladimir Kolosov

3. The Eurasian Economic Union: the geopolitics of authoritarian cooperation — 40
   Sean Roberts

4. The US Silk Road: geopolitical imaginary or the repackaging of strategic interests? — 64
   Marlene Laruelle

5. Benevolent hegemon, neighborhood bully, or regional security provider? Russia's efforts to promote regional integration after the 2013–2014 Ukraine crisis — 80
   Andrej Krickovic and Maxim Bratersky

6. (Dis-)integrating Ukraine? Domestic oligarchs, Russia, the EU, and the politics of economic integration — 103
   Julia Langbein

7. Building identities in post-Soviet "de facto states": cultural and political icons in Nagorno-Karabakh, South Ossetia, Transdniestria, and Abkhazia — 127
   John O'Loughlin and Vladimir Kolosov

8. The decline and shifting geography of violence in Russia's North Caucasus, 2010-2016 — 152
   Edward C. Holland, Frank D.W. Witmer and John O'Loughlin

9. Clearing the Fog of War: public versus official sources and geopolitical storylines in the Russia-Ukraine conflict — 181
   Ralph S. Clem

10. Cleavages, electoral geography, and the territorialization of political parties in the Republic of Georgia 202
David Sichinava

11. The political geographies of religious sites in Moscow's neighborhoods 223
Meagan Todd

Index 251

Citation Information

The chapters in this book were originally published in various issues of *Eurasian Geography and Economics*. When citing this material, please use the original page numbering for each article, as follows:

Chapter 1

Editorial: political geographies of the post-Soviet Union – a farewell to Eurasian Geography and Economics from the editors
John O'Loughlin and Ralph S. Clem
Whilst the book chapter has been updated, the original version was published in *Eurasian Geography and Economics*, volume 58, issue 6 (December 2017), pp. 587–591

Chapter 2

Who identifies with the "Russian World"? Geopolitical attitudes in southeastern Ukraine, Crimea, Abkhazia, South Ossetia, and Transnistria
John O'Loughlin, Gerard Toal and Vladimir Kolosov
Eurasian Geography and Economics, volume 57, issue 6 (December 2016), pp. 745–778

Chapter 3

The Eurasian Economic Union: the geopolitics of authoritarian cooperation
Sean Roberts
Eurasian Geography and Economics, volume 58, issue 4 (August 2017), pp. 418–441

Chapter 4

The US Silk Road: geopolitical imaginary or the repackaging of strategic interests?
Marlene Laruelle
Eurasian Geography and Economics, volume 56, issue 4 (August 2015), pp. 360–375

Chapter 5

Benevolent hegemon, neighborhood bully, or regional security provider? Russia's efforts to promote regional integration after the 2013–2014 Ukraine crisis
Andrej Krickovic and Maxim Bratersky
Eurasian Geography and Economics, volume 57, issue 2 (April 2016), pp. 180–202

Chapter 6
(Dis-)integrating Ukraine? Domestic oligarchs, Russia, the EU, and the politics of economic integration
Julia Langbein
Eurasian Geography and Economics, volume 57, issue 1 (February 2016),
pp. 19–42

Chapter 7
Building identities in post-Soviet "de facto states": cultural and political icons in Nagorno-Karabakh, South Ossetia, Transdniestria, and Abkhazia
John O'Loughlin and Vladimir Kolosov
Eurasian Geography and Economics, volume 58, issue 6 (December 2017),
pp. 691–715

Chapter 8
The decline and shifting geography of violence in Russia's North Caucasus, 2010-2016
Edward C. Holland, Frank D.W. Witmer and John O'Loughlin
Eurasian Geography and Economics, volume 58, issue 6 (December 2017),
pp. 613–641

Chapter 9
Clearing the Fog of War: public versus official sources and geopolitical storylines in the Russia-Ukraine conflict
Ralph S. Clem
Eurasian Geography and Economics, volume 58, issue 6 (December 2017),
pp. 592–612

Chapter 10
Cleavages, electoral geography, and the territorialization of political parties in the Republic of Georgia
David Sichinava
Eurasian Geography and Economics, volume 58, issue 6 (December 2017),
pp. 670–690

Chapter 11
The political geographies of religious sites in Moscow's neighborhoods
Meagan Todd
Eurasian Geography and Economics, volume 58, issue 6 (December 2017),
pp. 642–669

For any permission-related enquiries please visit:
http://www.tandfonline.com/page/help/permissions

Notes on Contributors

Maxim Bratersky is a Professor and Leading Research Fellow in the Faculty of World Economy and International Affairs at the National Research University Higher School of Economics, Russia. He is interested in Russia's foreign policy, US foreign policy, and international relations of Asia.

Ralph S. Clem is Emeritus Professor of Geography at Florida International University-Miami, USA. His research focuses on the interface between national security and geopolitics in the post-Soviet space.

Edward C. Holland is Assistant Professor of Geography in the Department of Geosciences at the University of Arkansas, USA. His research interests range across a variety of topics, including political violence, religion, and critical geopolitics, and are generally focused on the Russian Federation.

Vladimir Kolosov is a Professor, Deputy Director, and Head of Laboratory at the Institute of Geography at the Russian Academy of Sciences, Moscow, Russia. He is Past President of the International Geographical Union. His main research interests are political geography and geopolitics.

Andrej Krickovic is an Assistant Professor in the Faculty of World Economy and International Affairs at the National Research University Higher School of Economics, Russia. His research and teaching areas are international security, international relations theory, Russian foreign policy, Chinese foreign policy, regionalism, and research methods.

Julia Langbein is a Senior Researcher at the Otto-Suhr-Institute for Political Science, Freie Universität Berlin and at the Centre for East European and International Studies (ZOiS), Berlin, Germany. Her research focus lies in the field of comparative political economy (with a focus on Eastern Europe and the post-Soviet space), transnational economic integration, and institutional development.

Marlene Laruelle is Research Professor of International Affairs, Director of the Central Asia Program, and Associate Director of the Institute for European, Russian and Eurasian Studies (IERES) at George Washington University, USA. She works on political, social, and cultural changes in the post-Soviet space.

John O'Loughlin is College Professor of Distinction in Geography at the University of Colorado-Boulder, USA. He has conducted field work for over 25 years in the states of the former Soviet Union on state building, national conflicts, and emerging geopolitical orientations.

Sean Roberts is Senior Lecturer in Politics and International Relations at the University of Winchester, UK. His areas of expertise are Russian foreign and security policy, Russian domestic politics, authoritarian regime dynamics, and regional integration in the post-Soviet space.

David Sichinava is Assistant Professor in the Department of Human Geography at Tbilisi State University, Georgia, and Senior Policy Analyst at the Caucasus Research Resource Centres Georgia. He is interested in political and urban geography.

Gerard Toal is Professor of Government and International Affairs in the School of Public and International Affairs at Virginia Tech, USA. He writes about US foreign policy, geopolitics, and conflict regions.

Meagan Todd is a Research Affiliate in the School of Global and International Studies at Indiana University Bloomington, USA. She specializes in the political geography of religion in Russia by examining the rights of minorities to practice religion in public space in increasingly authoritarian Russia.

Frank D.W. Witmer is a computational geographer and Assistant Professor in the Department of Computer Science and Engineering at the University of Alaska Anchorage, USA. He conducts research in violent conflict and human-environment interactions using spatial statistical methods, remote sensing data, and simulation.

Introduction: political geographies of the post-Soviet Union

John O'Loughlin and Ralph S. Clem

When the Soviet Union dissolved in 1991 one-sixth of the world's land surface entered a state of geopolitical flux on a scale not seen anywhere since the end of World War II, with 15 new sovereign states emerging on the territory of the former USSR. This vast expanse at the heart of the Eurasian landmass has since witnessed societal unrest, insurrections, revolutions, civil wars, and *de facto* secessions, (several of which have resulted in massive casualties), human suffering and infrastructure damage on a vast scale, and the displacement of millions of people. With the collapse of the Soviet-led Warsaw Pact military alliance in Central and Eastern Europe and the subsequent eastward expansion of the North Atlantic Treaty Organization (NATO), heightened tensions between the West and a resurgent Russia now exist, catalyzed by Russia's seizure of Crimea in early 2014 and NATO's vigorous response thereto. Russia's somewhat problematic engagement with the European Union (EU), the creation of its own political and customs unions (especially the Eurasian Economic Union – EAEU), and the growing economic influence of China (especially in the former Soviet Central Asia) likewise complicate international relations in the post-Soviet space.

The chapters in this volume reflect a "new regional geography" that should shed light on these significant contemporary political subjects, dilemmas, and crises. Stated as "research whose scope extends beyond disciplinary boundaries to embrace current public and political debate" (Murphy and O'Loughlin 2009, 241), such work is now increasingly expected by the public as well as funding agencies and academic institutions as part of the responsibility of scholars. Presenting an interesting and rich-in-detail account is significantly enhanced if the author(s) also connect their message to two wider audiences: for the public-political where matters are debated and often hotly contested and for the academic where researchers seek empirical verification of (sometimes hyperbolic) theoretical declarations.

Toward that goal, these chapters also deal, in order of presentation, with geopolitical dynamics across the post-Soviet space through the intermediary level of individual states within that space through regions within those states and finally to specific locales within those regions. This is in keeping with our view that the best understanding of this strategically vital and complex area engages what Gerard Toal refers to as "thick geopolitics," a concept that "... strives to describe the geopolitical forces, networks, and interactions that configure places and states" where the many multi-layered influences of location, distance, and place come very much into play (2017, 279). A "thick geopolitics" that is aware of local, regional, and international scalar effects on specific developments can be a holy grail for political geographers and a welcome antidote to broad brush generalizations of other social sciences and anecdotal descriptions from journalists (i.e., "thin geopolitics").

In that spirit, the volume opens with four chapters dealing with macro-regional and international issues, another four focusing on individual states, and concluding with two that examine spatial patterns within states down to the provincial and neighborhood level. John O'Loughlin, Gerard Toal, and Vladimir Kolosov's opening piece discusses the "Russian World" (*Russkiy mir*) concept in the context of the ongoing geopolitical tensions in Russia's "Near Abroad", that area within the post-Soviet space that Moscow traditionally considers a zone in which it has not only longstanding historical and cultural ties, but one which also now abuts NATO's eastern frontier and thus is of vital national security interest to the Kremlin. Probing the views of persons living in strategically sensitive conflict areas by means of public opinion views and preferences within this crucial contact zone, the efficacy of geopolitically charged terms is found to vary among respondents in four different separatist regions in the Caucasus-Black Sea region, reflecting the enduring historical legacy of the Imperial and Soviet periods while underscoring the importance of place-specific factors, especially around the conduct of the 1990s wars, in matters geopolitical.

Sean Roberts in Chapter 3 explores the manner and extent to which authoritarian regimes within the post-Soviet space cooperate through regional economic integration. He finds that the imperatives of regime security often mitigate, or negate outright, the ability of these states to coordinate meaningfully on policy through the Eurasian Economic Union (EAEU). This is especially relevant in current discussions about the possibility of economic liberalization, which is viewed as antithetical to the survival of authoritarian rule. Geopolitically, the challenges of balancing regional economic integration with the perpetuation of authoritarian rule effectively preclude the formation of a stronger bloc of states that might play a more powerful role in international affairs. As a complement to this inter-state discussion which revolves around Russia, the traditional regional power, Marlene Laruelle follows in Chapter 4 with a fascinating analysis of how an outside actor, in this case the United States, seeking to build influence in the five former-Soviet Central Asia states (Kazakhstan, Tajikistan, Uzbekistan, Kyrgyzstan, and Turkmenistan) in the post-Cold War period through an ill-conceived "Silk Road" policy, failed to understand the drivers of regional geopolitics. The US was completely outmaneuvered by Russian and Chinese initiatives based on long-standing regional/historical ties and, in the Chinese case, a willingness to make a huge financial investment in the region.

The question of whether or not Russia is seeking to and is able to establish itself as a regional hegemon, taken up in Chapter 5 by Andrej Krickovic and Maxim Bratersky, continues the macro-level geopolitical analysis of Moscow's influence across the post-Soviet space. This chapter focusing on the use of "soft power" as a means of enticing other states within that space into its orbit (as opposed to "losing" them to Western interests). That strategy having failed in the case of Ukraine forced Russia to exert "hard power" (i.e., military force), but the authors see long term dangers in foreclosing Russian economic and soft power in the furtherance of Moscow's regional integration goals.

Julia Langbein in Chapter 6 investigates the interface between international economic integration on the part of the EU vis-à-vis an individual post-Soviet state and counter-efforts, or lack thereof, on the part of Russia by examining the impact of these competing influences on the automotive industry in Ukraine. As it turns out, for their own reasons, both the EU and Russia failed to engage in a meaningful way with this very important sector of Ukraine's economy, and as a consequence, this industry went through two major downturns and the ultimate result is missed opportunity for all parties concerned except Ukrainian oligarchs who manipulated the system for their own benefit.

Turning from the economic to the political-cultural domain, John O'Loughlin and Vladimir Kolosov in Chapter 7 return to an area of research in political geography that has not received adequate attention – that of the role of symbolism in promoting loyalty to, and identification with, new political units such as those that have emerged in the post-Soviet space. In the case of four *de facto* states that have emerged as a result of internal conflict and unrecognized by almost all members of the international community, ensuring the support of the existing populations through both provision of public goods and reliance on the continued support of the patron (Russia in these post-Soviet cases; Bakke et al. 2018) is matched by promotion of local icons. The article shows a mixed picture of success in this regard since political and cultural figures from earlier Tsarist and Soviet eras are still strongly present in all republics. Only Nagorno-Karabakh has successfully complemented state-building with recognition of locals as major symbolic players in their nation-building, while residents of Transdniestria (also known as Transnistria) at the opposite pole still identify strongly with the Soviet heritage and Abkhazia and South Ossetia show a mixed local and Russian/Soviet heritage and attachments.

The study of conflict in the North Caucasus region of Russia by Edward Holland, Frank Witmer, and John O'Loughlin in Chapter 8 uses information sources that appear in newspapers and press reports to document trends in and the spatiality of events. The North Caucasus region of Russia has been characterized by conflict since the early post-Soviet years of the 1990s, but political developments related to the Kremlin management of the region's economy and security that rely on the partnership with Ramzan Kadyrov, president of the Chechen Republic, has dampened down the conflicts considerably and pushed the flashpoints away from Chechnya to neighboring Dagestan. The level of attacks on Russian forces might be predicted to be negatively correlated with the amount of federal spending (subsidies) in the local area, but the analysis in this article does not support this expectation, as little of the significant Kremlin largess reaches the pocketbooks of the residents of the area.

Chapter 9 by Ralph Clem tackles the seemingly impossible task of parsing and evaluating the evidence from an active war zone, in this case the Donbas conflict in eastern Ukraine. With massive amounts of news, both real and fake, emanating from the various conflict protagonists and their supporters especially in the realm of social media, it can be extremely difficult to evaluate claims and counterclaims. But close examination of social media use by war participants can also produce clear and convincing evidence of military actions, as was documented in the use of a Russian missile in shooting down Malaysia Airlines flight 17 over the Donbas war zone in July 2014 (Toal and O'Loughlin 2018). Despite such evidence, television audiences differ greatly in their views and information sources about such spectacular events. Clem's article takes a broader view and examines a wider variety of the information on the Donbas conflict to show the evidence for Russian trans-border aggression despite official denials from the Kremlin. The conclusions of this and other works on the Ukraine conflict do not rely on official sources, nor even on journalists on the ground, thus presaging a development that marks the new era of conflict analysis in an information-overloaded world.

State-making and nation-building in the post Soviet years continues within both formal *de jure* and unrecognized (partially recognized) *de facto* republics. As David Sichinava shows in Chapter 10, the Republic of Georgia is one of the most democratic states (after the Baltic republics now in the European Union) that emerged after the 1991 Soviet implosion. Though separatist violence has resulted in an uneasy ceasefire and two *de facto* republics (Abkhazia and South Ossetia) in Georgia, its geopolitical orientation toward the West and away from Russia is now almost irreversible but its internal electoral map is quite unstable and unpredictable. The slow emergence of a Western-style electoral geography coincides with the beginnings of consistent (election to election) levels of support for parties that are still personalized but becoming more ideological. Building government structures and confidence in a democratic model not only requires fair electoral procedures but is also promoted by stable parties/coalitions and support bases. Existing ethnic and social cleavages in Georgia are reflected in electoral preferences and a resulting geographic polarization is visible on the maps included in the chapter.

In the last chapter, Meagan Todd's article takes the study of conflict geography down to the truly local layer by examining the potential for social protest escalating to violence in specific areas of Moscow over access to public spaces and the control of public expenditures. Both the Russian Orthodox church and the Muslim communities within the city have tried to build edifices to serve the burgeoning numbers who attend the respective services in the atmosphere of religious freedom that is now (since the 1997 Law on Religious Freedom) guaranteed by the Russian state. However, the tiny number of mosques vastly underserves the Muslim observant, and efforts to build new structures have been met with solid resistance from neighborhood groups and the official authorities. By contrast, the number of Orthodox churches is growing rapidly, but new building proposals sometimes run into competing claims on scarce public sites for parks and other uses. Using an ethnographic approach, Todd shows how citizen groups have differential access to the political process and achieve different rates of success in a tightly controlled political environment.

The discipline of political geography is characterized by a variety of methodologies and topical foci. Often divided by concentrations of research at the global (interstate), national and local scales, the discipline typically studies contemporary developments in evolving relations between states, the making of nations around mythical constructs and the consolidation of states, and local politics in the form of electoral contests and neighborhood politics. What holds the discipline together is a focus on "geo-politics" which can be defined in its broadest form as the intersection of political power and territory. Territory control is often the basis of power and the struggle for territorial control and influence is the most expressive form of political geography. In all of the chapters of this book, this territory-power nexus is evident, whether at the interstate scale as Russia seeks to continue its traditional influence in the post Soviet space through economic, political and military means, or at the national and local scales where territorial competition has devolved into war or struggles over access to public spaces in semi-authoritarian contexts. Scale linkages are critical in determining the outcomes as external actors (Russia and the West) remain deeply involved in continuing power struggles in countries like Ukraine, Georgia, and Kazakhstan. The outcomes of these contests is by no means determined yet. While commentators frequently assess the probable outcomes on the basis of the relative economic or military strengths of the contestants, a deeper understanding of the encompassing localized geopolitics is necessary to generate more precise answers. The chapters in this book, selected for their contemporary insights on the varied territorial-power disputes in the former Soviet Union at different scales, help in this inquiry as the geopolitical contexts continue to evolve nearly 30 years after the end of Communism.

References

Bakke, Kristin M., Andrew M. Linke, John O'Loughlin, and Gerard Toal. 2018. "Dynamics of State-Building after War: External-Internal Relations in Eurasian *De Facto* States." *Political Geography* 63: 159–173.

Murphy, Alexander B., and John O'Loughlin. 2009. "New Horizons for Regional Geography." *Eurasian Geography and Economics* 50: 241–251.

Toal, Gerard. 2017. *Near Abroad: Putin, the West and the Contest over Ukraine and the Caucasus.* New York: Oxford University Press.

Toal, Gerard, and John O'Loughlin. 2018. "'Why Did MH17 Crash?': Blame Attribution, Television News and Public Opinion in Southeastern Ukraine, Crimea and the *De Facto* States of Abkhazia, South Ossetia and Transnistria." *Geopolitics* 23: 882–916.

Who identifies with the "Russian World"? Geopolitical attitudes in southeastern Ukraine, Crimea, Abkhazia, South Ossetia, and Transnistria

John O'Loughlin, Gerard Toal and Vladimir Kolosov

ABSTRACT

The concept of the Russian world (*Russkii mir*) re-entered geopolitical discourse after the end of the Soviet Union. Though it has long historical roots, the practical definition and geopolitical framing of the term has been debated and refined in Russian political and cultural circles during the years of the Putin presidency. Having both linguistic-cultural and geopolitical meanings, the concept of the Russian world remains controversial, and outside Russia it is often associated with Russian foreign policy actions. Examination of official texts from Vladimir Putin and articles from three Russian newspapers indicate complicated and multifaceted views of the significance and usage of the *Russkii mir* concept. Surveys in December 2014 in five sites on the fringes of Russia – in southeastern Ukraine, Crimea, and three Russian-supported de facto states (Abkhazia, South Ossetia, and Transnistria) – show significant differences between the Ukrainian sample points and the other locations about whether respondents believe that they live in the Russian world. In Ukraine, nationality (Russian vs. Ukrainian) is aligned with the answers, while overall, attitudes toward Russian foreign policy, level of trust in the Russian president, trust of Vladimir Putin, and liking Russians are positively related to beliefs about living in the Russian world. In Ukraine, the negative reactions to geopolitical speech acts and suspicions about Russian government actions overlap with and confuse historical linguistic-cultural linkages with Russia, but in the other settings, close security and economic ties reinforce a sense of being in the Russian "world."

Introduction

The current geopolitical standoff between Russia and NATO on the European continent has been described as the worst security crisis since the end of the cold war (Wilson 2014; Sakwa 2015). While the origins of the crisis go back more than a decade, it escalated dramatically in early 2014 with the flight of the democratically elected president of Ukraine, Viktor Yanukovych, in face of violent protests against his rule. This turn of events precipitated a decision by Russian President Vladimir Putin to authorize a stealthy Russian invasion of Crimea. At the same time, pro-Russian activists across southeastern Ukraine, most successfully in the Donbas area within Donetsk and Lukansk oblasts (regions), sought to exploit the vacuum of power and legitimacy crisis created by Yanukovych's departure to advance their own autonomy and separatist aspirations. By May 2014 the Ukrainian military, and an assorted collection of pro-Kyiv militias, were at war with Russian-backed separatists. An imperfect ceasefire arrangement largely stabilized the frontlines in the Donbas in a second Minsk Agreement in February 2015.

In December 2014, we organized simultaneous social scientific surveys in five locations in Russia's near abroad – in six oblasts of southeastern Ukraine (not including the Donbas war zone), in Crimea (the territory annexed to Russia in March 2014), and in the three "de facto" states of Abkhazia, South Ossetia, and Transnistria. We sought to examine how the violent and tumultuous events that year affected the (geo)political attitudes of the various nationalities living in these areas. This paper examines one of these questions; a query that asked respondents whether they believed their current location – oblast or de facto state – was part of the Russian world (*Russkii mir*). Part one of this paper explains our understanding of the term as a geopolitical frame and provides a brief history of its evolution and operation within the Russian political establishment. Part two looks at some media representations of the term in Russia. Part three examines the results from the five surveys about the relevance of this geopolitical framing in each setting.

What and where is *Russkii mir*? A geopolitical frame in post-Soviet space

The study of geopolitical cultures is, in part, the study of the rhetorical framing practices that organize and delimit the world political map into convenient categories and recognizable spaces (Toal 2017). At the outset of the cold war, in a speech before Congress on 12 March 1947, President Truman framed world affairs as a struggle between two ways of life. "One way of life," Truman argued, "is based upon the will of the majority, and is distinguished by free institutions, representative government, free elections, guarantees of individual liberty, freedom of speech and religion, and freedom from political oppression." The other way of life is based "upon the will of a minority forcibly imposed upon the majority. It relies upon terror and oppression, a controlled press and radio; fixed elections, and the suppression of personal freedoms" (Truman 1947). The Truman Doctrine defined the cold war as a clash of opposing worlds. Geopolitical frames like "the free world" and "the West" gave this clash an abstract and moralized definition rather than geographic and material one (Craig and Logevall 2012). Cold war geopolitical culture was organized around a moralized dichotomy between free and enslaved states in Western geopolitical discourse and between capitalist imperialist states and worker republic states within Communist discourse (Westad 2005).

In rhetorical terms, geopolitical frames are performative (sentences that perform an action), although they may appear as merely constative (sentences that describe something as true or false) (Austin 1975). What this means is that their use helps create and constitute the very categorical scheme they proclaim. States become part of the "free world" or "the West" by being described as such, even if these states have characteristics that call such a designation into question (e.g. Turkey and Greece during the cold war as "the free world" or Japan as part of "the West"). Indeed, for many geopolitical frames, their definitional substance is fluid and ambivalent. It is the power of the act of categorizing and framing that matters, not whether it is accurate. Geopolitical frames work as protean signifiers that float above multiple categorical definitions, working largely as dichotomizing performatives; namely as speech acts that function to draw boundaries between "us" and "them." Mass media play a central role in the reproduction of historical myths and narratives in the public field. They legitimize particularistic interpretations of current events by representing them as an objective truth and mobilizing collective memory concerning historical injustices, thus strengthening a group identity by the confirmation of negative stereotypes (Laine 2017).

The term "Russian world" (*Russkii mir*) is one of many used historically to describe the Russian state and empire as a distinctive civilizational space. Medieval sources, for example, described the civilization of ancient Rus as a Russian world. The modern framing has a long genesis beginning with its initial promulgation in St. Petersburg in the 1870s. It was used by some intellectuals, including the classic Russian playwright, Alexander Ostrovskii (1823–1886), who understood it in spiritual terms, as a community of Orthodox Christians living in the unity of belief, rites, and traditions. His articulation distinguished the concept of the "Russian world" from the Pan-Slavic doctrine (MacKenzie 1964; *Khristianskaya* 2007). *Russkii mir* emerged anew as a gathering node for self-definition and meaning after the collapse of the Soviet Union. This renewed articulation has been conditioned by the geopolitical context of its emergence and by multiple overlapping understandings of the term. The collapse of the Soviet Union left Russia searching for a new "national idea" (Allensworth 1998). The socio-demographic realities of that collapse are well known. Up to 25 million ethnic Russians found themselves beyond the borders of the Russian Federation. Some, like Russians in Central Asia and Moldova, were caught up in ethno-territorial turmoil while others were soon on the receiving end of discriminatory citizenship laws in newly independent nationalizing states, such as Latvia and Estonia (Smith 1999). Belonging to the dominant titular group in the new states became a crucial advantage in competition for prestigious jobs, housing, and land. In many regions, like southeastern Ukraine, people with mixed ethnicities were labeled as Russian-speakers by Russian officials who advocated their language rights (Chinn and Kaiser 1996; Tishkov 1997).

Different factions across the Russian political spectrum worked out varied approaches to this geopolitical dilemma, some decidedly revisionist and imperialist in ambition. For example, in 1993 Dmitri Rogozin (currently Deputy Prime Minister of the Russian Federation) initiated the creation of a Congress of Russian Communities (KRO) to organize reuniting Russian communities and NGOs in former Soviet republics and autonomous republics of Russian Federation. It served as a vehicle for advocacy on behalf on "stranded Russians" as well a political project articulating revisionist geopolitical schemes (Ingram 1999). In 1995 a liberal faction around the well-connected Russian political technologist Gleb Pavlovskii established a consulting agency, Foundation for Efficient Politics, whose first project was planning the KRO's unsuccessful electoral campaign. Nevertheless, this agency was soon involved in a number of Kremlin's projects including Boris Yeltsin's electoral campaign and the implementation of the Kremlin initiative to formulate a new Russian idea (Shchuplenkov 2012). Articles by two employees in Pavlovskii's public relations consulting firm, Petr Shchedrovitskii and Efim Ostrovskii, promoted the elaboration of the idea of a "Russian World" in the post-Soviet context (Laruelle 2015). Following the ideas of Shchedrovitsky's father, the Soviet liberal philosopher Georgii Shchedrovitskii who specialized in the field of semiotics, they viewed the common language as a tool to bridge different communities spread across the world. They further wanted to maintain interaction between divergent post-Soviet states and to promote bonds with other countries using the associations of Russian speakers who had left the Soviet Union/Russia at different time periods. Shchedrovitskii speculated about "intellectual network structures" having no center and independent of politics, and even about a new cosmopolitan form of statehood based on networks of diasporas and the principle of multiculturalism, though he did not ignore the role of these communities as a resource for Russian foreign policy (Shchedrovitskii 2000; Petro 2015).

While the definition of the cultural sphere was Russophone, the implicit definition was of a Russian ethno-scape (Appadurai 1996). The term Russian/*Russkii* (ethnic form) and not Rossian/*rossiiskii* (civic form) enjoyed greater emphasis. This reflected a conscious effort to rehabilitate pride in matters explicitly Russian and move beyond what some perceived as the tarnished Westernizing of Rossian/*rossiiskii* as liberal multiculturalism. Large numbers of scholars and politicians were involved in debate about how to understand Rossian/*rossiiskii* political identity. Using the results of numerous surveys, authors claim that the Rossian/*rossiiskii* political nation – integrating citizens of different ethnic background, living together for centuries and sharing the same political, economic, and informational space – is a sociological fact, though a few authors associate it only with the Russian ethnic group (Drobizheva 2009, 2011, 2013; Semenenko

2010; Tishkov 2010, 2011, 2013). Others believe that the Russian/*rossiiskii* political nation can emerge only in a truly democratic society (Pain 2003, 2013). Pavlovskii and colleagues likewise explicitly rejected the exclusivist ethno-nationalism of those on the political right in Russia (Clover 2016). The tension within the Russian world concept between a broad attempt to encompass all Russian speakers and a narrower ethnic Russian understanding of the term has thus never been fully resolved. Russian serves as the language of communication and social promotion for most ethnic groups and thus, it is very difficult to clearly separate Russian ethnic culture from the Rossian political definition.

Russkii mir is polysemous, a catchphrase that is sufficiently fluid, vague, and empty in substance, a sound bite with the useful quality of being ambiguous in substance but clear in its broad boundary – drawing identity-defining function. As used by Putin and other officials, the term *"Russkii mir"* has three interconnected sets of meaning: linguistic, biopolitical, and civilizational.

The first meaning as a cultural and linguistic definition ostensibly has little to do with politics. The Russian world is the cultural sphere of the Russian language and its productions. Just as there is a Francophone world well beyond France's borders, so also is there a Russian world, a community of a shared spoken language and culture. A language-centric definition, however, is inevitably entangled with the biopolitical and geopolitical situation of Russian "compatriots" (*sootechestvennik*) abroad (Zevelev 2001, 2014). Recognizing that Russian (national) culture was also being developed abroad, the first congress of compatriots was held exactly on the same days of the coup against Mikhail Gorbachev in August 1991. "Compatriot" was a coinage of the late Soviet period as the Russian state came to terms with the new geopolitical order and its sense of responsibility to those beyond Russia's borders who looked to the country as a cultural hearth and for protection. "Compatriot" is what Foucault scholars view as a biopolitical term: it concerns the organization, management, and security of populations (Lemke 2011). All states, to different extents and degrees, use biopolitical techniques to manage and monitor their diasporas. For example, Croatia, Hungary, and Poland use biopolitical criteria for maintaining relations with their compatriots by issuing cards to them that provide the holders with access to considerable privileges. This support is inevitably geo-political since these populations reside in the territory beyond "their homeland" within its present boundaries.

The notion of "compatriot" was legally defined in Russia in the federal law "On the State Policy toward Compatriots Abroad" adopted on 24 May 1999 and completed in 2002–2003 by a number of amendments. The Putin administration facilitated a Congress of Compatriots in October 2001 to address what Putin saw as a neglect of this diaspora. In his address to the Congress, Putin gave voice to what would be a consistent theme during his rule – the injustice of the sudden collapse of the Soviet Union. On the one hand, Putin regularly acknowledged a firm intention to support "compatriots" in their struggle for civil rights and against discrimination and their aspiration to keep alive their language and culture. On the other hand, he stated that as Russia was recovering its power, it needed the help of its diaspora. Putin explained that a compatriot is defined not by a legal category or status. Rather it is something that involves spirit and personal choice: it is a "question of self-determination. I would say even more precisely, spiritual self-determination" (Putin 2001).[1] From the beginning, this second biopolitical understanding of *Ruskii mir* is framed in a language of spirituality, of community that transcends the materiality of actually existing political borders. In one of his 2014 speeches on the Ukraine crisis, Putin noted "When I speak of Russians and Russian-speaking citizens I am referring to those people who consider themselves part of the broad Russian community, they may not necessarily be ethnic Russians, but they consider themselves Russian people." (Putin 2014b). The notion of the Russian world, Putin explained, "from time immemorial went far beyond the geographical boundaries of Russia and even far beyond the boundaries of the Russian ethnos." A common cultural and information space is a key instrument of the interaction between the state and fellow countrymen abroad. "What matters is not where you live geographically, what matters is your mentality, your aspirations and, as I said, the person's self-identification" (Putin 2001).

Over the years, Putin's understanding of compatriots did not change. More than a decade later he defined compatriots as those who "share a common concern for Russia's future and its people, a commitment to be useful to your historical homeland, to promote its socioeconomic development and strengthen its international authority and prestige" (Putin 2012). Compatriots were the biopolitical substance – a substance, ironically, defined in terms of its spirit – of the idea of the Russian world. Russian leaders observe enormous diversity across the Russian world but always stress the need for its consolidation and solidarity. They consider relations with compatriots as an intrinsic element of the country's soft power (Ministry of Foreign Affairs 2013).

The two ostensibly separate understandings of the Russian World, linguistic and biopolitical, were officially bound together in June 2007 when Putin signed a decree establishing an organization called the Russian World Foundation (Gorham 2011). Its institutional mission was to promote the Russian language within Russia and abroad and to encourage interest in Russian history and culture. The establishment of the Foundation was inspired by the experience of organizations like *Alliance Française* or the British Council (Russkii mir 2007) and contribute to keeping or to strengthening Russian cultural influence. The Russian World Foundation established programs in 80 countries across the world. It has held international congresses in Russia annually for the last decade. At each Congress, the government regularly confirmed the commitment to support compatriots through the Russian World Foundation. In 2015 the Russian government adopted a new federal program, "Russian language," with a budget of about USD 100 million. There are also federal programs of compatriots' voluntary resettlement to Russia (though to peripheral regions) and a quota for them in Russian higher educational institutions.

As institutionalized in the Russian World Foundation, *Russki mir* is linguistic/cultural, biopolitical, and spiritual. The Russian Orthodox Church (ROC) is the only faith to have representatives on the Foundation's governing board – so the notion is sometimes considered equivalent to the community of believers of Russian Orthodoxy and a basic element on the ground of *Russkii mir*. (The ROC has myriad close ties to the Putin presidential administration and Russian state writ large.)

Critics of Russian foreign policy like Van Herpen (2016, 149) claim that *Russkii mir* is "part of a much more ambitious project that aims to give the Kremlin – again – the global ideological influence it had lost with the end of Communism." The term "Russian world" was part of the whole constellation of close notions designating the post-Soviet space as an area of particularly important to Russia's interests, such as "the near abroad," "historical space of Russia" (*istoricheskoe prostranstvo Rossii*), "the space of Russian language" (*prostranstvo russkogo yazyka*), and "the territory of Russia's responsibility" (*territoria rossiiskoi otvetstvennosti*). A number of authors insist upon Russophonie's apolitical and anti-colonial underpinnings and that the objective of this policy is to reunite, to integrate, and in some cases, to reconcile people of different ethnic, national, social, and ideological backgrounds (see the review in Gorham [2011]).

After Putin's return to the Russian presidency in 2012, the notion of a Russian world took on a more pronounced civilizational meaning (Laruelle 2015). In this sense, the term was a "global signifier" constituting Russia as a distinctive world power with its own civilizational space. Here the term functioned in opposition to competing global metageographic concepts like "the West" or "Atlanticism." "Russian World" is therefore an ideological foundation of the multi-polar world's concept as a cornerstone of Russian foreign policy. The official Foreign Policy Concept adopted by Putin soon after he came to power in 2000 proclaimed the need to oppose the establishment of the unipolar global structure and American hegemony. The most recent Concept included the mention of soft power, including the Russian World Foundation as an instrument of foreign policy (Grigas 2016), which should be backed by the potential of civil society, human communications, and contemporary IT and "other methods and technologies alternative to classic diplomacy." The Foreign Policy Concept stressed the growing importance of civilizational identity as an immanent manifestation of globalization (Ministry of Foreign Affairs 2013). Even if it is not directly used in the Foreign Policy Concept, the notion of the Russian world became a justification of Russia's specific place and influence in the world, an organic and natural element of its soft power and a confirmation of the world's diversity and multi-polar structure.

All of these competing understandings of the term have been over-determined in recent years by Russia's actions in Ukraine. Taking a critical perspective in relation to the Ukraine crisis, Wawrzonek (2014, 760) stated that "The neo-imperialist goals of Russian policy toward Ukraine in recent years have received a doctrinal foundation – the concept of the Orthodox civilizational community – the Russkiy mir." Further, he claimed that "Russkiy mir … should be considered a pretext for Russian political, economic, or 'security' policies toward Ukraine" (776). Kuzio (2015, 159) also accepts this geopolitical strategy behind *Russkii mir*, writing that "As Russian and Soviet identities were irrevocably intertwined in the Soviet Union, it is not surprising that the Russkii mir also mythologized the Soviet past." Saari (2014, 63) thinks that the *Russkii mir* project has inherited Soviet tactics and methods "without any ideology, values, long-term commitment of resources or any degree of responsibility." In a counteraction to Russia's perceived soft power action, in November 2016 the European Parliament supported a bill that listed and condemned the Russkii Mir Foundation as a propaganda arm of the Russian government along with RT television and the Sputnik news agency (Tass 2016).

In justifying soft and hard power actions in a set-piece speech on 18 March 2014, Vladimir Putin returned to the collapse of the Soviet Union as a trauma of sudden fragmentation for the Russian nation and those who identified with it.

> Millions of people went to bed in one country and awoke in different ones, overnight becoming ethnic minorities in former Union republics, while the Russian nation became one of the biggest, if not the biggest ethnic group in the world to be divided by borders.

Putin compared the situation of Russia to that of Germany, a nation divided after World War II. Russia, he pointed out, "unequivocally supported the sincere, unstoppable desire of the Germans for national unity." (Putin 2014a).

What Putin meant by the Russian world was given an environmental, biological, and super-cultural meaning a month later in his annual marathon *Direct Line* television event. Responding to a question – "what is the Russian people to you?" – he developed a distinction between Russia and the West that is worth quoting at length because of the biological and environmental determinism that it revealed:

> As for our people, our country, like a magnet, has attracted representatives of different ethnic groups, nations and nationalities. Incidentally, this has become the backbone not only for our common cultural code but also a very powerful genetic code, because genes have been exchanged during all these centuries and even millennia as a result of mixed marriages. And this genetic code of ours is probably, and in fact almost certainly, one of our main competitive advantages in today's world. This code is very flexible and enduring. We don't even feel it but it is certainly there. So what are our particular features? We do have them, of course, and I think they rely on values. It seems to me that the Russian person or, on a broader scale, a person of the Russian world, primarily thinks about his or her highest moral designation, some highest moral truths. This is why the Russian person, or a person of the Russian world, does not concentrate on his or her own precious personality … Western values are different and are focused on one's inner self. Personal success is the yardstick of success in life and this is acknowledged by society. The more successful a man is, the better he is. This is not enough for us in this country. … Death is horrible, isn't it? But no, it appears it may be beautiful if it serves the people: death for one's friends, one's people or for the homeland, to use a modern word. These are the deep roots of our patriotism. They explain mass heroism during armed conflicts and wars and even sacrifice in peacetime. Hence there is a feeling of fellowship and family values. Of course, we are less pragmatic, less calculating than representatives of other peoples, and we have bigger hearts. Maybe this is a reflection of the grandeur of our country and its boundless expanses. Our people have a more generous spirit. (Putin 2014c)

In these sentences, Putin seems to echo the idea of a "super-ethnos" (an alliance of different ethnic groups) championed by Lev Gumilev, a well-known historian and Eurasianist, although without using the term (Bassin 2016; Clover 2016). His argument, however, is squarely within the mainstream of Russian Slavophile discourse where dichotomies between "spiritual Slavdom" and an "individualistic West" are common (a dichotomy, ironically, that is transposed from nineteenth-century German Romantic thought). His understanding of patriotism (willingness to die for the homeland) is also recognizably Western. Indeed it perfectly illustrates Max Weber's classic definition of the nation as subjective belief in common descent, an affective community where the individual is expected ultimately to face death in the group interest (see Vujačić 2015).

To speak of the Russian world in this way is to potentially mobilize a variety of possible meanings in combination and interconnected. The survey research question we analyze here does not presume or test whether respondents know the different contextual meanings of the term. Rather it is a prompt that asks respondents to accept or refuse an act of identification with *Russkii mir*. In the context of 2014, a year of deepened polarization around geopolitics, it is understandable that many interpret this as a question asking them if they identify with the Russian state under Vladimir Putin as opposed to a Russo-phonic cultural sphere that does not involve endorsing one state or one ruler.

Russkii mir and three Russian newspapers

We have elaborated above the development and the use of the *Russkii mir* concept at the highest levels of the Russian government and the suspicions about its advocacy in a foreign policy sense. However, the public reception of the term and its salience in Russian media is quite uncertain. Is the term only a political ruse by the Kremlin and the subject of a scholarly discussion or does it resonate with the public? We examine three Russian newspapers that span the political spectrum for the years 2014–2016. A liberal newspaper *Nezavissimaya Gazeta* (Independent Newspaper) is compared to two further from the mainstream, *Zavtra* (Tomorrow), a national-patriotic outlet, and *Sovetskaya Rossiya* (Soviet Russia), which is close to the Communist party. As Laruelle (2015) noted regarding the term "Novorossiya," the political discourses of *Zavtra* and *Sovetskaya Rossiya* about geopolitical concepts surprisingly converge: their authors develop the same ideas, use the same arguments, and even the same specific colorful, threatening, and belligerent language. Both newspapers assert Russia's great-power status and messianic role in the world, the values of ultra-conservative Orthodoxy, and treat *Russkii mir* as a synonym for an alternative civilization at war with a hostile "Western civilization."

A common and often repeated point of the two newspapers is that current events in southeastern Ukraine are critically important for Russia: it is not only about this specific conflict but also about the future of Russia and *Russkii mir* as a civilizational space under constant threat of fragmentation and destruction by the West and the US. "Donbas and Crimea have turned out to be on the edge of the clash of civilizations; they felt themselves the bang of the Western machine ready to absorb and to dig into Slavic lands" (Averyanov 2014). Igor Girkin (*nom-de-guerre* Strelkov, a field commander during the first stage of the war in Donbas) clearly explains this idea in *Sovetskaya Rossiya*:

> War is declared on Russia, and if it was not unleashed in the Donbas, it would begin in Crimea or somewhere else. This is a stage of the big war between the West aspiring to global dominance and the Russian world. (Samelyuk 2014)

Stories in *Sovetskaya Rossiya* are targeted at elderly Communist party voters nostalgic for the Soviet Union. They are imbued by an anti-Western and particularly anti-American stance (Boldyrev 2014), the psychology of the "besieged fortress" fighting against the American *diktat* for sovereignty, territorial integrity, and independence: "the language of Russian national philosophy is the language of resistance." (Nikitin 2007). *Sovetskaya Rossiya* repeatedly emphasizes the role of the "core," "state-shaping" Russian nation and declares that Russia is a nation-state "but not from the perspective of the ethnic composition of population" (Bobrov 2014).

In May 2014, at a critical moment in the development of the Ukrainian crisis, *Sovetskaya Rossiya* published a long article where it was noted that the ideologists of Russian nationalism who had been until recently *personae non gratae* in federal media were now regularly invited to talk shows on the main television stations. While recognizing the dangers of nationalism, Kirillov (2014) opted for the recreation of an empire as a means to avoid the risk of outbursts of Russian and other nationalisms. Like other newspapers, *Sovetskaya Rossiya* has never delimited the boundaries of the desirable hypothetical empire, or the Russian world. On the one hand, *Sovetskaya Rossiya* accepts the official concept of the Russian world as a community of all those who speak Russian and identify with Russian culture. They also define the Russian world as a network of large and small communities, defending Russian civilization and its spiritual culture without the single coordinating center, ready to act for the sake of the historical motherland. On the other hand, the newspaper reacted to the creation of the Russkii Mir Foundation by declaring that its official concept cannot be applied in practice. They suggested instead two options: to associate the Russian world with the national political space or to include "compatriots" abroad in it (Zakhar'yin 2007). The newspaper blames the Putin government for the "purposeful politics of compatriots' alienation from Russia, while at the same time opening the doors for immigrants" (Nikitin 2005; Anuchkin-Timofeev 2011). In other words, the paper includes in the Russian world former parts of the Soviet Union, neighboring republics, and territories with an important share of Russian speakers in population. According to *Sovetskaya Rossiya*, Russians, Ukrainians, and Belarussians are a single people, and the paper separates the inhabitants of Western regions from the rest of Ukraine as "professional traitors of Slavs since ancient times" (Zadornov 2014). In the paper's opinion, the organizers of the coup in Kyiv wanted not only control over Ukraine but to involve Russia into the conflict.

Zavtra, in contrast to *Sovetskaya Rossiya*, rarely uses the term "Russian world" but its content clearly shows how its authors understand its extension and borders and what role this concept plays in their understanding of the 2014–2015 events in Ukraine. The overriding preoccupation of *Zavtra* is that Russians should strengthen the state (*gosudarstvennichestvo*) and maintain its sovereignty. They believe that the vocation of Russia is to be the core of an empire interpreted as a voluntary association of Russians and "small" neighboring peoples under the umbrella of a powerful common great power state. The founder and the editor-in-chief of *Zavtra*, the writer Alexander Prokhanov, emphasizes that "the idea that we live in a great power has always accompanied Russian history." The Russian (Tsarist) empire and the Soviet Union as its successor have never been a nation-state, their roots are in a "symphony of cultures" (Glushik 2014).

"Patriotic" ideologists dream about a restoration – at least, partly – of the borders of the former Soviet Union. Quite naturally, like Communists, they firmly support all forms of integration between post-Soviet countries and, in particular, its latest form – the Eurasian Economic Union. However, the patriots always stress the role of Russians as the "state-shaping" people, as "the spiritual, cultural, economic center of new integration's structures in the post-Soviet space." According to Prokhanov (2014), the Russian state is based on

> four powerful forces, four beliefs: Orthodoxy opening the way to endless azure from where divine paradise senses, the light of justice and love. ... Russian culture, our great language and music that God awarded us, reuniting us through music and lyrics with the Divine mystery. ... Oh God, what a happiness is to be Russian!

Russian civilization is thus a "big system based on the values which are radically different from the Western ones" (Prokhanov 2014).

A favorite xenophobic theme of *Zavtra*, as well as of *Sovetskaya Rossiya*, is the decrease in the number of ethnic Russians and the inflow of "Turkic-Muslim" migrants to Russian cities. An author in *Zavtra* declares that the Slavic civilization, despite "fragmentation," is a single civilization "sharing a common understanding of its destination" and totaling 200 million people (Vinnikov and Nagornyi 2014). "Ukraine and Russia are the same ... it is a historical part of Big Russia under the form of the Soviet Union destroyed in 1991 as an integrated geopolitical space" (Nagornyi 2014).

By contrast to these two organs, *Nezavissimaya Gazeta* is a more high-brow newspaper with strong readership in Moscow and St. Petersburg and among the educated and liberal sectors of society. It is often critical of Putin government actions. A screening of the term *"Russkii mir"* shows its almost-complete absence until 2007–2008, when the Russkii Mir Foundation was established. Even then, the term is found in only 24 stories until 2013. With 11 occurrences in 2013, 48 in 2014, 52 in 2015, and 27 in 2016, *Russkii mir*'s frequency in the newspaper reflects the events in Ukraine and the discussion about the extent to which Russians and Ukrainians share an identity and the increased use of the term by government officials. Op-eds by academics and politicians debate the meaning of the *Russkii mir* term and how it might be translated into concrete political actions. Authors are agreed about the political use of the term by the Putin government and the main elements of the concept, though much speculation is evident about its geopolitical implications.

Nezavissimaya Gazeta's (2016) editorial at the end of 2016 recognized the growing importance of the *Russkii mir* term, from an abstract historical term to "a support mechanism of the Russian diaspora," but worried that the public did not understand the implications of the (geo)political use of the term nor indeed its specific elements.

> In fact, (*Russkii Mir*) advocates promoting the "Russian mentality" abroad. And it already means Christian values, and sports achievements, and musical, technical, or scientific accomplishments. This new approach requires the involvement not only of the Foreign Ministry in the ideological struggle, but also the wider public.

Other 2015–2016 articles about *Russkii mir* probed the support for the concept by the Orthodox church (Lunkin 2016), the critique of the *Russkii mir* project by the European parliament and Russia's possible reaction (Gorbachev 2016), and the deeper meaning of Russian identity (Malinova 2015). A prominent opposition figure, Vladimir Ryzhkov (2015), bemoaned the hijacking of the *Russkii mir* term by the Putin government and its oppositional framing to international values:

> Making the first post-Soviet territorial increment of Russian territory from the territory of another state (Crimea), and having widely used nationalist rhetoric ("Russian values", "Russian World"), Vladimir Putin has radically changed the ideological nature of the Russian state. Earlier, Russia appealed to the values of development, modernization, respect for international law, broad international cooperation, and human rights.

In sum, the term "*Russkii mir*" is not one that was particularly central to Russian geopolitical culture until 2014; it was known but not widely debated. That it has become a matter of controversy and debate is a function of how it has been used by the Putin administration to frame Russia's foreign policy vision and interests, particularly from early 2014 onward.

The relative silences about and the uncertainty around the term "*Russkii mir*" at the highest levels of the Russian state are reflected in a variety of media outlets. Ranging from a vague cultural-language promotion project like the *Alliance Française* to a conspiratorial view from outside Russia of a geopolitical project to extend Russian territorial control, *Russkii mir's* meaning remains highly controversial. By asking about the salience of the term in a variety of settings outside Russia but which are heavily involved with Russia, we can probe the relevance of the concept in the context of the renewed confrontation between Russia and the West in the Black Sea-Caucasus region.

Survey data and definitions

Shortly after the annexation of Crimea into Russia and during a relative lull in the conflict in the Donbas after the intense August 2014 fighting, we conducted a comparative and representative public opinion survey in five locations in the Black Sea and the south Caucasus region. The overall project was set in the context of the changing geopolitical relations between Russia and the West that featured the 2014 Ukrainian crisis as its fulcrum. What the "Russian world" means in this hostile political environment and where its margins are lie at the center of our interest in mapping and analyzing the new realities of the near abroad. By asking about the salience of the *Russkii mir* term in a variety of settings outside Russia but which are heavily involved with Russia, we can probe the relevance of the concept in the context of the renewed confrontation between Russia and the West in the Black Sea-Caucasus region.

In 2010–2011 we examined the beliefs and attitudes of residents in the four Russian-supported de facto states of Abkhazia, South Ossetia, Transnistria, and Nagorny-Karabakh (O'Loughlin, Kolosov, and Toal 2014). The later project from which this paper emanates examined three of these regions (Nagorny-Karabakh was excluded since Armenia is the patron state and Russia is a more remote supporter), the annexed peninsula (Crimea), and the broader swath of contested territory in the south and east of Ukraine. This large area, which revisionists in Russia labeled as "Novorossiya," consists of eight oblasts, including the two heavily involved in the Donbas war, those of Luhansk and Donetsk (see O'Loughlin, Toal, and Kolosov 2017). Because of the ongoing conflict in these two oblasts, we were only able to conduct the survey in a reliable manner in six of the eight oblasts that comprise southeastern Ukraine. (The oblasts are listed in Table 1). Consequently, we use the term SE6 rather than "southeastern Ukraine" throughout the text to indicate this substantial but not complete coverage.

The five research sites offer an exploratory and useful *tour d'horizon* of the resonance of the concept of *Russkii mir* on the borders of Russia itself. The sites include two regions with large ethnic Russian minorities (Transnistria and SE6), a region with a large Russian majority (Crimea), and two regions with small (Abkhazia) or negligible Russian (South Ossetia) populations. In language terms, all of the regions have a strong Russian linguistic presence, and Russian is normally the language of "inter-ethnic communication." All sites were formerly in the Soviet Union and all have experienced disputes over the eventual disposition of the respective territories. In the case of Abkhazia, South Ossetia, Transnistria, and parts of southeastern Ukraine, the territorial dispute was violent, while in the case of Crimea, the annexation of March 2014 was peaceful. Most significantly, Russian geopolitical interests and foreign policy actions in the near abroad over the past 25 years since the

Table 1. Responses to the question "Does your region belong to *Russkii mir?*" by oblast in Ukraine.

	Dnipro	Zaporizhzhia	Mykolaiv	Odesa	Kharkiv	Kherson
Strongly agree	3.17	6.12	0.00	11.92	11.67	5.75
Agree	13.99	6.80	20.63	13.21	16.96	5.17
Disagree	19.22	9.18	17.46	18.65	14.32	7.47
Strongly disagree	51.12	39.46	35.98	36.79	31.28	62.07
Don't know	8.77	38.10	14.81	15.28	23.35	11.49
Refuse	3.73	0.34	11.12	4.15	2.42	8.05

collapse of the Soviet Union have involved these five regions. (For comparisons of other geopolitical attitudes in the de facto states, see Toal and O'Loughlin [2016]).

Critical to the project was that the timing of the interviews in the five settings should be exactly the same. In an environment where the geopolitical atmosphere was changing rapidly and where the conflict in the Donbas region was not stable, comparison of the results would have been jeopardized if the interviews were staggered over time as contextual circumstances changed. All interviews were conducted in the last two weeks of December 2014 in the same manner of face-to-face doorstep interviews. Interviews were conducted in Russian (or Ukrainian, by respondent choice, in the SE6 Ukraine oblasts). The average interview lasted 52 min. Sample sizes were 2033 in SE6 Ukraine, 750 in Crimea, 800 in Abkhazia, 500 in South Ossetia, and 750 in Transnistria. The question about *Russkii mir* was the same in all settings and the predictor variables analyzing the responses to the *Russkii mir* question are also directly comparable across the samples.

The survey of about 127 individual questions was organized into three sections. The demographic section of 29 questions and the generic section of 80 questions about contemporary geopolitical developments were exactly the same in all settings. The third section of each survey was specifically oriented to the local conditions and consisted of about 18 questions. The sampling procedure design was a four-step process with random selection at each stage. First, the sample was divided by each of the main regions in each location proportionate to the most recent census population over 18. For each district, all settlements were stratified by size and type (village, small town, town/city), and the probability of each settlement being included in the sample is proportional to its size. Next, for each settlement or group of settlements, a random selection of voting precincts was made, and for each precinct, the initial address was selected with street, house, and apartment chosen randomly. Starting with the initial address, respondents were selected by the method of the modified route sample. Lastly, in the selection procedure for respondents, after getting the initial address, the interviewer made a list of potential respondents ("chain"), who lived in sequential apartments and questioned every third or fifth respondent from the list. Follow-up checks by supervisors were completed for 10% of the completed questionnaires. The response rate varied from 41% in SE6 Ukraine to over 75% in Crimea.

As we have indicated, the conception of *Russkii mir* can take on different hues from a vague and unthreatening promotion of Russian culture, language, and literature to a scripted and aggressive geopolitical vision of in-gathering of ethnic Russians and Russian-speakers as well as unity with the territories where they reside. In this latter interpretation, the Crimean annexation was seen as the first of several such appropriations planned in the Kremlin. In the survey question, we did not specify a particular notion nor exclude others; each individual respondent answered based on his/her own understanding of what the term meant. Vladimir Putin's multiple uses of the term during 2014 and the highly divergent opinions of his actions during the year likely colored some if not most responses.

In our survey, we simply asked, "Do you believe that your region (oblast, republic) is part of *Russkii mir* (the Russian world)?" Respondents could give an answer that ranged from "strongly agree" to "strongly disagree" or could reply with a "don't know" answer. The ratio of refusals was low in all locations, but the rate of "don't know" at 20% in the SE6 oblasts differs dramatically from the very low rate in other locations. Unlike other parts in the survey where high rates of "don't knows" indicate a high level of sensitivity to a particular question, there is no evidence that respondents considered the *Russkii mir* subject as highly sensitive. In the Ukrainian case, the one-in-five response of "don't know" is more likely a reflection of a sense of uncertainty about the meaning of the term (a political vs. a cultural meaning) and the mismatch between the political boundary and the cultural one with Russia. Living in SE6 Ukraine where Russian is commonly spoken, where Russian television was available on cable (but is now banned), and where Russian history and culture have been prominent and widely studied can easily generate a sense of living in the *Russkii mir*, especially for older citizens. A rejection of this background can be made for political reasons and nationalistic motivations that privileges Ukrainian independence and wishes to erect a cultural barrier to the strong Russian influence in the region.

Summary results for the five study sites

It was in the six oblasts of southeastern Ukraine that the *Russkii mir* concept garnered the most varied responses due to the intensity of political discussions after the beginning of an extraordinary political mobilization and after an anti-Russian government came to power in Kyiv. A clear majority of respondents in the SE6 did not believe that their oblasts were part of the Russian world. However, the ratio of respondents (27.4%) who hold the opposite view varied considerably across the six oblasts (Table 1). The relatively low ratio of respondents in Ukraine who believe that they are in an area that could properly be labeled as *Russkii mir* could be due to a number of factors, including resentment at Putin's capture of the term and concern that it might be used to acquire evidence for a possible Russian occupation. With more than half of the respondents using Russian as the home language, a rejection of the *Russkii mir* appellation is probably a response to the contentious geopolitical circumstances of SE6 in late 2014.

Kharkiv, located close to the border with Russia, registered the highest ratio of respondents who feel that their oblast belongs to the Russian world. The 1989 Soviet census indicated that 53% of the city population (more than half of the oblast) were ethnic Russians. Since the disintegration of the Soviet Union this ethnic Russia ratio fell to 33%; Ukrainians are 63% (Ukraina Segodnya 2016), but many of them are of mixed Russian-Ukrainian background (Kolosov and Vendina 2011). The city's population remains largely Russian-speaking, and as one of the largest city and industrial centers in the former USSR, its economy was based on large plants directed by all-union ministries. Until 2014 its economy remained closely tied to the Russian market (Kolosov and Vendina 2011). In Odesa the situation is to some extent similar, with most inhabitants living in the Russian-speaking capital city. By contrast with Kharkiv and Odesa oblasts, Dnipro (petrovsk)[2] oblast, also containing a predominantly Russian-speaking city of about one million inhabitants, contained fewer respondents that think that their region is part of the Russian world. As a polycentric region, its urban population was recruited from the surrounding densely populated countryside and more broadly, from central Ukraine. As in most cities of SE6, ethnic Russians dominated in the city's population (they were 42% in 1926; Argument 2013), but unlike Donbas cities, their ratio decreased and now amounts to only 18%. Though 90% of the population speaks Russian, the city is to a large extent bilingual: about 40% speak Ukrainian fluently and another 30% believe that they speak the state language well. Over half of local printed media is in Ukrainian (Dnipro Gorod 2016). Since March 2014 the city has been dominated by the tycoon Igor' Kolomoisky, appointed as mayor by the new Kyiv authorities. Controlling the largest Ukrainian private bank and a large part of industrial activity, he quickly took the Kyiv government's side in the conflict, funding his own paramilitary units to protect the region and fight Russian-backed separatists in the Donbas. These rapid forceful moves in the oblast, close to the Donbas front line, determined the resulting political environment in the city.

In our survey, the concept of *Russkii mir* was not popular in Zaporizhzhia, Kherson, and Mykolaiv oblasts, despite sizable ethnic Russian populations living mostly in cities (about 30% in Zaporizhzhia oblast and 20% in two other regions). Russian is widely spoken, and a possible explanation of the apparent lack of correlation with the sense of belonging to the Russian world is the origin of population. In the last decades of the USSR, it was growing due to migration from the Ukrainian-populated rural hinterland. A high percentage of "don't know" answers and refusals is noticeable, particularly in Zaporizhzhia oblast (38%, compared to only 12% in Dnipro), indicating a hesitation or an uncertainty about the cultural or political meaning of the question.

The graphs in Figures 1–5 display the respective ratios across the five research sites for the level of agreement with the prompt about the region belonging to *Russkii mir*. The values for each location are compared for key socio-demographic groups – the main nationalities, four age groups, highest and lowest educational levels, and for one key ideological orientation – whether the respondent believed that the end of the Soviet Union was a "right step." We have shown before in multiple studies (e.g. O'Loughlin, Kolosov, and Toal 2014) that the choice by a respondent on the question about whether the end of the Soviet Union was a right or a wrong step is a powerful predictor of a wider set of beliefs about Russia and Russians, about President Vladimir Putin's motivations and actions, about geopolitical opinions on broader issues like NATO expansion, and about the fairness of the post-Soviet liberal economic order as it operates in the respective regions. Generally, individuals who judge that the end of the USSR was a positive development are pro-West, support a democratic political system, are fearful or

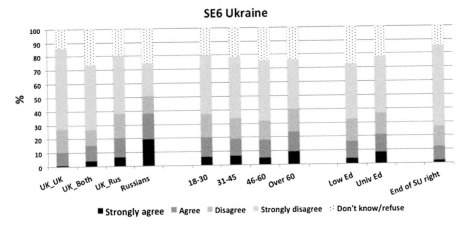

Figure 1. Ratio of respondents in southeastern Ukraine agreeing that their oblast is part of "*Russkii mir*," by national and language groups, age, education, and attitudes to the end of the Soviet Union. December 2014 survey. Source: Authors' data.

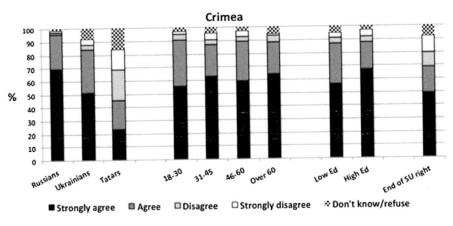

Figure 2. Ratio of respondents in Crimea agreeing that their region is part of "*Russkii mir*," by national groups, age, education, and attitudes to the end of the Soviet Union. December 2014 survey. Source: Authors' data.

suspicious about Putin's intentions, and have seen a positive uptick or, at least, not a dramatic fall in living standards over the past quarter-century.

Respondents in SE6 Ukraine show the lowest overall rate of agreement (combined ratios of strongly agree and agree) at about 30% (Figure 1). The most visible features of the graphs in this region are the large differences between Ukrainian-speakers and Russians. The differences between the language groups within the Ukrainian population on this *Russkii mir* question are notable since our analyses of other political controversial questions – such as the support for political leaders (Toal and O'Loughlin 2015a) – shows no significant differences within the Ukrainian nationality based on home language. The 15-point differences between Ukrainian Russian-speakers and Ukrainians who speak Ukrainian or who speak

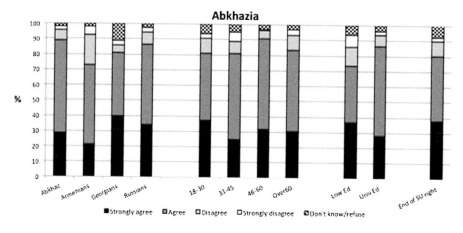

Figure 3. Ratio of respondents in Abkhazia agreeing that their republic is part of "*Russkii mir*," by national groups, age, education, and attitudes to the end of the Soviet Union. December 2014 survey. Source: Authors' data.

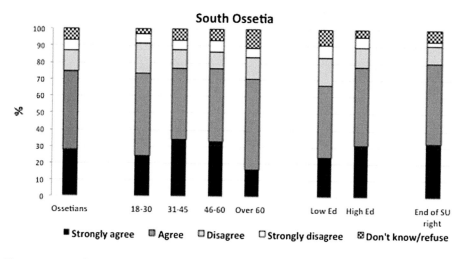

Figure 4. Ratio of respondents in South Ossetia agreeing that their republic is part of "*Russkii mir*," by Ossetian population, age, education, and attitudes to the end of the Soviet Union. December 2014 survey. Source: Authors' data.

both languages suggest that there is a sizable number of respondents who interpreted the question in a cultural-linguistic sense. Because the "don't know" ratio does not vary by linguistic or national group, it is thus unlikely to be politically motivated. As expected, just over 10% of those who believe that the end of the Soviet Union was a correct move agree that their oblast is part of *Russkii mir*. These respondents are most determined to put the Soviet past behind them and move away from the Russian orbit. On the other hand, it is rather surprising that there are no sizable differences between age or educational groups.

In stark contrast to SE6 Ukraine, the ratio that believes that their region is part of *Russkii mir* is three times higher, at over 90%, in Crimea (Figure 2). The level of

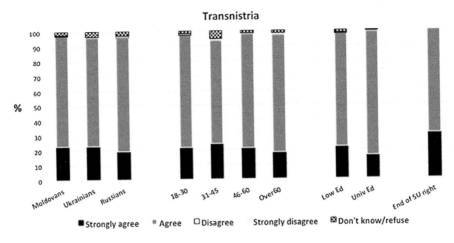

Figure 5. Ratio of respondents in Transnistria agreeing that their republic is part of "*Russkii mir*," by national groups, age, education, and attitudes to the end of the Soviet Union. December 2014 survey. Source: Authors' data.

the strongest support (over two-thirds strongly agree) was the highest among the set of five survey respondents. The survey took place about nine months after the March 2014 annexation to Russia and only the Tatar minority (about 12% of the population) disagrees with the interpretation that the peninsula was part of the Russian world. The support for the annexation in the survey showed huge support for the Russian majority population and the Ukrainian minority, though Tatars showed a high ratio (over 30%) of "don't knows," indicating the sensitivity of the opinions about the annexation and the marginal position of this minority in the new political environment. The responses on the question about *Russkii mir* align well with the responses on the annexation. No differences between age or educational groups are noteworthy. Over two-thirds of people in Crimea believe that the end of the Soviet Union was a mistake; the minority who think it was a right step show about 20% less agreement that the peninsula belongs to *Russkii mir*.

Abkhazia has the most diverse nationality mix of the five study sites and the most complex politics based both on the different group interests and a significant split within the Abkhaz establishment about the strength of security, economic, and political associations with Russia. Like the other de facto states and Crimea, over 80% agreed that the republic is part of *Russkii mir*, with Armenians less likely than the other three groups to agree with the sentiment (Figure 3). As with other questions we have asked in Abkhazia in both 2010 and 2014, the ratio of Georgians saying that they were unable to give an opinion is higher than the other groups. At 12% in this instance, the "don't know" ratio was much lower than other questions about the direction of the republic or relations with Russia where the ratio approaches one in three respondents. No differences by age or educational level or by those who believe that the end of the Soviet Union was a correct move are evident in the graphs.

South Ossetia is now populated almost completely by Ossetians after the displacements of ethnic Georgians in the wake of the 2008 five-day war with Georgia. While there is a small Georgian minority in Akhalgori (Leningor) rayon, we did not sample this isolated population due to concerns about the reliability of responses and interviewer accessibility. About 75% of respondents in all demographic and other categories agreed that South Ossetia is part of *Russkii mir* (Figure 4), but like Abkhazia, this high level of agreement with the prompt hides differences about the scale of integration into Russia. After the 2008 war, Russia recognized both Abkhazia and South Ossetia as independent states, but a strong minority in both republics prefers annexation to Russia. Both republics are heavily dependent on Russian economic and security guarantees.

Transnistria vies with Crimea as the region with the highest level of agreement with the *Russkii mir* statement, with over two-thirds of respondents accepting the "agree" option (Figure 5). This very high ratio is not surprising, as the republic is heavily Russified and since the brief war in 1992, most residents wish to be part of the Russian Federation (O'Loughlin, Toal, and Chamberlain-Creanga 2013). A referendum in 2006 confirmed that orientation with 98% support for annexation. After the Maidan revolt in Kyiv and the ratcheting of tensions between Russia and the Western alliance, the Transnistrian government tried to position the republic as a region of the Russian Federation. That is the line that is heavily pushed by local media and by all segments of the political establishment. Differences among groups are insignificant in the face of a widespread consensus that the republic's best option for economic development in the face of a crisis in the domestic employment sectors and a fast-declining population is full integration into Russia.

Modeling the responses to the *Russkii mir* question

Previous work by the authors and numerous other students of post-Soviet political attitudes have shown that a relatively small set of key predictors is consistently related to significant cleavages in the countries that emerged from the USSR. In the 1990s a sizable literature emerged on the social and regional cleavages that underlie the post-Communist party formations and the bases of their support. Countries such as Poland, where few questions about the survivability of the state or the extent of its borders emerged in the post-Communist period, stand in contrast with countries such as Ukraine, Georgia, and Russia, where geographic cleavages are as important as socio-demographic ones. In these latter countries, center-periphery differences related to both distance from the metropolis and regional ethnic and cultural dissimilarities from the majority populations – a geographic cleavage – need to be considered in addition to the usual compositional political factors. Building on the much-debated but widely-used Lipset-Rokkan (1967) cleavages for Western European politics, Kitschelt (1995) extended and modified the approach for the new electoral politics in the post-Communist states. Zarycki (2000) blended the Lipset-Rokkan and Kitschelt models to inform his analysis of Polish electoral choices by studying axes of political alignment. We follow his model to understand the answers to our question about whether the region in which the respondent lives is part of *Russkii mir*. The theoretical axes motivate the selection of predictors. To allow proper comparison across the five research sites, all predictors are included in the model, whether significant or not.

The first axis, named the "citizenship" axis by Zarycki following Kitschelt, defines the elements of belonging to the state with full inclusive rights. It arranges inclusiveness from a sense of universal citizenship regardless of demographic characteristics to one that is based on ethnic identification where minorities are marginalized. In mixed ethnic societies like Ukraine, Moldova, Georgia, and Russia, the majority-minorities question motivated fundamental political questions about the direction of the state. To animate the "citizenship" axis, we need to consider the nature of ethnic differences within the study sites. Due to population displacements at the end of the respective post-Soviet conflicts, South Ossetia is essentially homogenously Ossetian. In the other four sites, Russians are a majority (Crimea) or a large minority (Transnistria, Abkhazia, and SE6 Ukraine). We include ethnic Russian status as a predictor that helps to account for beliefs of belonging to the Russian world. However, where Russians are a minority, two predictors that measure the strength of attachment to and support of the state are potentially useful in the models. A measure of ethnic pride (respondents who reported that they were "very proud" of their ethnic group) and a willingness to defend the state (affirmative responses to the question about whether the respondent or a family member would take up arms to oppose an invasion) are direct and effective measures of attachment to the state. Those who answered these questions in the negative are likely to be alienated from the state. In our model, we expect Russian ethnicity to be positively related to the *Russkii mir* question while ethnic pride and propensity to fight are expected to be negatively related, except in Crimea, where Russians are a majority and the region is heavily Russified. We include a fourth predictor on this dimension – the level of ethnic distancing from Russians. Since all sites except South Ossetia have sizable numbers of Russians, since Russia is a powerful and omnipresent geopolitical power in the region, and since Russian media is readily accessible, personal attitudes to Russians ("warm to Russians") are expected to be important in influencing opinions about residence in *Russkii mir*.

The second axis is a "values" one. For Kitschelt and Zarycki, values relate to the political dimension of authoritarianism versus liberalism and is also connected to the conservative religious beliefs (the collective) opposed to Western ideals of individual liberties and practices. Our discussion earlier of the concept of *Russkii mir* stressed its communal appeal that is contrary to Western individualism. For post-Communist societies, one can substitute the continued appeal of communist ideologies and authoritarian principles that guarantee a modicum of material well-being against the Western liberal and political model. Our predictors emanating from this axis are four in number. As we have described before, the collapse of the Soviet Union strongly affects the range of opinions about current affairs, both domestically and about relations with the other post-Soviet states, especially Russia. In the face of perceived Russian actions in its neighborhood – the near abroad – opinions about the supposed aims of President Putin are paramount. For this reason, we include a variable that measures whether the respondent trusts Putin (a binary measure). The other three measures are motivated by attitudes to the collapse of the USSR. The first asks whether it was a right move (a binary measure), and whether the respondent self-identifies a left-of center political preference (a binary measure). Lastly, we include a sense of the direction of the current prospects for the respondent's family, asking if they believe that they will be better off two years after the survey (a binary measure). We expect trust in Putin and a left of center ideology to be positively related and the other two measures to be negatively related to the sense of being included in part of the Russian world.

The third axis – the "interests" one – is generated by the uneven distribution of resources after the collapse of Communism. As has been well documented, a significant divide emerged between younger and older generations, and between status groups based on educational levels (Kolosov 1993; Pavlovskaya 2004; Kolosov and O'Loughlin 2011). The social fabric of Communist times was severely damaged by a capitalist model that provided little support to those whose livelihoods became tenuous with the end of the centrally planned model. The "interests" axis ranges from those who support the post-1991 economic model to those who oppose it because of its deleterious effects on their material status. In our models, we expect those who grew up and were socialized in the Soviet times (age over 65), those with a low level of education (less than high school), those with low incomes (can only afford food or worse off), and those who self-identified as being in a bad mood to have higher levels of agreement that they reside in the *Russkii mir*. Many theorists point to the centrality of emotions in conflictual environments. How to operationalize and measure this, however, is contentious. The current political situation is expected to influence a respondent's self-reported affective disposition – measured by the current mood (happy, normal, sad, anxious).

In addition to the three sets of variables based on the axes/cleavages visible in the post-Communist years, we also include four controls in the models. These variables are not of intrinsic predictive value in themselves since we have no theoretical justification for their inclusion. However, they might have confounding effects in the models. We include a gender control and one based on the respondent's self-reported interest (or lack of interest) in politics. We also include two controls based on television viewing habits, which we have demonstrated are significant in the information war regarding recent developments in Ukraine and Russian geopolitical actions since 2014 (Toal and O'Loughlin 2015b). Television as the main source of news and watching more than 20 h of television a week are expected to influence a respondent's view of Russia and the Russian world, but since the television sources are multiple, it is difficult to specify a particular relationship with a sense of being in the Russian world.

Models

The results of the modeling of belonging to *Russkii mir* for the five research sites are presented in graphical form in Figures 6–10. The lengths of the bars correspond to odds ratios from logit models where the outcome variable is agreement (combined strongly agree and agree) with the statement that the respondent lives in a region that belongs to *Russkii mir*. The odds ratios give the chances that a particular respondent of a certain group (e.g. ethnic Russians) will agree with the statement. A value of 1.75 indicates a 75% increased likelihood for this group (compared to the others, in this case non-Russians) holding other factors constant. Values less than 1.0, such as 0.75, indicate a 25% decreased likelihood of agreement. Significant values are shown as hatched bars.

Of the seventeen predictors (including the four control variables), only three of them show significant relationships in three of the sites. Attitudes toward Russia and level of trust in the Russian president are significantly positive (expected direction) in the SE6 Ukraine and Transnistria, while trusting Putin in Crimea and liking Russians in South Ossetia are positively related to the outcome variable (again in

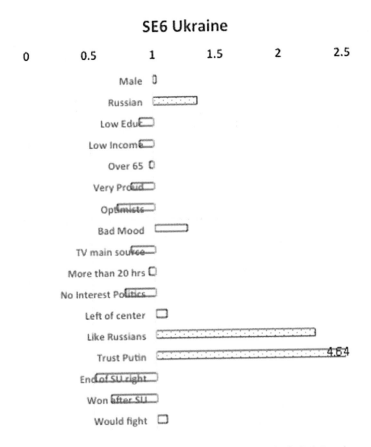

Figure 6. Odds ratio plot of predictive model of agreement with the belief that the respondent's oblast in southeastern Ukraine is part of "*Russkii mir*." Data from December 2014 survey. Source: Authors' data.

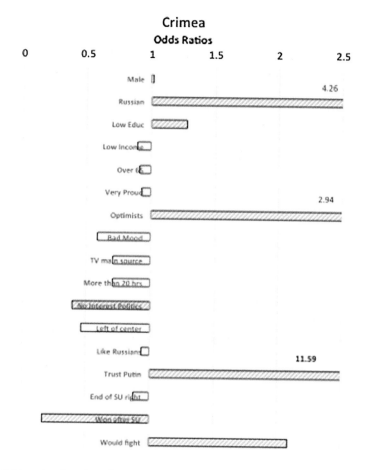

Figure 7. Odds ratio plot of predictive model of agreement by respondent with the belief that Crimea is part of "*Russkii mir.*" Data from December 2014 survey. Source: Authors' data.

the expected direction). The other predictor that is significant in three of the sites is that for respondents who believe that they will be better off in two years (optimists). The distinction between SE6 (where ethnic Russians and pro-Russia attitudes are in a minority) and the other sites is seen in this indicator that is negatively related to the outcome variable but is positively related in Crimea and Abkhazia. These relationships are expected and provide further evidence of the different political and nationality contexts of the survey. In general, most respondents in SE6 are suspicious of Russian government intentions while elsewhere, respondents want closer relations with Russia and deeper incorporation into the Russian world.

The significant negative relationship for optimists in SE6 matches the same relationship for two predictors that measure attitudes toward the end of the Soviet Union. As hypothesized, respondents in this region who thought it was a right move and those who rate an improvement in their family material status after the end of the Soviet Union show a significant negative relationship with a sense of belonging to *Russkii mir* (Figure 6). In a sense, these relationships reflect a desire

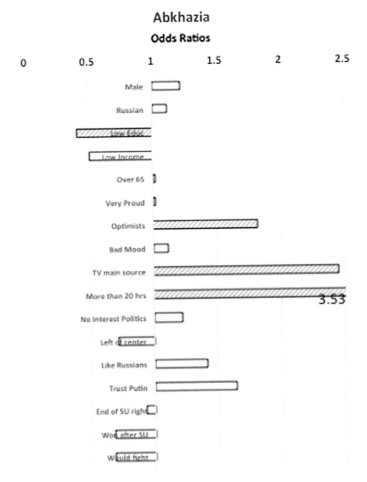

Figure 8. Odds ratio plot of predictive model of agreement by respondent with the belief that Abkhazia is part of "*Russkii mir.*" Data from December 2014 survey. Source: Authors' data.

to move away from the Russian world and for a stronger level of interaction with the West, including NATO and the European Union. The divide in SE6 is strongly along nationality lines, with Russians showing a significant positive value there. To some extent, the ethnic cleavage in SE6 spills over into the annexed territory of Crimea, where the same predictors (Russians with a positive value and respondents who report winning after the end of the Soviet Union with a negative value) show similar strength and direction (Figure 7). Similar groups line up in the same way about whether their region belongs to *Russkii mir*, though now they live in different countries after the March 2014 annexation.

Those who perceive that they have lost out since the end of the Soviet Union are expected to want to be part of the Russian world since for them it offers a nostalgic memory of social protection and a possible integration into Russia's security orbit. This feeling is evident in the significant positive value for those with low education in Crimea (Figure 7) and for those who report a bad mood in SE6 (Figure 6) and

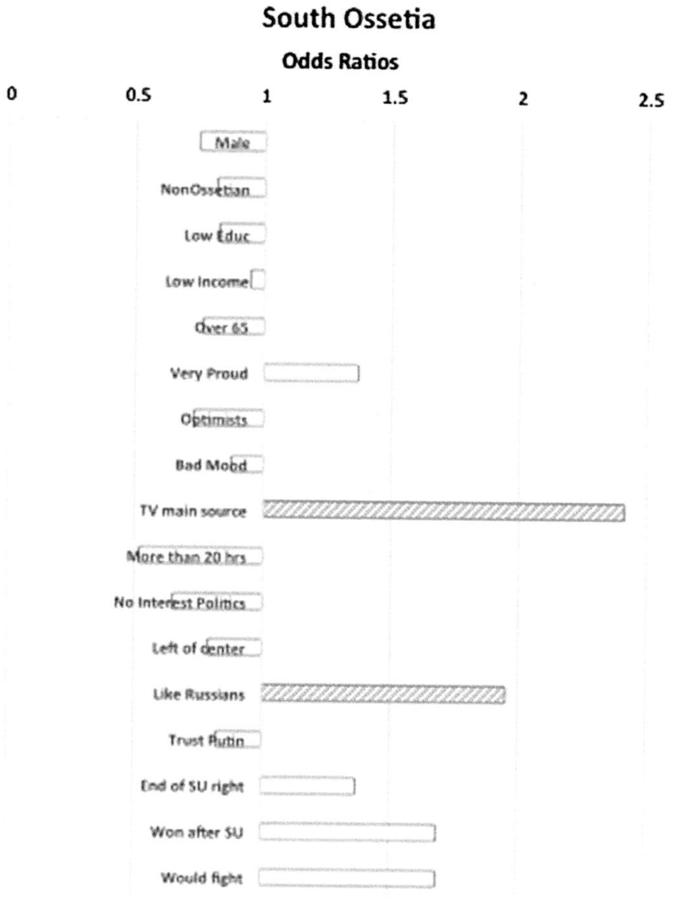

South Ossetia

Odds Ratios

Figure 9. Odds ratio plot of predictive model of agreement by respondent with the belief that South Ossetia is part of "*Russkii mir.*" Data from December 2014 survey. Source: Authors' data.

Transnistria (Figure 10). The significant negative value for poorly educated individuals in Abkhazia (Figure 8) is related to the ethnic division of labor and in-out group status in the political arena, with Georgians forming the "out group" in the republic. In the latter case, the opinion that they do not live in a region of the Russian world is a statement about their geopolitical orientation toward Georgia and the Western alliance.

Crimea is the only site where those who said that they would be willing to take up arms to defend their territory show a significant positive value (Figure 7). Most Crimeans remained in favor of the unification with Russia as reported elsewhere in our survey and a high level of political mobilization and pro-Russian sentiment is evident in other questions. Those who affirmed that they were willing to fight to defend their territory are a small minority – less than 25% in all sites – and in the Crimean case, their views confirm their advocacy of their new status as a subject of the Russian Federation and their opposition to a return to Ukraine. The control variables are generally unimportant in the models except in Abkhazia and South

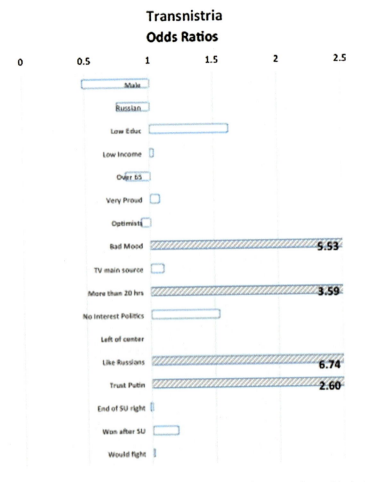

Figure 10. Odds ratio plot of predictive model of agreement by respondent with the belief that Transnistria is part of "*Russkii mir*." Data from December 2014 survey. Source: Authors' data.

Ossetia (Figure 9), where television viewing is significant and positively related to a sense of belonging to the Russian world, an anticipated finding since Russian television stations dominate the viewership.

Of the five sites, both SE6 and Crimea show seven significant predictors, Abkhazia and Transnistria have four each, while South Ossetia only has two significant variables. These comparisons reflect the divided societies in Ukraine and Crimea containing both substantial nationality differences and differing views about Russian geopolitical motivations and actions. South Ossetia is now quite homogenous and therefore few divides mark its society. While Abkhazia and Transnistria have heterogeneous ethnic mixes, both republics are closely aligned to Russia and depend on the Putin government for financial subventions that keep state services running.

In regard to expected relationships, all of the significant variables aligned as the hypotheses anticipated. But overall, only 22 of the possible 85 relationships (including the control variables) show a significant value. The cleavages that have emerged in post-Communist societies about domestic political choices, still evident a quarter-century after the events of 1989–1991, are also relevant to foreign relations. In our case, the question about belonging to the Russian world is both a domestic and a foreign policy one. Its wording suggests a geopolitical orientation as well as a possible cultural direction. On the domestic political scene, opinion about the Russian world aligns with political party choices since they typically conform to the classifications from Lipset and Rokkan (1967) and Kitschelt (1995) about resources distributions (left-right choice), governance rules (liberal vs. authoritarians), and citizenship policies (a civic vs. an ethnic definition). While the settings of the surveys are quite diverse in ethnic makeup and international legal status – only SE6 is part of an internationally recognized entity, the responses to the question about belonging to the Russian world are consistent across the sites given the contextual political circumstances in each.

Conclusions

Geopolitical framing practices are part of the everyday operation and practice of geopolitics. In 2014 the storyline developed by the Putin administration to justify its annexation of Crimea and interventionism in southeastern Ukraine framed the conflict in Ukraine as result of a "fascist coup" backed by the United States and NATO in order to wrest Ukraine from the longstanding partner nation that is Russia. According to this storyline, in history and demographics Ukraine was part of the Russian world, a Russophonic and cultural sphere at the least, a separate civilizational space in its grandest articulations. Large segments of the Ukrainian population were Russian speakers and ethnic Russians. Russia, in this storyline, had a responsibility to protect and defend the Russian world in its near abroad against resurgent fascism and NATO encroachment (Toal 2017).

This paper has sought to measure the degree to which this storyline of cross-border common community resonates in contested regions of Russia's near abroad. Reviewing differing understandings of the term *"Russkii mir,"* it underscores that the term has no fixed and essential meaning. Rather, it is a geopolitical speech act that either works or does not for individuals, both inside Russia and outside its borders. Drawing upon a simultaneous survey in five contested territorial locations beyond Russia's border, we presented evidence that select populations do indeed identify with the notion of a Russian world and see it as something to which they belong.

Unlike other questions about geopolitical matters in the neighborhood of Russia where strong correlations reflect divides along political, nationality, and ideological preferences, the answers about whether the respondent lives in the Russian world were not as predictable. A deep divide along ethnic lines among the residents of six oblasts in southeastern Ukraine about the direction of the country, about responsibility for the war, and about the orientation of the country to the EU/US/West or to the Eurasian Customs Union is not replicated in the Russian world question.

The partly cultural, partly geopolitical meanings embedded in the term *"Russkii mir"* generate a murkier picture of preferences. For a geographically wider set of locations on the borders of Russia, the de facto states, and the annexed territory of Crimea, less doubt is visible since Russia dominates these territories economically and in security terms. Becoming more embedded in the Russian world is the clear preference of the vast majority of their residents.

The term and concept *"Russkii mir"* has achieved a growing presence in political discourse at the highest levels in Moscow, and its implications for the support of compatriots abroad has alarmed many in the countries adjoining Russia. Our survey shows varying acceptance of the belief among respondents that they live in the Russian world, despite their location in heavily Russified areas. This reluctance to accept the moniker for their region indicates a high level of suspicion of the geopolitical implications in the term.

Notes

1. The Kremlin's English language translation of this speech is incomplete. The citations here are translations of the Russian text of the speech.
2. On May 19, 2016, Dnipropetrovsk was officially renamed Dnipro.

Acknowledgments

In conducting this project, we are grateful for the care and attention of the survey organizations and their key personnel, Natalia Kharchenko and Volodymyr Paniotto (Kyiv), Alexei Grazhdankin (Moscow) and Khasan Dzutsev (Vladikavkaz). We are thankful for the cooperation of the 4783 respondents in the five locations who filled out the respective questionnaires. A preliminary version of this paper was presented at the Association for Slavic, East European and Eurasian Studies annual meeting in Philadelphia in November 2015. Thanks to Lucan Way and other panelists for their comments on that version. Close readings of the paper by Ralph Clem and Nancy Place improved the text.

Disclosure statement

No potential conflict of interest was reported by the authors.

Funding

The research is supported by a grant from the U.S. National Science Foundation, Political Science program through the RAPID initiative [grant number 1442646].

References

Allensworth, Wayne. 1998. *The Russian Question: Nationalism, Modernization, and Post-communist Russia.* Lanham, MD: Rowman and Littlefield.

Anuchkin-Timofeev, A. 2011. "Drama Russkogo mira." [Drama of the Russian World.] *Sovetskaya Rossiya*, August 16. http://www.sovross.ru/articles/651/10580/comments/7.

Appadurai, Arjun. 1996. *Modernity at Large: Cultural Dimensions of Globalization*. Minneapolis: University of Minnesota Press.

Argument. 2013. *Etnoyazychnyy sostav naseleniya oblastnykh tsentrov Ukrainy v 1926 godu* [Ethnic Composition of Regional Centers of Ukraine in 1926]. http://argumentua.com/stati/etnoyazychnyi-sostav-naseleniya-oblastnykh-tsentrov-ukrainy-v-1926-godu.

Austin, John L. 1975. *How to Do Things With Words*. Cambridge, MA: Harvard University Press.

Averyanov, Vitalii. 2014. "Novorossia – nash avangard." [Novorossia – Our Vanguard.] *Zavtra* 23 (1072), June 5. http://old.zavtra.ru/content/archive/?number=23&year=2014&month=6.

Bassin, Mark. 2016. *The Gumilev Mystique: Biopolitics, Eurasianism and the Construction of Community in Modern Russia*. Ithaca, NY: Cornell University Press.

Bobrov, Alexander. 2014. "Yanvar' Maidana i obmana." [January of Maidan and Fraud.] *Sovetskaya Rossiya* 10 (13958), January 30. http://www.sovross.ru/articles/1022/17457.

Boldyrev, Yurii. 2014. "Voina vse spishet? Ili vpravit nam mozgi?" [The War will Write Off Everything? Or will it Set it Straight?] *Sovetskaya Rossiya* 29 (13977), March 15. http://www.sovross.ru/articles/1039/17818.

Chinn, Jeff, and Robert Kaiser. 1996. *Russians as the New Minority: Ethnicity and Nationalism in the Soviet Successor States*. Boulder, CO: Westview Press.

Clover, Charles. 2016. *Black Wind, White Snow: The Rise of Russia's New Nationalism*. New Haven, CT: Yale University Press.

Craig, Campbell, and Fredrik Logevall. 2012. *America's Cold War: The Politics of Insecurity*. Cambridge, MA: Harvard University Press.

Dnipro Gorod. 2016. *Naselenie goroda* [Population of the City]. http://gorod.dp.ua/inf/geo/?pageid=109.

Drobizheva, Leokadia M. 2009. *Rossiiskaya identichnost' v Moskve i regionakh* [Russian Identity in Moscow and the Regions]. Moscow: MAKS Press.

Drobizheva, Leokadia M. 2011. "Rossiiskaya identichnost' i tendentsii v mezhetnicheskikh ustanovkakh za 20 let reform." [Russian Identity and Tendencies in Ethnic Stereotypes During 20 Years of Reforms.] In *Rossia reformiruyushayasia: Ezhegodnik 2011* [Russia in Reforms; Yearbook 2011], edited by Mikhail K. Gorshkov, Issue 10, 72–85. Moscow: Nestor-Istoria.

Drobizheva, Leokadia M., ed. 2013. *Grazhdanskaya, etnicheskaya i regionalnaya identichnost: vchera, segodnia, zavtra* [Civil, Ethnic and Regional Identity: Yesterday, Today and Tomorrow]. Moscow: Rossiiskaya Entsiklopedia.

Glushik, Ekaterina. 2014. "Ukrainskaya rana." [Ukrainian Wound.] *Zavtra* 27 (1076), July 3. http://old.zavtra.ru/content/view/ukrainskaya-rana/.

Gorbachev, Alex. 2016. "Rossiyu postavili v odin ryad s 'Islamskim gosudarstvom: Evropa ob'yavila Moskve informatsionnuyu voynu." [Russia Put on a Par with the 'Islamic State' – Europe Announced an Information War on Moscow.] *Nezavissimaya Gazeta*, November 21. http://www.ng.ru/politics/2016-11-21/1_6865_kontr.html.

Gorham, M. 2011. "Virtual Russophonia: Language Policy as 'Soft Power' in the New Media Age." *Digital Icons: Studies in Russian, Eurasian and Central European New Media* 5: 23–48.

Grigas, Agnes. 2016. *Beyond Crimea: The New Russian Empire*. New Haven, CT: Yale University Press.

Ingram, Alan. 1999. "'A Nation Split into Fragments': The Congress of Russian Communities and Russian Nationalist Ideology." *Europe-Asia Studies* 51: 687–704.

Khristianskaya tsivilisatsia: sistema osnovnykh tsennostei. mirovoi opyt i rossiiskaya situatsia. Materialy nauchnogo seminara [Christian Civilization: The System of Basic Values. World Experience and the Russian Situation.] 2007. Proceedings of a Scientific Seminar, Issue 3, 113–114. Moscow: Nauchnyi expert.

Kirillov, Valery. 2014. "Russkii mir: Pravo na budushchee." [Russian World: The Right to the Future.] *Sovetskaya Rossia*, May14. http://www.sovross.ru/articles/1064/18331.

Kitschelt, Herbert. 1995. "Formation of Party Cleavages in Post-Communist Democracies." *Party Politics* 1: 447–472.

Kolosov, Vladimir, and Olga Vendina, eds. 2011. *Rossiiko-ukrainskoe pogranich'e: Dvadtsat' let razdelennogo edinstva* [The Russian-Ukrainian Borderland: Twenty Years of Separated Unity]. Moscow: New Chronograph.

Kolosov, Vladimir. 1993. "The Electoral Geography of the Former Soviet Union, 1989–1991: Retrospective Comparisons and Theoretical Issues." In *The New Political Geography of Eastern Europe*, edited by John O'Loughlin and Herman van der Wusten, 189–216. London: Belhaven Press.

Kolosov, Vladimir, and John O'Loughlin. 2011. "After the Wars in the South Caucasus State of Georgia: Economic Insecurities and Migration in the 'De Facto' States of Abkhazia and South Ossetia." *Eurasian Geography and Economics* 52: 631–654.

Kuzio, Taras. 2015. "Competing Nationalisms, Euromaidan, and the Russian-Ukrainian Conflict." *Studies in Ethnicity and Nationalism* 15: 157–169.

Laine, Jussi. 2017. "Vospriatie Rossii v obshchestvennom soznanii Finlandii." [Perception of Russia in the Finnish Public Consciousness.] *Comparative Politics Russia* 8: 123–139.

Laruelle, Marlene. 2015. *The "Russian World": Russia's Soft Power and Geopolitical Imagination*. Washington, DC: Center on Global Interests.

Lemke, Thomas. 2011. *Bio-politics: An Advanced Introduction*. New York: New York University Press.

Lipset, Seymour Martin, and Stein Rokkan. 1967. "Cleavage Structures, Party Systems, and Voter Alignments: An Introduction." In *Party Systems and Voter Alignments: Cross-National Perspectives*, edited by Seymour Martin Lipset and Stein Rokkan, 1–64. New York: Free Press.

Lunkin, Roman N. 2016. "Antizapadnichestvo pobezhdayet russkii mir: iz soyuznikov 'moskovskoye' pravoslaviye ostavilo sebe tol'ko islamskii Vostok" [Anti-Westernism Wins the Russian World: Of 'Moscow's' Allies, Orthodoxy has Only the Islamic East Left.] *Nezavissimaya Gazeta*, May 18. http://www.ng.ru/ng_religii/2016-05-18/3_rusmir.html.

MacKenzie, David. 1964. "Panslavism in Practice: Cherniaev in Serbia (1876)." *The Journal of Modern History* 36: 279–297.

Malinova, Olga. 2015. "V poiskakh samikh sebya: rossiyskaya identichnost' kak ustoychivaya neopredelennost'." [Finding Themselves: Russian Identity as a Stable Uncertainty.] *Nezavissimaya Gazeta*, November 24. http://www.ng.ru/stsenarii/2015-11-24/9_search.html.

Ministry of Foreign Affairs, Russian Federation. 2013. *The Foreign Policy Concept of the Russian Federation*, February 12. http://www.rusemb.org.uk/in1/.

Nagornyi, Alexander. 2014. "Zashchitit li Rossia russkikh?" [Will Russia Protect Russians?] *Zavtra* 28 (1077), July 10. http://old.zavtra.ru/content/view/zaschitit-li-rossiya-russkih/.

Nezavissimaya Gazeta. 2016. "Nedootsenennyye vozmozhnosti proyekta 'Russkii mir': Novyye aspekty bor'by za vliyaniye v mirovoy politike." [The Possibility of the Project 'Russian World' is Underestimated: New Aspects of the Struggle for Influence in World Politics.] December 21. http://www.ng.ru/editorial/2016-12-21/2_6890_red.html.

Nikitin, Vladimir S. 2005. "Sozdadim Russkii mir." [Let's Create the Russian World.] *Sovetskaya Rossiya* 142 (12753), October 27. http://www.sovross.ru/old/2005/142/142_3_4.htm.

Nikitin, Vladimir S. 2007. "Rossia: ot bedy k pobede." [Russia: from Adversity to Victory.] *Sovetskaya Rossiya* 11–12 (12933), January 30. http://www.sovross.ru/old/2007/11/11_3_3.htm.

O'Loughlin, John, Vladimir Kolosov, and Gerard Toal. 2014. "Inside the Post-Soviet De Facto States: A Comparison of Attitudes in Abkhazia, Nagorny Karabakh, South Ossetia, and Transnistria." *Eurasian Geography and Economics* 55: 423–456.

O'Loughlin, John, Gerard Toal, and Rebecca Chamberlain-Creanga. 2013. "Divided Space, Divided Attitudes? Comparing the Republics of Moldova and Pridnestrovie (Transnistria) using Simultaneous Surveys." *Eurasian Geography and Economics* 54: 227–258.

O'Loughlin, John, Gerard Toal, and Vladimir Kolosov. 2017. "The Rise and Fall of 'Novorossiya': Examining Support for a Separatist Geopolitical Imaginary in Southeastern Ukraine." *Post-Soviet Affairs* 33: 124–144.

Pain, Emil A. 2003. *Mezhdu imperiei i natsiei: Modernistskii proekti i ego traditsionalistskaya alternativa v natsionalnoi politike Rossii* [Between Empire and Nation: A Modernizing Project and its Traditionalist Alternative in Russian Ethnic Policy]. Moscow: Liberal Mission.

Pain, Emil A. 2013. "Budeshchee postimperskikh obshchetsv XXI veka." [The Future of Post-imperial Societies of the XXI Century.] *Russia in Global Affairs* 3. http://www.globalaffairs.ru/number/Buduschee-postimperskikh-obschestv-XXI-veka–16008.

Pavlovskaya, Marianna. 2004. "Other Transitions: Multiple Economies of Moscow Households in the 1990s." *Annals of the Association of American Geographers* 94: 329–351.

Petro, Nicolai N. 2015. *Russia's Orthodox Soft Power.* Washington, DC: Carnegie Council. http://www.carnegiecouncil.org/publications/articles_papers_reports/727.

Prokhanov, Alexander. 2014. "Schastie byt' russkim." [Happiness to be Russian.] *Zavtra* 1 (1050), January 2. http://old.zavtra.ru/content/view/schaste-byit-russkim/.

Putin, Vladimir V. 2001. "Vystuplenie Presidenta Rossiiskoi Federatsii V. V. Putina na Kongresse sootechestvennikov." [Speech of the Russian President V.V. Putin at the Congress of Compatriots.] October 11. http://old.nasledie.ru/politvnt/19_44/article.php?art=24.

Putin, Vladimir V. 2012. "Vladimir Putin's Address to the Participants of the Fourth World Conference of Compatriots," October 26. http://en.kremlin.ru/events/president/news/16719.

Putin, Vladimir V. 2014a. "Address by President of the Russian Federation," March 18. http://eng.kremlin.ru/news/6889.

Putin, Vladimir V. 2014b. "Address to Conference of Russian Ambassadors and Permanent Representatives," July 1. http://en.kremlin.ru/events/president/news/46131.

Putin, Vladimir V. 2014c. "Direct Line with Vladimir Putin," April 17. http://en.kremlin.ru/events/president/news/20796.

Russkii mir. 2007. *Mezhdunarodnyi opyt podderzhki sootechestvennikov za rubezhom* [International Experience of Support of Compatriots Abroad]. Moscow: Russkii mir.

Ryzhkov, Vladimir. 2015. "Nostal'giya po imperii: Chto budet, yesli Rossiya po ushi vtyanetsya v voyny na postsovetskom prostranstve." [Nostalgia for Empire: What will Happen if Russia is Involved up to its Ears in War in the Former Soviet Union.] March 3. *Nezavissimaya Gazeta* www.ng.ru/ng_politics/2015-03-03/10_nostalgy.html.

Saari, Sinukukka. 2014. "Russia's Post-orange Revolution Strategies to Increase its Influence in Former Soviet Republics: Public Diplomacy po russkii." *Europe-Asia Studies* 66: 50–66.

Sakwa, Richard. 2015. *Frontline Ukraine: Crisis in the Borderlands*. London: I.B. Tauris.

Samelyuk, Anna. 2014. "Atakuyut Novorossiu: Interv'yu Igorya Strelkova." [Attacking Novorossiya: Interview with Igor' Strelkov.] *Sovetskaya Rossiya* 112 (14060), October 7. http://www.sovross.ru/articles/1125/19532/comments/4.

Semenenko, Irina S. 2010. "Rossiiskaya identichnost' pered vyzovami XXI veka: problemy i riski na puti formirovania grazhdanskoi natsii." [Russian Identity Before the Challenges of the 21st Century: Challenges and Risks in the Way of the Formation of a Civic Nation.] In *Destabilizstsia mirovogo poriadka i politicheskie riski razvitia Rossii* (Destabilization of the World Order and Political Risks for Russia's Development), edited by Vladimir I. Pantin and Vladimir V. Lapkin, 37–51. Moscow: Institute of World Economy and International Relations of Russian Academy of Sciences.

Shchedrovitskii, Petr. 2000. "Kto i schto stoit za doctrinoi Russkogo mira?" [Who and What is Behind the Doctrine of the Russian World?] *Russkii Archipelag*. http://www.archipelag.ru/ru_mir/history/history99-00/shedrovicky-doctrina/.

Shchuplenkov, Oleg. 2012. *Natsionalnaya ideya Rossii* [National Idea of Russia]. 6 vols. Moscow: Nauchnyi Expert.

Smith, Graham. 1999. *The Post-Soviet States: Mapping the Politics of Transition*. London: Edward Arnold.

Tass. 2016. "RF mozhet prinyat' mery v svyazi s rezolyutriyey Evroparlamenta protiv rossiiyskikh SMI." [Russia may take Measures in Connection with the Resolution of the European Parliament Against the Russian Media.] *Nezavissimaya Gazeta*, November 29. http://www.ng.ru/news/563842.html.

Tishkov, Valery A. 1997. *Ethnicity, Nationalism and Conflict in and after the Soviet Union: The Mind Aflame*. London: Sage.

Tishkov, Valery A. 2010. *Rossiiskii narod: kniga dlya uchitelya* [Russian People: A Book for Teachers]. Moscow: Prosveshchenie.

Tishkov, Valery A. 2011. *Edinstvo v mnogoobrazii* [Unity in Diversity]. Orenburg: Editorial Center of Orenburg State Agrarian University.

Tishkov, Valery A. 2013. *Rossiiskii narod i smysl natsionalnogo samosoznania* [Russian People and the Sense of National Self-consciousness]. Moscow: Nauka.

Toal, Gerard. 2017. *Near Abroad: Putin, the West and the Contest for Ukraine and the Caucasus*. New York: Oxford University Press.

Toal, Gerard, and John O'Loughlin. 2015a. "How Popular are Putin and Obama in Crimea and Eastern Ukraine. " *Monkey Cage blog, Washington Post*, January 22. https://www.washingtonpost.com/news/monkey-cage/wp/2015/01/22/how-popular-are-putin-and-obama-in-crimea-and-eastern-ukraine/.

Toal, Gerard, and John O'Loughlin. 2015b. "Russian and Ukrainian TV Viewers Live on Different Planets." *Monkey Cage Blog, Washington Post*, February 26. https://www.washingtonpost.com/news/monkey-cage/wp/2015/02/26/russian-and-ukrainian-tv-viewers-live-on-different-planets/.

Toal, Gerard, and John O'Loughlin. 2016. "Frozen Fragments, Simmering Spaces: The Post-Soviet De Facto States." In *Questioning Post-Soviet*, edited by Edward C. Holland and Matthew Derrick, 103–126. Washington, DC: Wilson Center Press.

Truman, Harry S. 1947. "Address before a Joint Session of Congress, Recommending Assistance to Greece and Turkey," March 12. http://www.americanrhetoric.com/speeches/harrystrumantrumandoctrine.html.

Ukraina Segodnya. 2016. *Katalog vedushchikh predpriyatii Ukrainy – Kharkiv* [Catalog of Leading Enterprises of Ukraine – Kharkiv]. http://www.rada.com.ua/rus/RegionsPotential/Kharkiv/.

Van Herpen, Marcel. 2016. *Putin's Propaganda Machine: Soft Power and Russian Foreign Policy*. Lanham, MD: Rowman and Littlefield.

Vinnikov, Vladimir, and Alexander Nagornyi. 2014. "Ukraina kak nesostoyavsheesia gosudarstvo." [Ukraine as a Failed State.] *Zavtra* 28 (1077), July 10. http://old.zavtra.ru/content/archive/?number=28&year=2014&month=7.

Vujačić, Veljko. 2015. *Nationalism, Myth, and the State in Russia and Serbia*. Cambridge: Cambridge University Press.

Wawrzonek, Michał. 2014. "Ukraine in the 'Gray Zone': Between the 'Russkiy mir' and Europe." *East European Politics & Societies* 28: 758–780.

Westad, Odd Arne. 2005. *The Global Cold War*. New York: Cambridge University Press.

Wilson, Andrew. 2014. *Ukraine Crisis: What it Means for the West*. New Haven, CT: Yale University Press.

Zadornov, Mikhail. 2014. "Nado li vvodit' rossiiskie voiska?." [Should There be Russian Troops?] *Sovetskaya Rossiya* 24 (13972), March 5. http://www.sovross.ru/articles/1036/17744.

Zakhar'yin, V. R. 2007. "Monopolia na ekstremizm." [Monopoly on Extremism.] *Sovetskaya Rossiya* 19–20 (12939), February 13. http://www.sovross.ru/old/2007/19/19_2_5.htm.

Zarycki, Tomasz. 2000. "Politics in the Periphery: Political Cleavages in Poland Interpreted in their Historical and International Context." *Europe-Asia Studies* 52: 851–873.

Zevelev, Igor. 2001. *Russia and its New Diasporas*. Washington, DC: US Institute of Peace Press.

Zevelev, Igor. 2014. "The Russian World Boundaries: Russia's National Identity Transformation and New Foreign Policy Doctrine." *Russia in Global Affairs*, no. 2. http://eng.globalaffairs.ru/number/The-Russian-World-Boundaries-16707.

The Eurasian Economic Union: the geopolitics of authoritarian cooperation

Sean Roberts

ABSTRACT

Understanding cooperation among authoritarian regimes remains a puzzle for researchers; in particular, those working in post-Soviet Eurasia. Research suggests that autocrats are becoming increasingly coordinated in their efforts to thwart democracy, with authoritarian-led regional organizations offering an effective vehicle to extend autocrat time horizons. In contrast, older studies, including insights from failed regional integration among former Soviet states, suggest that the absence of democracy limits cooperation, although in both cases there is a lack of detail on the mechanisms enabling or constraining relations between autocrats. This article addresses this shortcoming by developing a theoretical framework based around autocrat survivability or "regime security" and applying it to the important case of the newly formed Eurasian Economic Union (EAEU), drawing on original interview data with experts and stake-holders in Belarus, Kazakhstan, and Russia. The argument forwarded in this article is that concerns over regime security create antagonistic cooperation drivers. In the case of the EAEU, regime security provides a strong explanation for the inability of member states to coordinate policy. The implication is that future studies should pay close attention to the way the material and ideational aspects of authoritarian rule combine to drive, but also limit relations between autocrats.

Introduction

Research on the international dimensions of authoritarian rule suggests that autocrats are becoming increasingly coordinated in their efforts to thwart democracy, and that authoritarian-led regional organizations offer an effective vehicle to bolster non-democracies and extend autocrat time-horizons (Ambrosio 2009; Bader, Grävingholt, and Kästner 2010; Börzel, van Hullen and Lohaus 2013; Libman 2015; Silitski 2010;

von Soest 2015; Tansey 2016). With the fragile status of democracy in many parts of the world, authoritarian cooperation has the potential to alter the political balance and to create a zone of illiberal states united in mutually reinforcing institutions.

Indeed, authoritarian cooperation is not static, but appears increasingly ambitious, in particular in the post-Soviet space. Literature has already noted the expanding remit of the Russia–China led Shanghai Cooperation Organization (SCO), which has grown to include election monitoring and formal cooperation with other regional organizations (Ambrosio 2009, 182). But of potentially greater significance is the creation of a qualitatively new kind of authoritarian integration project in the form of the Eurasian Economic Union (EAEU).

This important regional organization was established in 2015 and represents an alternative to the EU, but in a de-politicized institutional framework, offering current and prospective members a path to economic modernization without democratic conditionality. Hailed by its founders as a means to achieve deep integration in a short period of time and to change the geopolitical configuration of the continent (Putin 2011), the EAEU has already raised concerns in some quarters of a "re-Sovietization" of the region (FT 2012). With what we know about the geographical limits of democratic diffusion (Kopstein and Reilly 2000) and with the backdrop of the 2008 Global Financial Crisis and the EU's internal problems, the EAEU may well find opportunities to expand its influence. To an extent, this is already happening. The EAEU incorporated Kyrgyzstan and Armenia as new members in 2015 in regions (Central Asia and the South Caucasus) where the EU's influence is under pressure.

However, while recent studies suggest that authoritarian regimes are becoming increasingly integrated, older studies, including insights from failed integration among former Soviet states present a conflicting image of authoritarian cooperation, suggesting that the absence of democracy acts as a significant centrifugal force (Allison 2008; Bohr 2004; Collins 2009; Haas 1966). Rather than facilitate cooperation, similar authoritarian political systems act as a barrier, supported by the poor results of post-Soviet regional integration since 1991 (Kobrinskaya 2007; Olcott, Åslund and Garnett 1999; Vinokurov and Libman 2012).

In fact, the EAEU, despite its relative novelty, appears to reflect this puzzle, with scholars unsure if the EAEU is seeking EU-style "deep integration" or something more modest and in line with the general experience of post-Soviet integration (Dragneva and Wolczuk 2012, 220). In short, there appear to be conflicting images of authoritarian cooperation – one where autocrats are increasingly coordinated and united in their dealings with each other, and one where authoritarian regimes are greatly restricted in their capacity for international cooperation. Clearly, these images are of great significance for both regional and international politics, in particular in the context of democratic rollback in post-Soviet Eurasia and other regions. But, as discussed in this article, part of problem lies in the fragmented and implicit treatment of the dynamics of authoritarian cooperation in existing studies.

This article addresses this shortcoming by revisiting the extant regime studies and regionalism literature to elaborate a framework to explain authoritarian cooperation, based on the importance of autocrat survivability or "regime security," in contrast to mainstream theories emphasizing state survival and state security and which includes both material and ideational components. The remainder of this article then applies this framework to the case of the EAEU, utilizing primary and secondary sources as well as original qualitative data from fieldwork conducted in Belarus, Kazakhstan, and Russia in 2014 and Russia in 2016. This fieldwork included semi-structured interviews with thirty stakeholders, policymakers, and experts in these three founding member states, including representatives from foreign and economic ministries and the nascent Eurasian Economic Commission – the supranational component of the EAEU (Appendix 1), offering a unique opportunity to explore the dynamics of authoritarian cooperation from the perspective of those states involved.

This article proceeds as follows. The first two parts outline the puzzle of authoritarian cooperation and develop the regime security framework. The remainder of this article then applies this framework to consider the case of the EAEU, first in terms of regime security as a driver for cooperation and then as a brake, inhibiting cooperation. The final section considers the way regime security and regime identity combine to limit deeper cooperation, in particular sovereignty pooling. The argument forwarded in this article is that concerns over regime security create antagonistic drivers for cooperation. In the case of the EAEU, regime security offers some explanation for cooperation between member states, while providing a stronger explanation for their inability to coordinate policy. The implication is that future studies should pay close attention to the way the material and ideational aspects of authoritarian rule combine to drive, but also limit relations between autocrats.

The puzzle of authoritarian cooperation

Coinciding with the perceived retreat of global democracy, scholars have turned their attention to the international dimensions of authoritarian rule, including authoritarian cooperation, in an attempt to better understand this emerging trend. Although under-theorized (von Soest 2015, 628) with research said to be in its early stages of development (Tansey 2016, 201), existing studies suggest that authoritarian regimes are becoming increasingly coordinated in pursuit of a common agenda that includes the establishment of regional orders and the replacement of democratic norms with their own non-democratic values (Ambrosio 24, 2009). Some scholars have even hypothesized an emerging "authoritarian international" united in a common objective of thwarting unwelcome international influences, notably the spread of democracy (Silitski 2010, 341). Although evidence is far from conclusive, the appearance of relatively dynamic authoritarian-led regional organizations, such as the SCO and EAEU, supports the notion that authoritarian cooperation is evolving, if not deepening.

According to this literature, regional organizations potentially represent a significant "next step" in authoritarian cooperation, offering an effective means to stabilize and bolster autocrats with the potential to create an expanding zone of illiberal coordination. Libman (2015), for example, theorizes a number of regime-boosting functions performed by authoritarian-led regional organizations, including legitimacy provision, economic support, governance transfer, mutual learning, and the socialization/financial support of private business interests. Tansey (2016, 68) highlights the importance authoritarian sponsorship via regional organizations, including money and weapon transfers and diplomatic support. At the micro level, it has even been suggested that the freedom of movement associated with regional integration may help autocrats redirect opponents away from activism in their home country toward politically or economically motivated migration (Obydenkova and Libman 2015, 21).

Overall, this increasingly complex authoritarian cooperation has a "common sense" logic. It has long been noted by first-wave integration scholars that shared values and beliefs facilitate regional integration (Nye 1968, 423), while more recent literature has also indicated that autocrats prefer cooperating with other autocrats in "similar systems" (Bader, Grävingholt, and Kästner 2010, 96). In the case of post-Soviet Eurasia, there is no shortage of commonalities that may be expected to drive cooperation among authoritarian regimes, some of which are noted in existing studies (Lane 2015, 14; Libman and Vinokurov 2012, 14). Beyond authoritarianism, these commonalities include similar economic systems (relatively high levels of state intervention and a high importance of hydrocarbons), high levels of economic interdependence, and similar security and modernization concerns, but also a range of cultural, historical, and linguistic features (notably the prevalence of Russian language) that go some way to explaining the close alignment among certain post-Soviet states following the collapse of the USSR.

However, the relatively recent interest in the dynamics of authoritarian cooperation and its potential to redefine regions stands at odds with existing research detailing the problems of cooperation, including integration, in the post-Soviet space. While it is not inconceivable that autocrats in the region have gradually grown more united in their opposition to a single issue (e.g. the spread of democracy) or that some galvanizing event has focused states on the need for deeper cooperation (e.g. the Global Financial Crisis), when viewed from the perspective of previous studies, this apparent increase in authoritarian coordination is puzzling.

The literature detailing post-Soviet regionalism is illustrative of this point. Between 1991 and 2010, 36 regional organizations were established in the post-Soviet space, making it one of the most active regions in the world in terms of integration (Keukeleire and Petrova 2016, 270). While the participating states were not always clear-cut autocracies, the turnover of regional organizations combined with the prevalence of non-democratic regimes in the region points to a significant coordination problem. In fact, scholars have resorted to "adjective regionalism" to capture this reality, characterizing integration as "ink-on-paper" (Vinokurov and

Libman 2012, 53) "stalled" (Collins 2009, 254) or "virtual" (Allison 2008, 185), pointing to a circumstance where high-level statements of intent are rarely matched by corresponding levels of commitment.

At a generalized level, a number of explanations have been forwarded to explain these poor results, such as the fear of economic and military asymmetry that exits between Russia and all other post-Soviet states (Libman and Vinokurov 2012, 192), elite preoccupation with defending sovereignty (Olcott, Åslund and Garnett 1999, 22–28), and a general predilection to ignore sovereignty pooling commitments and to by-pass the rules (Kobrinskaya 2007, 14, 15). However, in many cases, the prevalence of non-democratic regimes was also viewed as a significant barrier to effective cooperation (Allison 2008; Bohr 2004; Collins 2009), notably in the way that national interests are subordinated to the interests of the ruling group. This chimes with the work of early integration scholars who noted the importance of democracy as a pre-requisite for successful integration, highlighting the difficulty of authoritarian regimes in meeting their integration commitments (Haas 1966, 106).

From this perspective, the puzzle of authoritarian cooperation is that similarity, in particular similar political systems, is theorized as both a push and pull factor in relations between autocrats. While these conflicting images of authoritarian cooperation are not necessarily mutually exclusive, the puzzle is magnified by the absence of a clear framework that can serve as an entry point to better identify the drivers behind these processes and how they work.

The framework: regime security as a key variable in authoritarian cooperation

The approach taken in this article is to reconcile these competing images by revisiting the regime studies and regionalism literature in order to find common ground between regime type and cooperation drivers. As discussed below, the desire of autocrats to retain power makes concerns over regime security a key variable for understanding authoritarian cooperation – a point that is often implicit and fragmented in existing literature. Regime security is defined as "the condition where governing elites are secure from violent challenges to their rule" (Jackson 2013, 162) and stands in contrast to mainstream theories that emphasize state security/ survival as a foreign policy driver (see below). For the purposes of this article, cooperation occurs when actors adjust their behavior through a process of policy coordination so that partners view them as facilitating the realization of their own objectives (Keohane 2005, 50, 51).

Despite the recent focus on the democracy-thwarting potential of authoritarian cooperation, existing literature gives a strong hint that regime security is an underlying motivation. There is an acknowledgment, for example, that the ultimate goal of autocrats is to extend time horizons or to ensure what Ambrosio terms "regime survivability" (2009, 19). As Von Soest notes; regimes collaborate when faced with an existential threat (2015, 624), and it is the threat to regime survival more than any other that drives authoritarian foreign policies, in particular in the post-Soviet space (White and Feklyunina 2014, 239). Overall, cooperation is seen as a logical security strategy in the sense that "isolated regimes are vulnerable" and that authoritarian leaders who sit back and watch allies "founder" magnify threats to their own rule (Whitehead 2014, 9).

As detailed in the case of the EAEU, authoritarian cooperation designed to bolster regime security does not necessarily take the form of collective security or defense organizations, although this sometimes is the case. We know from comparative experience that economic cooperation among autocrats often carries a distinct security aspect. The Gulf Cooperation Council (GCC), for example, was formed as a vehicle for economic and security cooperation in response to the dual threats of the USSR and Iran (Walt 1987, 270) but also internal dissent and internal pressures (Priess 1996). But, this is by no means an indication of "authoritarian exceptionalism." State security has long been theorized as a central foreign policy driver (Waltz 1979) and an important component of economic integration per se (Buzan 2003), including European economic integration (Moravcsik 1998, 5, 6), although for authoritarian leaders, survival becomes a first-order interest to which all others are subordinate – a point supported in post-Soviet area studies literature (Allison 2004, 469; Collins 2009, 251; Miller and Toritsyn 2005, 332).

Regime security as a brake on cooperation

In sum, the hypothesis that regime security drives authoritarian cooperation is both testable and derived from existing literature and provides an entry point to better understand relations between authoritarian regimes. However, a corollary of this is that cooperation that is deemed to endanger regime security is likely to be avoided or curtailed by the autocrats in question. Again, existing regime studies and regionalism literature support this assertion, although again it is fragmented and often implicit.

For example, from what we know of the ability of intergovernmental organizations to promote democracy, there are obvious reasons why autocrats may be reluctant to deepen interdependence with other states through authoritarian-led equivalents. International organizations are seen to be effective democracy promoters because they provide increased leverage over the domestic affairs of member states, opening the way for targeted sanctions if a member strays from the collective norms of that organization. Sanctions such as de-legitimization through diplomatic pressure, economic sanctions, and even the negative conditionality of expulsion from the organization are powerful mechanisms to affect domestic political outcomes (Pevehouse 2002, 522). Some scholars have already asked if autocrats prefer weaker forms of cooperation (von Soest 2015, 633), presumably in acknowledgment of this point.

As a result, and under certain circumstances, cooperation may enhance but also weaken regime security. Cooperation may gain autocrats powerful patrons able to provide security guarantees and bolster incumbents, vis-à-vis domestic and international opposition – as the regime studies literature suggests. But, cooperation may afford the same partner states an undesirable level of influence over domestic affairs, which may undermine key pillars of regime control and stability. Even without the threat of direct interference, cooperation commitments that involve sustained policy coordination may require significant changes to existing socio-economic relations within and between states, which may directly challenge the ruling group.

This point, on the delicate multi-balancing that autocrats must engage in, is a little ambiguous in existing studies. Tansey (2016, 4), for example, mentions that for autocrats to stay in power, they must consider the balance of international pressure and support bearing on them, to maximize the latter and minimize the former. But for any regime, the option of maximizing international support, in particular through deeper forms of cooperation, must be weighed against domestic pressures. Close cooperation with other states may accrue direct material benefits for the ruling group but unsettle the delicate balance among domestic constituencies.

This suggests the presence of tipping points, based around the balance between the internal and external aspects of regime security and which potentially limit the extent of authoritarian cooperation. This, in many ways, is unsurprising. Existing literature has long emphasized the way that foreign policy in democratic states, including the decision to cooperate with other states, is an outcome of bargaining between domestic groups and the economic costs and benefits they anticipate. Frameworks that combine domestic- and international-level factors to explain foreign policy decisions, such as Putnam's dual-level game (Putnam 1988) or more sophisticated versions (Milner 1997), usefully capture the tension between domestic constituencies. The limited pluralism found in most authoritarian regimes reduces scope for bargaining, but concerns over regime security mean that autocrats must consider the domestic impact of their international cooperation.

Regime security and regime identity

This leads to the final point that the nature of regime security depends in no small part on the characteristics of the regime in question. The authoritarian regime-type is hardly a residual category, and existing literature has long highlighted variations between authoritarian regimes, notably their differing institutions and locales of power (Geddes 2003; Huntington and Moore 1970). However, alongside these aforementioned material variations, careful attention needs to be given to ideational differences in order to avoid a narrow materialist ontology that may limit our understanding of authoritarian cooperation.

The literature on regional integration is quite illustrative. We know, for example, that the degree of commitment democratic states are able to make to each other depends on the compatibility of internal and external norms (Schimmelfennig and Sedelmeier 2005). The new regionalism literature has also highlighted the importance of identity for integration processes (Soderbaum and Shaw 2003), including "Othering" (Neumann 2003), while theorists have made an explicit connection between identities, institutions, and democratic regime stability. Moravcsik, for example, notes that social actors provide support to government in return for institutions that accord with their "identity-based preferences," institutions which then become "legitimate" in their eyes (Moravcsik 1997, 525).

In terms of authoritarian regimes and in particular for the post-Soviet space, there have been attempts to highlight the role of identity in explaining foreign policy alignment (Ambrosio 2006), balancing behavior (Gvalia et al. 2013) and the failure of Central Asian regionalism (Rosset and Svarin 2014). Elsewhere, literature has highlighted the role of national consciousness as an influence on integration (Hale 2008, 191) and the general ability of identity politics to "trump" material factors when it comes to foreign policy (White and Feklyunina 2014, 238).

While the emerging literature on authoritarian cooperation has acknowledged that authoritarian regimes do not share a "common identity" (von Soest 2015, 626), there are grounds to go one step further and state that authoritarian regimes have unique identities, which are often cultivated as a component of the ruling group's legitimacy. This identity may take a number of forms, notably ethnonational, but it is nonetheless an important non-material aspect of regime security that incumbents must take into account when choosing international partners and in deciding the degree of commitment they make. There is some evidence that autocrats may forfeit a materially advantageous alignment with a foreign partner because of domestic opposition and the perceived threat to identity-based legitimacy (Ehteshami et al. 2013, 227), meaning autocrats must not only balance internal and external pressures, but also material and ideational threats.

In this way, regime legitimacy is a function not only of regime performance (e.g. the ability to provide security to citizens, economic prosperity, etc. [Miller 2006, 26–29]), but also the ability of incumbents to safeguard particular values, a way of life, or set of relations between society and the state. As regime identity is defined by the ruling group but also contested (other voices and identities exist, but are often repressed), any international cooperation that generates sustained contradictions to the regime's identity risks encouraging and enabling challengers to "outflank" incumbents with broad-based appeals for political change. This is particularly important in electoral authoritarian regimes. As most authoritarian regimes invest in a nominal "democratic identity," elections are important events, but so too the ability of the ruling group to appeal to a core electorate and, if needed, mobilize pro-regime counter-protests. Both are a function of the material but also ideational capacity of incumbents.

In sum, the consideration of complex inter- and intra-regime balancing and ideational and material security provides a multi-level, multi-factorial framework to explore the dynamics of cooperation among autocrats. The next sections develop these ideas further through examination of the EAEU, where evidence suggests that concerns over regime security do indeed create antagonistic drivers, facilitating but also inhibiting policy coordination.

The case of Eurasian economic integration

The EAEU was formally signed into existence by the leaders of Belarus, Kazakhstan, and Russia in January 2015 and joined shortly afterward by Armenia (January 2015) and Kyrgyzstan (August 2015). It builds on an older idea of creating a customs union among newly independent states (Vinokurov 2007, 26) that found expression in the CIS Customs Union (1993), Eurasian Economic Community (2000), Eurasian Customs Union (2010), and Single Economic Space (2012). Despite a bright start that saw trade volumes among Customs Union members increase in 2011–2012, the combination of sanctions and counter-sanctions surrounding the ongoing Ukraine crisis and drop in oil prices has seen a consistent decrease in mutual trade, 2013–2016. Total exports (millions, USD) by EAEU member states to EAEU partners declined from a high of 67,856 in 2012 to 42,536 in 2016 (Eurasian Commission 2017).

In terms of suitability, the EAEU has high value as a case study of authoritarian cooperation. First, the EAEU is a good example of an authoritarian-led regional organization. The EAEU founding members (Belarus, Kazakhstan, and Russia) are clear non-democracies ("not free") with little to suggest they are marginal cases (Freedom House 2016) – despite attempts by ruling groups in each to sustain a façade of democracy. Second, the fact that the EAEU is an economic and not a security organization provides an ideal opportunity to test the claim that regime security drives cooperation between autocratic states. Finally, the EAEU is an important case because of its implications for the region, but also the way it continues to draw divergent accounts from the literature, in particular if the EAEU is seeking EU-style deep integration or something more modest (Dragneva and Wolczuk 2012). As such, this article is able to contribute to this debate by exploring EAEU drivers and their strength and direction.

The following discussion draws on primary and secondary sources, as well as in-depth interviews conducted in 2014 just before the official unveiling of the EAEU, with follow-up interviews conducted in 2016 (Appendix 1). Each interview focused on the motivations and perspectives of integration as seen by experts and stakeholders in each state. Overall, the choice of interviews is well-suited for this research, offering insider views from the three founding member states and a unique opportunity to explore the dynamics of authoritarian cooperation from the perspectives of those states involved. For ethical reasons interviewees are anonymized and cited by country (Belarus, Kazakhstan, and Russia) or institutions (Eurasian Economic Commission) only.

Regime security as a driver for cooperation?

From the outset, the appearance and development of the EAEU has been accompanied by an obvious economic logic; meaning that, in the official discourse, economic rather than security factors present themselves as "first order" drivers. With the inclusion of Armenia and Kyrgyzstan in 2015, the EAEU comprises a huge market of over 170 million people with a combined GDP of more than USD 2 trillion, creating substantial opportunities for domestic producers and for attracting foreign investment. With an explicit aim of realizing the four economic freedoms and of coordinating economic policy among member states, the EAEU offers a long list of advantages for member states.

For Belarus and Kazakhstan, unfettered access to the huge Russian market offers immediate benefits, not least the opportunity to modernize their respective economies, without the risk of full exposure to the globalized economy. Companies in Belarus and Kazakhstan have the opportunity to become part of the Russian production chain and to benefit from Russian investment, as well as external investment from non-member states wishing to access the common market. For Russia, improved access to markets in Belarus and Kazakhstan (9 million and 17 million people, respectively) is not insignificant, in particular for key manufacturing sectors such as the automotive industry. Business also has the opportunity to relocate to member states to take advantage of differing tax regimes, if and when issues with residency are resolved.

In addition and at a macro level, there is also an underlying assumption in each member state that the global economy is "evolving in the direction of regional economic blocks and alliances" and that the next stage of development will likely see the creation of transregional alliances, like the Transatlantic Trade and Investment Partnership agreement (T-IPP) (Interview, Russia, 2014). As such, there is a consensus that the Eurasian region risks being left behind if it continues to hesitate with its own integration projects. In fact, it is here at the macro level that we also find evidence that integration has a more tangible regime security component alongside the obvious economic drivers.

The first indication that the EAEU is a reaction to a perceived threat to regime security is seen in the form of the ongoing effects of the 2008 Global Financial Crisis. These effects were quick to materialize in the region, seen in Russia's economic slump in 2009 and the financial crisis in Belarus in 2011, but also in the longer-term drop in the price of oil – a key cog in the political economy of post-Soviet authoritarianism. Falling demand for oil without a corresponding drop in production has seen oil prices decline by two-thirds in the period 2008–2016, affecting Russia and Kazakhstan as major oil exporters but also Belarus, whose economy derives significant revenue from refining oil. The removal of tariffs and a commitment to create common access to pipeline infrastructure are part of an attempt to mitigate the reliance that each member state has on international commodity markets.

In addition, the economic slowdown that followed the 2008 financial crisis refocused regional leaders on the need to diversify their economies (Putin 2014a), with the removal of tariffs seen as an important step in stimulating other sectors of the economy and raising competitiveness in a region that lags behind others on most comparative measures. As such, the ongoing effects of the 2008 financial crisis have led to an understanding that cooperation is essential, or as one interviewee remarked, "keeping together gives us added value" (Interview, Russia, 2014). In this sense, the EAEU serves as a vehicle for economic modernization among states with similar levels of development.

In fact, this idea of economic vulnerability ties into the second threat to regime security that underpins Eurasian economic integration – the growing influence of the "democratic" European Union (EU), but also "authoritarian" China in the region. In many ways, this is unsurprising, as customs unions are designed to create but also divert trade (Mattli 1999, 11), and without trade diversion the presence of two huge economies on the western and eastern flanks poses a longer-term threat to the security of ruling groups in the region. Even though the economies of Belarus, Kazakhstan, and Russia benefited from rising oil prices and a general economic upturn in the period 2000–2008, the expansion of EU and Chinese economies is hard to ignore. In the period 1991 up the creation of the EAEU in 2015, the EU's GDP grew from USD 7.8 to USD 16.3 trillion, while China's GDP increased from USD 3.8 billion to USD 11 trillion (World Bank EU 2017a; World Bank China 2017b) – in both cases dwarfing the USD 2 trillion combined GDP of EAEU member states.

For Belarus and Russia, the implications of Chinese investment and the growing trade imbalance have yet to receive public attention, but it is Kazakhstan where a growing nationalist sentiment combines with a fear of China (Interview, Kazakhstan, 2014) and where the nature of Kazakhstan–China relations are openly discussed in public. Here, the economic threat posed by China is seen in terms of cheap goods undercutting domestic producers, but also the unease felt in some quarters at growing Chinese influence in the country. As such, the EAEU is viewed as a necessary step to shield Kazakhstan from China's presence in the region.

Like China, the EU is also viewed as an economic threat, but here the boundary between economics and politics becomes blurred. For some, the new generation of European Union Association Agreements, drafted in 2009, served as a direct stimulus for Eurasian economic integration, as Russia, along with other states in the region, was forced to respond (Adomeit 2012). This corresponds to a general understanding in Moscow at least, that if Russia is not pushing a cooperation agenda in the region, then others will fill the vacuum: "if Russia is not actively promoting this concept [integration], then neighboring countries will get engaged in other integration projects" (Interview, Russia, 2014).

More importantly, there is also a belief among member states that the EU is heavily politicized, meaning that political expansion accompanies economic expansion – something viewed as intrinsically threatening. This links to the third regime security threat that the EAEU is designed to counter – externally provoked regime change or a Ukrainian-style "Maidan" uprising, spreading as a contagion to neighboring states, either through a spontaneous demonstration effect or directed by hostile third parties.

In both Russia and Belarus, notions of a fifth column are frequently articulated by the regime. In Russia, this is a fully fledged public discourse focused on the entire spectrum of regime opponents (Lipman 2015), while in Belarus the Polish minority has been targeted, in particular following the introduction of a Polish Ethnicity Card for citizens of the former USSR in 2007 (BTI 2014, 5). Kazakhstan also has significant issues with domestic opposition, at times drawing direct criticism from the EU (Savchenko 2015) and exacerbated by ongoing concerns surrounding political stability and leadership succession. The EAEU provides an opportunity to stabilize regimes through gradual economic development, as well as offering the prospect of collective action and external assistance from partners, if and when needed.

In this sense, there is some evidence that the role of the EAEU is to complement other security-focused integration projects to resist the spread of color revolution. In June 2012 Belarus leader Aleksandr Lukashenko met with Vladimir Putin to discuss the threat of sanctions and Western-sponsored regime change (Lukashenko 2012). Belarus, like Russia, is no stranger to EU economic sanctions. Following a brief period of détente with the West that saw Belarus sign up to the EU's Eastern Partnership in 2009, the EU imposed rolling sanctions in the wake of Lukashenko's December 2010 presidential election victory and an accompanying crackdown on regime opposition figures (Benzow 2011). The EAEU, alongside other structures, such as the Belarus-Russia Union State Treaty and the Shanghai Cooperation Organization, provides an extra layer of support, including financial assistance, in circumstances when other sources of credit may become unavailable. In this regard, the EAEU is an important source of regime credit, acting through the financial mechanisms of the Eurasian Development Bank (EDB) and the Eurasian Stabilization and Development Fund (ESDF). Since its creation in 2006, the EDB has invested around USD 4.85 billion in the region, while the ESDF, with capital of USD 8.5 billion, acts a regional IMF and a "lender of last resort" (Vinokurov 2017, 58)

In addition, membership of the EAEU increases the prospects of direct support from Russia, both in terms of hard security guarantees but also direct economic support. Again, Belarus is illustrative of the economic benefit of closely aligning with Russia, not only through Moscow-approved bailouts from the EDB in 2011, but also energy subsidies. Following a deal with Russia in 2014, Belarus pays no export duties on the oil it imports from Russia, refines, and then sells on international markets (Interfax 2014), and overall it is estimated that Russian subsidies account for up to 20% of Belarus's GDP, with half of all that country's exports going to Russia (Bentzen and Dietrich 2016). As one interviewee noted, "Eurasian integration carries with it opportunities for us to keep on getting energy subsidies and rebates from Russia" (Interview, Belarus, 2014). This tallies with other studies of the EAEU that emphasize the importance of "extractive relations" and the ability

Regime security as a brake on cooperation?

In the case of the EAEU there is no shortage of evidence of coordination problems between member states, although these problems are in some ways exacerbated by international conditions. This is particularly evident in terms of foreign policy coordination against the backdrop of the Ukraine crisis and Russia's unilateral counter sanctions introduced in August 2014, which have proved a point of contention for both Belarus and Kazakhstan. In fact, the unwillingness of other EAEU member states to support Russian sanctions against Ukraine in itself shows the very real limits of authoritarian cooperation.

Elsewhere, the relatively high number of coordination problems seen in the period 2015–2016 does not necessarily reflect deep-seated regime security issues, but general problems of trust-building familiar to most regional integration projects. In particular, the use of non-tariff barriers remains a problem, with each member state accusing partners of selectively applying them to protect domestic producers from competition.

For just one member state – Kazakhstan – the period 2015–2016 witnessed numerous trade disruptions with EAEU partners. In 2015 and 2016, Kazakhstan temporarily banned Russian food imports, citing health concerns and the need to protect consumers from poor-quality products (gov.Kz. 2016). In March 2015, Kazakhstan imposed a 45-day ban on Russian oil imports to protect domestic producers, and in May 2015, the government approved new standards limiting the import of Russian and Belarusian vehicles (Rodeheffer 2015). This comes against a backdrop of currency devaluation and persistent accusations that member states are trying to boost the competitiveness of domestic producers.

In most cases, the numerous disputes among EAEU members in the period 2015–2017 and following the official unveiling of the EAEU reflect the weakness of institutions and by extension the union's enforcement mechanisms. The supranational component of the EAEU – the Eurasian Economic Commission – has little power to influence domestic institutions beyond urging compliance with regulations. In 2017, for example, the commission acknowledged its inability to intervene in the so-called "meat war" between Belarus and Russia and to force the Russian food inspection agency to remove punitive measures on Belarussian meat (Wolczuk and Dragneva 2017, 14). Likewise, the judicial body of the Union – the EAEU court – was deemed an ineffective mechanism for settling disputes by legal experts even before the EAEU came online in 2015 (Checkalov 2014, 12). Unsurprisingly, the Supreme Council, which consists of each head of state, is the highest body of the EAEU, meaning member state leaders retain high levels of "manual control" over the integration process as a whole (Roberts and Moshes 2016, 9).

At a deeper level, the choice of weak institutions and the selective enforcement of integration commitments reflect the underlying relationship between regime security and policy coordination. For example, while the problem of non-tariff barriers is not unique to the post-Soviet space or other regional integration projects, there are fears among EAEU member states that too much integration will mobilize domestic opposition and increase the political and economic leverage of partner states to an unacceptable level. This is particularly so for Russia's smaller partner states, where leaders understand the advantages of close cooperation with their larger neighbor but also the risks involved.

Every post-Soviet state has economic relations with Russia and so some degree of interdependence, typically asymmetrical. The economies of Belarus and Kazakhstan and new EAEU members Armenia and Kyrgyzstan are reliant on Russian energy and energy providers in various forms, including the supply of oil and gas and transit of oil and gas, but also the operation of electricity and nuclear power sectors. In addition, most states in the region rely on access to the Russian market, not least in the form of worker remittances sent back from Russia. But for both Belarus and Kazakhstan, there are specific regime security issues at stake.

For Belarus, relations with Russia continue to operate on two levels. At an official level, Belarus and Russia are the closest-aligned states in the post-Soviet space; an alignment underpinned by the 1999 bilateral Union State Treaty and joint membership of a host of multilateral regional organizations. However, both states have a long history of coordination problems, including numerous oil and gas pricing conflicts and a host of other disagreements both before and after the official unveiling of the EAEU in 2015. In 2016, Moscow and Minsk clashed over plans to build a Russian airbase in Belarus and over the latter's introduction of border regulations affecting Russians traveling from the Kaliningrad enclave. Although Vladimir Putin identified Belarus as Russia's closest "strategic partner" during a meeting with the Belarus President in July 2015 (Kremlin 2015) at an unofficial level, Aleksandr Lukashenko is viewed by Russia as a "situational partner" – unreliable and willing to support Russian initiatives, but only if the price is right.

However, the ability of the Lukashenko regime to extract concessions from Russia is tempered by the fact that the regime's existing limited leverage will diminish if the Belarus economy is liberalized, as envisaged under the aegis of the EAEU. There is an acknowledgment within Belarus that reform is dangerous, not least because Russia is best placed to take advantage. Russia has greater bureaucratic resources and expertise than any other member state, which can be used to lobby and determine the path of integration. Moreover, it is Russian companies and Russian capital that are best poised to take advantage of new openings in neighboring markets if and when liberalizing reforms happen.

In terms of regime security, there is a fear that Russian criminal organizations or the Russian state may use economic reforms in Belarus to take a controlling stake in key industries, possibly through manipulating shareholders (Interview, Belarus, 2014) and from this position attempt to remove Lukashenko and replace him with a more malleable partner. Lukashenko himself has repeatedly claimed that Russia is seeking to undermine him, even going so far as to identify Russia as an existential threat to Belarus, alongside NATO (Moscow Times 2014). Following yet another gas pricing dispute in May 2016, Russian experts did not rule out a Moscow-backed attempt to remove Lukashenko in the near future (Hodasevich 2016a).

In comparison to Belarus, Kazakhstan's economic reliance on Russia is significantly less, but as interviewees noted, similar concerns exist that liberalizing reforms will see Russian companies overrun domestic competition (Interview, Kazakhstan 2014). Moreover, Russia's increasing unpredictability as a partner is also creating concerns. In economic terms, Moscow's unilateral decision to impose counter-sanctions in 2014 dismayed many within Kazakhstan, as well as figures within the Eurasian Economic Commission (Interview, Eurasian Commission, 2016). In foreign policy terms, there is concern that Russia is pulling Kazakhstan into an isolationist project, leading some to question the motivations behind Russia's integration drive. As such, Russia's motives are increasingly viewed as geopolitical, in sharp contrast to Astana's stringent desire for a "de-politicized economic-only project" (Interview, Kazakhstan 2014).

However, a qualitatively different issue for Kazakhstan is Russia's unpredictable use of force. The annexation of Crimea in 2014 and subsequent emergence of Putin's "Crimea Doctrine" (Putin 2014b), and the readiness to intervene beyond the territory of the Russian Federation to defend ethnic Russians is a significant development. The combination of a large ethnic Russian population in the north of Kazakhstan and uncertainty concerning the post-Nazarbaev power succession opens up the possibility of Russian intervention in the future. While rarely mentioned in high-level statements within Kazakhstan, this hypothetical situation is acknowledged in Russian circles as a likely source of consternation in Kazakhstan, as well as other neighboring states (interview, Russia, 2014). For some, there is little doubt that any political instability in Kazakhstan that threatens Moscow's interests will be met with military intervention: "Russia will use force in Kazakhstan as a last resort to restore order" (Interview, Russia, 2016).

In addition, for both Belarus and Kazakhstan, there are clear indications that integration must be balanced with external security concerns but also internal tensions. In Belarus, attempts by the government to enforce EAEU regulations that small traders selling imported goods provide details of their origin have already resulted in protests. The so-called edict 222 came into force in January 2016 and resulted in several localized demonstrations in February and March 2016, with protestors and the opposition United Civic Party demanding the government's dismissal. Although relatively small in size (they numbered several hundred), this law may affect up to 120,000 small businesses (Marples 2016), but more importantly these protests were explicitly anti-Eurasian Union in nature (Hodasevich 2016b) and provide an early indication of the problems of enforcing integration commitments.

In Kazakhstan, moves to amend a law allowing foreign citizens to own land sparked widespread protests in April 2016, with tens of thousands taking to the streets. Eventually, Kazakhstan's leadership was forced to introduce a moratorium in August 2016 (RIA Novosti 2016). This comes against a backdrop of persistent negative media commentary on the EAEU within Kazakhstan (Likhachev 2015) and a growing dissatisfaction among business elites in the country at the poor results of integration to date. Throughout 2015, the government came under sustained (and successful) pressure from domestic producers to limit Russian imports. The pro-business "loyal" opposition party, Ak Zhol, which supports protectionism, has repeatedly cautioned against Kazakhstan's involvement in the EAEU and any temptation to sacrifice the nation's sovereignty and independence.

Regime identity and Eurasian economic integration

What the cases of Belarus and Kazakhstan show is the delicate balance between the costs and benefits of policy coordination in political systems that lack the institutional means to absorb and accommodate popular dissatisfaction. In line with the regime security framework, both cases also reveal the ideational challenges that authoritarian cooperation poses, in particular, sustained policy coordination and sovereignty pooling, when the adjustments needed to meet commitments to partner states threatens to undermine the regime identity that ruling groups have cultivated over a long period of rule.

In the case of the post-Soviet space, the breakup of the Soviet Union initiated a process of nation building in newly independent states, a process in which ruling groups became embedded first as initiators and then as defenders of an emerging national identity (Fawn 2004, 20). As such, for many post-Soviet states, regime identity is closely entwined with national identity, and with one or two notable exceptions (Belarus) this identity is based around ethnicity. In these terms, the resilience of many authoritarian regimes in the region is based on the (continuing) support of a core ethnic group which confers additional legitimacy on rulers, who promote themselves as "defenders" of particularistic group interests.

One exception is Belarus. Unlike most post-Soviet states, Belarusian national identity is weakly developed, although both Russia and Europe are often cast as significant "Others" (White and Feklyunina 2014, 163, 165), with the regime shifting the state's official discourse between the two. In the place of nationalism, long-standing leader Aleksandr Lukashenko has built regime identity around a commitment to a socially oriented state that sees relatively high levels of social welfare (albeit low quality) compared to Russia and Kazakhstan (Cook 2007) and relatively high levels of state ownership in the economy.

For Belarus, and for Lukashenko in particular, any economic reform must be gradual in order to preserve the regime's material control over the state but also the regime's credentials as a defender of a particular way of life that has changed little since 1991. In order for Lukashenko to retain power, he must carefully coordinate his policies in accordance with the preferences of his core electorate and supress market forces. As Zlotnikov notes, it would be difficult for Lukashenko to change his negative attitude toward the private sector because it would go against his own convictions, but also that of his electoral base (Zlotnikov 2004, 138).

However, this places Belarus in a paradoxical situation in the context of EAEU and the requirements of sustained policy coordination and economic integration. There is an open acknowledgment within the country and within the other EAEU member states that the Belarus economy lags in terms of levels of economic liberalization. In 2011, following the Belarus financial crisis, loans from the EDB were conditioned on economic reform, including privatization, but so far the Belarus leadership has continued to stall. The government's social and economic plan for the period 2016–2020 promises to "modernize property relations" (Belarus 2016) but is at odds with other high-level statements on economic development. Lukashenko's address to the People's Assembly in July 2016 ruled out any deviation from the established "gradualist" approach to modernization (Lukashenko 2016). Within Belarus there is an opinion that the country will never meet all its integration commitments: "Belarus will never adopt the economic rules and the principles of liberalism which prevail in Russia" (Interview Belarus 2014).

In Kazakhstan, under the long-standing leadership of Nursultan Nazarbaev, the regime has built an identity around the values of sovereignty and political and economic independence but underpinned by support for ethnic Kazakh nationalism. So, while Nazarbaev has been a consistent champion of Eurasian integration as a necessary tool to modernize the country, the regime's ethnonationalism implies a circumscribed limit to the extent of Kazakhstan's relations with Russia. Among the more controversial moves the regime has made in recent years include a program for increasing the number of Kazakh-speakers in the country to 95% by 2025 as well as a plan to "modernize" the Kazakh alphabet by abandoning the existing Cyrillic in favor of a Latin script (Roberts and Moshes 2016, 13). In addition, there are indications that identity is hindering foreign policy coordination with Moscow and contributing to "civilizational conflicts" as the regime gravitates toward ethnically similar Turkey and Azerbaijan (Galstyan 2017).

The problem for the ruling group in Kazakhstan is that Eurasian integration is increasingly challenging the regime's identity as a defender of the key values of sovereignty and independence, as well as the interests of ethnic Kazakhs. While the focus of land reform protests in April 2016 carried a distinct anti-Chinese character (Solov'eva 2016), there exists a significant anti-Russian sentiment among Kazakh nationalists (Interview, Kazakhstan, 2014). This sentiment could create problems for the regime if integration is perceived to disproportionally benefit Russia – as indeed the current media discourse within the country attests. Available opinion-poll data from 2014 just prior to the formation of the EAEU showed that ethnic Kazakhs (13.1%) were almost three times more likely to oppose integration than ethnic Russians (4.9%) living in Kazakhstan (KISI 2014, 47). In terms of regime security, failure to protect the right flank and to defend the values of sovereignty and the interests of ethnic Kazakhs opens opportunities for challengers to appropriate the ideological space vacated by the regime as defenders of the nation's key constituencies and values.

Conclusions

This article contributes to the emerging literature on the international dimensions of authoritarian rule, but also the nascent Eurasian Economic Union (EAEU). The argument forwarded in this article is that concerns over regime security create antagonistic cooperation drivers. In the case of the EAEU, regime security offers some explanation for cooperation between member states while providing a stronger explanation for their inability to coordinate policy. Rather than united in the perception of a common threat, EAEU member states are busy balancing multiple challenges to regime security, including those resulting from cooperation with their authoritarian partners. At the same time, rather than offer a problem-free route to "regime upgrading" or modernization, ambitious cooperation projects, notably those involving economic integration, appear to contain tipping points and diminishing regime security returns. Despite the appearance of the ambitious EAEU, this article confirms the older regionalism literature that highlighted the poor results of post-Soviet integration since 1991.

In terms of better understanding the dynamics of authoritarian cooperation, there are a number of avenues that present themselves. The first is the difficult task of specifying tipping points in authoritarian cooperation. A second, related question is the compatibility of authoritarianism with any form of economic integration involving liberalizing reforms. There is an argument that economic liberalism helps produce democracy by separating economic and political power (Ulfelder 2008, 274). Economic integration is also assumed to go "hand in hand with the development of a democratic identity" (Van der Vleuten and Hoffmann 2010, 739). From these perspectives, it is questionable if anything but token economic integration is compatible with authoritarian resilience.

A third and potentially more intriguing research avenue relates to authoritarian strategies for balancing internal and external pressures within regional organizations. This may involve a closer consideration of the malleability of regime identity in the hands of long-serving autocrats and the possibility that regimes may develop multiple identities to ease integration concerns. Cross-regional research would be best suited for observing the range of strategies designed to limit the impact of integration commitments while still extracting benefits. As in the case of the EAEU, this may include the use of non-tariff barriers, the creation of weak supra-national institutions, and general "feet-dragging." Another feature of the EAEU is the prevalence of individual state opt-outs and multi-speed integration that typically involve concessions in certain areas (e.g. regarding the Union's expansion) in order to buy time in others. In addition, the whole issue of multi-vectored foreign policies and an emerging discourse surrounding "big Eurasia," including open regionalism, in particular in Belarus and Kazakhstan, suggest a mild attempt or strategy of "soft balancing" Russia. Examining these strategies and others in the context of regional integration, both in and beyond the post-Soviet space, would serve to deepen our understanding of the dynamics of authoritarian cooperation.

Acknowledgments

The author would like to thank the Finnish Institute of International Affairs and the Finnish Ministry for Foreign Affairs for making this research possible. In particular, the author would like to thank Arkady Moshes; Katri Pynnöniemi, and Anaïs Marin for their contribution to the initial research conducted in 2014. All views expressed in this article are those of the author.

Disclosure statement

No potential conflict of interest was reported by the author.

References

Adomeit, Hannes. 2012. "Putin's Eurasian Union: Russia's Integration Project and Polices on Post-Soviet Space". *Neighborhood Policy Paper*. Zurich: Center for International and European Studies.

Allison, Roy. 2004. "Regionalism, Regional Structures, and Security Management in Central Asia." *International Affairs* 80: 463–483.

Allison, Roy. 2008. "Virtual Regionalism, Regional Structures and Regime Security in Central Asia." *Central Asian Survey* 27: 185–202.

Ambrosio, Thomas. 2006. "The Non-material Cost of Bandwagoning: The Yugoslav Crisis and the Transformation of Russian Security Policy." *Contemporary Security Policy* 27: 258–281.

Ambrosio, Thomas. 2009. *Authoritarian Backlash: Russian Resistance to Democratization in the Former Soviet Union*. Ashgate: Farnham.

Bader, Julia, Jörn Grävingholt, and Antje Kästner. 2010. "Would Autocracies Promote Autocracy? A Political Economy Perspective on Regime-type Export in Regional Neighbourhoods." *Contemporary Politics* 16: 81–100.

Belarus. 2016. "Proekt programmy social'no-ekonomicheskogo razvitiya Respubliki Belarus' na 2016–2020 gody." [The Socio-economic Development Programme of the Republic of Belarus, 2016–2020.] Accessed 2 July 2017. http://www.president.gov.by/ru/sobranie/

Bentzen, Naja, and Christian Dietrich. 2016. "Belarus: A Repressed Economy." At a Glance, European Parliament. Accessed 2 July 2017. http://www.europarl.europa.eu/thinktank/en/search.html?keywords=001345

Benzow, Gregg. 2011. "EU Imposes Sanctions against Belarus." *DW News*. Accessed 2 July 2017. http://www.dw.com/en/eu-imposes-sanctions-against-belarus/a-14807433

Bohr, Annette. 2004. "Regionalism in Central Asia: New Geopolitics, Old Regional Order." *International Affairs* 80: 485–502.

Börzel, Tanja, Vera van Hüllen and Mathis Lohaus. 2013. "Governance Transfer by Regional Organizations: Following a Global Script?" SFB Governance Working Paper Series 43. Accessed 2 July 2017. http://edoc.vifapol.de/opus/volltexte/2015/5726/

BTI. 2014. *BTI Belarus Country Report*. Accessed 2 July 2017. http://www.bti-project.org/fileadmin/files/BTI/Downloads/Reports/2014/pdf/BTI_2014_Belarus.pdf

Buzan, Barry. 2003. "Regional Security Complex Theory in the Post-cold War World." In *Theories of New Regionalism*, edited by F. Soderbaum and T. Shaw, 140–159. New York: Palgrave Macmillan.

Checkalov, Dmitriy. 2014. "Poryadok Razresheniya Sporov v Evraziiskom Ekonomicheskom Soyuze." *Integrites Law Firm Kazakhstan* Accessed 2 November 2017. http://www.integrites.com/ru/publication/792

Collins, Kathleen. 2009. "Economic and Security Regionalism among Patrimonial Authoritarian Regimes: The Case of Central Asia." *Europe-Asia Studies* 61: 249–281.

Cook, Linda. 2007. *Postcommunist Welfare States: Reform Politics in Russia and Eastern Europe.* Ithaca, NY: Cornell University Press.

Dragneva, Rilka, and Kataryna Wolczuk. 2012. "Commitment, Asymmetry, and Flexibility: Making Sense of Eurasian Economic Integration." In *Eurasian Economic Integration: Law, Policy, and Politics*, edited by R. Dragneva and K. Wolczuk, 204–221. Cheltenham: Edward Elgar.

Ehteshami, Anoushiravan, Raymond Hinnebusch, Heidi Huuhtanen, Paola Raunio, Maaike Warnaar, and Tina Zintl. 2013. "Authoritarian Linkage and International Linkages in Iran and Syria." In *Middle East Authoritarianisms*, edited by S. Heydemann and R. Leenders, 222–242. Stanford: Stanford University Press.

Eurasian Commission. 2017. "Vneshnyaya i vzaimnaya torgovlya tovarami Evraziiskogo ekonomicheskogo soyuza." [External and Mutual Trade in Goods by the EAEU.] Accessed 2 July 2017. http://eec.eaeunion.org/ru/act/integr_i_makroec/dep_stat/tradestat/time_series/Pages/default.aspx

Fawn, Rick. 2004. "Ideology and National Identity in Post-communist Foreign Polices." In *Ideology and National Identity in Post-communist Foreign Polices*, edited by R. Fawn, 1–42. London: Frank Cass.

Freedom House. 2016. "Eurasia." Accessed 2 July 2017. https://freedomhouse.org/regions/eurasia

FT (Financial Times). 2012. "Clinton Vows to Thwart New Soviet Union." *The Financial Times* Accessed 2 July 2017. https://www.ft.com/content/a5b15b14-3fcf-11e2-9f71-00144feabdc0

Galstyan, Areg. 2017. "Is the Eurasian Economic Union Slowly Coming Apart?" *The National Interest*. Accessed 2 November 2017. http://nationalinterest.org/feature/the-eurasian-economic-union-slowing-coming-apart-19947

Geddes, Barbara. 2003. *Paradigms and Sand Castles: Theory Building and Research Design in Comparative Politics*. Ann Arbor, MI: University of Michigan Press.

Gvalia, Giorgi, David Siroky, Bidzina Lebanidze, and Zurab Iashvili. 2013. "Thinking outside the Bloc: Explaining the Foreign Policies of Small States." *Security Studies* 22: 98–131.

Haas, Ernst. 1966. *International Political Communities*. New York: Anchor Books.

Hale, Henry. 2008. *The Foundation of Ethnic Politics: Separatism of States and Nations in Eurasia and the World*. Cambridge: Cambridge University Press.

Hodasevich, Anton. 2016. "Lukashenko davit na Moskvu." [Lukashenko Pressures Moscow.] *Nevazimaya gazeta*. Accessed 2 July 2017. http://www.ng.ru/cis/2016-05-16/6_lukashenko.html

Hodasevich, Anton. 2016. "Biznes obeshchaet Lukashenko bol'shie problemy: Predprinimateli Belorussii trebuyut otstavki pravitel'stva." [Business Promises Lukashenko a Big Problem: Entrepreneurs Demand the Government's Resignation.] *Nevazimaya Gazeta*. Accessed 2 July 2017. http://www.ng.ru/cis/2016-02-29/1_lukashenko.html

Huntington, Samuel, and Clement Moore. 1970. "Conclusion: Authoritarianism, Democracy and One-party Politics." In *Authoritarian Politics in Modern Society: The Dynamics of Established One-party Systems*, edited by S. Huntington and C. Moore, 509–517. New York: Basic Books.

Interfax. 2014. "Ves' ob"em eksportnykh poshlin ot prodannykh RB v 2015 godu nefteproduktov poidet v belorusskii byudzhet." [All Export Duties on the Sale of Belarus Oil Products in 2015 to Go to the Belarus Budget.] Accessed 2 July 2017. http://www.interfax.by/news/belarus/1168549

Jackson, Richard. 2013. "Regime Security." In *Contemporary Security Studies*, edited by A. Collins, 161–175. Oxford: Oxford University Press.

Keohane, R. 2005. *After Hegemony: Cooperation and Discord in the World Political Economy*. Princeton, NJ: Princeton University Press.

Keukeleire, Stephan, and Irina Petrova. 2016. "The European Union, the Eastern Neighbourhood and Russia: Competing Regionalisms." In *European Union and New Regionalism*, edited by M. Telo, 263–279. New York: Routledge.

KISI. 2014. *Vospriyatie grazhdanami voprosov evraziiskoi integratsii i uchastiya kazakhstana v evraziiskom soyuze* [Citizen Perceptions toward the Question of Eurasian Integration and the Participation of Kazakhstan in the EAEU]. Astana: Kazakhstanskii institut strategicheskikh issledovanii pri Prezidente Respubliki Kazahstana [The Kazakhstan Institute for Strategic Research under the President of the Republic of Kazakhstan].

Kobrinskaya, Irina. 2007. "The Post-Soviet Space: From the USSR to the Commonwealth of Independent States and Beyond." In *The CIS, the EU and Russia*, edited by K. Malfliet, L. Verpoest and E. Vinokurov, 13–21. Hampshire: PalgraveMacmillan.

Kopstein, Jeffrey, and David Reilly. 2000. "Geographic Diffusion and the Transformation of the Postcommunist World." *World Politics* 53: 1–37.

Kremlin. 2015. *Vstrecha s Prezidentom Belorussii Aleksandrom Lukashenko* [Meeting with the Belarus President Aleksandr Lukashenko]. Accessed 2 July 2017. http://kremlin.ru/events/president/news/49888

gov.Kz. 2016. "Kazakhstan vvel vremennye ogranicheniya na postavku krupno-rogatogo skota iz Rossiiskoi Federatsii." [Kazakhstan Introduced Temporary Restrictions on the Supply of Cattle to Russia.] Accessed 2 July 2017. http://mgov.kz/ru/aza-stan-respublikasy-resej-federatsiyasynan-iri-ara-mal-kirgizuine-ua-ytsha-shekteuler-engizdi/.

Lane, David. 2015. "Introduction: Eurasian Integration as a Response to Neo-liberal Globalization." In *The Eurasian Project and Europe*, edited by D. Lane and V. Samokhvalov, 3–22. New York: Palgrave Macmillan.

Libman, Aleksandr. 2015. "Supranational Organizations: Russian and the Eurasian Economic Union." In *Autocratic and Democratic External Influences in Post-Soviet Eurasia*, edited by A. Obydenkova and A. Libman, 133–158. Ashgate: Farnham.

Libman, Aleksandr, and Evgeny Vinokurov. 2012. *Holding-together Regionalism: Twenty Years of Post-soviet Integration*. Hampshire: Palgrave Macmillan.

Likhachev, Maksim. 2015. "Obraz EAES v mediaprostranstve Kazakhstana, Rossiiskii institut strategicheskikh issledovanii." [The EAEU in the Kazakhstani Media.] Accessed 2 July 2017. https://riss.ru/analitycs/22235/

Lipman, Maria. 2015. "Putin's Enemy Within: Demonising the 'Fifth Column.'" *European Council on Foreign Relations*. Accessed 2 July 2017. http://www.ecfr.eu/article/commentary_putins_enemy_within_demonising_the_fifth_column311513

Lukashenko, Aleksandr. 2012. "Lukashenko i Putin dogovorilis' sovmestno borot'sya protiv sanktsii Zapada." [Lukashenko and Putin Agreed on a Join Fight against Western Sanctions] Accessed 2 July 2017. http://telegraf.by/2012/06/lukashenko-i-putin-dogovorilis-sovmestno-borotsya-protiv-sankcii-zapada

Lukashenko, Aleksandr. 2016. "Doklad Prezidenta Belarusi na Pyatom Vsebelorusskom Narodnom Sobranii." [Report of the President.] Accessed 2 July 2017. http://president.gov.by/ru/news_ru/view/uchastie-v-pjatom-vsebelorusskom-narodnom-sobranii-13867/

Marples, David. 2016. "What is at Stake in the Small Traders Protests?" *Belarus Digest*. Accessed 2 July 2017. http://belarusdigest.com/story/what-stake-small-traders-protests-24292

Mattli, Walter. 1999. *The Logic of Regional Integration*. New York: Cambridge University Press.

Miller, Eric. 2006. *To Balance or Not to Balance: Alignment Theory and the Commonwealth of Independent States*. Aldershot: Ashgate.

Miller, Eric, and Arkady Toritsyn. 2005. "Bringing the Leader Back in: Internal Threats and Alignment Theory in the Commonwealth of Independent States." *Security Studies* 14: 325–363.

Milner, Helen. 1997. *Interests, Institutions and Information*. Princeton, NJ: Princeton University Press.

Moravcsik, Andrew. 1997. "Taking Preferences Seriously: A Liberal Theory of International Politics." *International Organization* 51: 513–553.

Moravcsik, Andrew. 1998. *The Choice for Europe*. Ithaca, NY: Cornell University Press.

Moscow Times. 2014. "Belarus' Lukashenko: Russia's Behaviour Arouses Suspicion." Accessed 2 July 2017. http://www.themoscowtimes.com/news/article/belarus-lukashenko-says-russia-s-behavior-arouses-suspicion/513476.html

Neumann, Iver. 2003. "A Region-building Approach." In *Theories of New Regionalism*, edited by F. Soderbaum and T. Shaw, 160–178. New York: Palgrave Macmillan.

Nye, Joseph. 1968. "Central American Regional Integration." In *International Regionalism: Readings*, edited by J. Nye, 377–427. Boston, MA: Little, Brown and Company.

Obydenkova, Anastassia, and Aleksandr Libman. 2015. "Modern External Influences and the Multilevel Regime Transition: Theory Building." In *Autocratic and Democratic External Influences in Post-Soviet Eurasia*, edited by A. Obydenkova and A. Libman, 7–47. Ashgate: Farnham.

Olcott, Martha, Anders Åslund, and Sherman Garnett. 1999. *Getting It Wrong: Regional Cooperation and the Commonwealth of Independent States*, Washington, DC: The Brookings Institution Press.

Pevehouse, Jon. 2002. "Democracy from the Outside-in?-International Organizations and Democratization." *International Organization* 56: 515–549.

Priess, David. 1996. "Balance-of-threat Theory and the Genesis of the Gulf Cooperation Council." *Security Studies* 5: 143–171.

Putin, Vladimir. 2011. "Novyi integratsionnyi proekt dlya Evrazii – budyshchee, kotoroe rozhdaetsya segodnya." [a New Integration Project for Eurasia-the Future Which is Being Conceived Today.] *Izvestiya*, October 3. Accessed 2 July 2017. http://izvestia.ru/news/502761

Putin, Vladimir. 2014a. "Soveshchanie s chlenami pravitel'stva. 12 fevralya 2014 goda, Moskovskaya oblast', Novo-Ogarevo." [Meeting with Members of the Government.] *Kremlin. Ru*. Accessed 2 July 2017. http://special.kremlin.ru/events/president/news/20217

Putin, Vladimir. 2014b. "Obrashchenie Prezidenta Rossiiskoi Federatsii." [Address of the President to the Russian Federation.] *Kremlin.Ru*. Accessed 2 December 2017. http://www.kremlin.ru/transcripts/20603

Putnam, Robert. 1988. "Diplomacy and Domestic Politics: The Logic of Two-level Games." *International Organization* 42: 427–460.

RIA Novosti. 2016. "V Kazakhstane prodlili moratorii na popravki v Zemel'nyi kodeks." [The Moratorium on Changes to the Land Code Continue in Kazakhstan.] Accessed 2 July 2017. https://ria.ru/world/20160819/1474754617.html

Roberts, Sean, and Arkady Moshes. 2016. "The Eurasian Economic Union: A Case of Reproductive Integration?" *Post-Soviet Affairs* 32: 542–565.

Rodeheffer, Luke. 2015. "Is a Trade War Brewing between Russia and Kazakhstan? Global Risks Insights." Accessed 2 July 2017. http://globalriskinsights.com/2015/06/is-a-trade-war-brewing-between-russia-and-kazakhstan/

Rosset, Damian, and David Svarin. 2014. "The Constraints of the past and the Failure of Central Asian Regionalism, 1991–2004." *Region: Regional Studies of Russia, Eastern Europe, and Central Asia* 3: 245–266.

Savchenko, I. 2015. "Kazakhstan Rejects Statements of the UN, EU and OSCE on Violations of Human Rights." *Open Dialogue*. Accessed 2 July 2017. http://en.odfoundation.eu/a/7073,kazakhstan-rejects-statements-of-the-un-eu-and-osce-on-violations-of-human-rights

Schimmelfennig, Frank, and Ulrich Sedelmeier. 2005. "Introduction: Conceptualizing the Europeanization of Central and Eastern Europe." In *The Europeanization of Central and Eastern*

Europe, edited by F. Schimmelfennig and U. Sedelmeier, 1–28. Ithaca, NY: Cornell University Press.

Silitski, Vitali. 2010. "Survival of the Fittest: Domestic and International Dimensions of the Authoritarian Reaction in the Former Soviet Union following the Colored Revolutions." *Communist and Post-Communist Studies* 43: 339–350.

Soderbaum, Frank, and Thomas Shaw. 2003. *Theories of New Regionalism*. New York: Palgrave Macmillan.

von Soest, Christian. 2015. "Democracy Prevention: The International Collaboration of Authoritarian Regimes." *European Journal of Political Research* 54: 623–638.

Solov'eva, Olga. 2016. "Moskva i Astana podelili krizis: Za poslednie dva goda ekonomiki Rossii i Kazakhstana sokratilis' v dollarovom izmerenii vdvoe." [Moscow and Astana Share a Crisis.] Accessed 2 July 2017. http://www.ng.ru/economics/2016-05-31/4_astana.html

Tansey, Oisn. 2016. *The International Politics of Authoritarian Rule*. Oxford: Oxford University Press.

Ter-Matevosyan, V., A. Drnoian, N. Mkrtchyan, and T. Yepremyan. 2017. "Armenia in the Eurasian Economic Union: Reasons for Joining and Its Consequences." *Eurasian Geography and Economics* 58: 340–360.

Ulfelder, Jay. 2008. "International Integration and Democratization: An Event History Analysis." *Democratization* 15: 272–296.

Van der Vleuten, Anna, and Andrea Hoffmann. 2010. "Explaining the Enforcement of Democracy by Regional Organizations: Comparing EU, Mercosur, and SADC." *JCMS: Journal of Common Market Studies* 48: 737–758.

Vinokurov, Evgeni. 2007. "Russian Approaches to Integration in the Post-Soviet Space in the 2000s." In *The CIS, the EU and Russia*, edited by K. Malfliet, L. Verpoest and E. Vinokurov, 22–46. Hampshire: Palgrave Macmillan.

Vinokurov, Evgeni. 2017. "Eurasian Economic Union: Current State and Preliminary Results." *Russian Journal of Economics* 3: 54–70.

Vinokurov, Evgeni, and Aleksandr Libman. 2012. *Eurasian Integration: Challenges of Transcontinental Regionalism*. Hampshire: Palgrave Macmillan.

Walt, Stephen. 1987. *The Origins of Alliances*. Ithaca, NY: Cornell University Press.

Waltz, Kenneth. 1979. *Theory of International Politics*. Long Grove, IL: Waveland Press.

White, Stephen, and Valentina Feklyunina. 2014. *Identities and Foreign Policies in Russia, Ukraine and Belarus*. New York: Palgrave Macmillan.

Whitehead, Laurence. 2014. "Anti-democracy Promotion: Four Strategies in Search of a Framework." *Taiwan Journal of Democracy* 10: 1–24.

Wolczuk, Kataryna, and Rilka Dragneva. 2017. "The Eurasian Economic Union: Deals, Rules and the Exercise of Power. Chatham House Research Paper." Accessed 2 November 2017. https://www.chathamhouse.org/publication/eurasian-economic-union-deals-rules-and-exercise-power

World Bank. 2017a. *EU*. Accessed 2 July 2017. http://data.worldbank.org/region/european-union

World Bank. 2017b. "*China*." Accessed 2 July 2017. http://data.worldbank.org/country/china

Zlotnikov, L. 2004. "In the Noose of Populism: Eleven Years of the Belarusian Economic Model (1991–2001)." In *The EU and Belarus: Between Moscow and Brussels*, edited by A. Lewis, 127–155. London: The Federal Trust for Education and Research.

Appendix 1. Interviews (alphabetical)

Akberdin, Rustam. Director of the Department for Development Entrepreneurship, Eurasian Economic Commission, (headquarters in Moscow).

Amrebayev, Aidar. Head of the First Kazakhstani President Centre, Institute of World Economics and Politics (IWEP), Kazakhstan.

Askarov, Tulegen. President of the Centre for Business Journalism BizMedia, Kazakhstan.

Bakenov, Ernar. Director of the Department of International Economic Integration, Kazakhstan.

Brodov, Roman. Head of Foreign Economic Policy Division, Ministry for the Economy, Belarus.

Busko, Vitaly. Member of the House of Representatives, Deputy Chairman of the Standing Commission on International Affairs, Belarus.

Evseev, Vladimir. Deputy Director, The CIS Institute, Russia.

Filippov, Alexander. Head of the Youth Affairs Department, Belarusian State University of Culture and Arts; previously expert with the Information-Analytical Centre of the Administration of the President (2010–2013), Belarus.

Guryanov, Alexander. Deputy-Minister for Foreign Affairs, Belarus.

Kariagin, Vladimir. Chairman of Belarusian Republican Confederation of Entrepreneurship, Belarus.

Karimsakov, Murat. President of the Eurasian Economic Club of Scientists, Kazakhstan.

Kassenova, Nargis. Director of Central Asian Studies, KIMEP University, Kazakhstan

Kishkembayev, Askar. Head of the Secretariat of the Minister for Economy and Financial Policy, Eurasian Economic Commission, (headquarters in Moscow).

Knobel, Aleksandr Laboratory of International Trade, Gaidar Institute for Economic Policy, Russia.

Kortunov, Andrei. President of the New Eurasia foundation, Russia.

Kozhakov, Asan. Ambassador-at-Large, Ministry of Foreign Affairs, Kazakhstan.

Kravchenko, Valentina. Deputy Director of the Department for Financial Policy, Eurasian Economic Commission, (headquarters in Moscow).

Krishtapovich, Lev. Deputy Director of the Information-Analytical Center of the Administration of the President, Belarus.

Mukhamedjanova, Darya. Chief Research Fellow, Economic Studies Department, Institute for Strategic Studies, Kazakhstan.

Nursha, Askar. Coordinator of Projects on Foreign Policy Issues, Institute of World Economics and Politics (IWEP), Kazakhstan.

Polyanski, Dmitry. Deputy Director, First Department of CIS countries, Ministry for Foreign Affairs, Russia.

Postnikova, Natalya. Senior Research Fellow, Gaidar Institute for Economic Policy, Russia.

Rakhmatullina, Gulnar. Minister for Economic and Financial Policy, Eurasian Economic Commission, (headquarters in Moscow).

Rusakovich, Andrei. Head of the Foreign and Security Policy Studies Centre, Belarus.

Sankubayev, Amirbek. Head of the Financial Market Division, Eurasian Economic Commission, (headquarters in Moscow).

Satpayev, Dosym. Director of the Risk Assessment Group, Kazakhstan.

Sultangalieva, Alma. Advisor to the Director, Institute of World Economics and Politics (IWEP), Kazakhstan.

Survillo, Vitali. Vice President of All-Russia Public Organization Delovaya Rossiya, Russia.

Ulakhovich, Vladimir. Deputy Chair of the Chamber of Trade and Commerce, Belarus.

Volchkova, Natalya. Lead Economist, Centre for Economic and Financial Research, Russia.

The US Silk Road: geopolitical imaginary or the repackaging of strategic interests?

Marlene Laruelle

Central Asia appears a highly fertile region for producing inflated imaginaries aimed at both domestic and external actors. Since the 1990s and more openly even since 2011, official Washington embraced the evocative and romantic concept of the Silk Road in formulating US policy for Central Asia. This article uses the critical geopolitics approach to understand US foreign policy assumptions and projections about post-Soviet Central Asia and its broader environment. I argue that the US version of the Silk Road can be interpreted as a geopolitical imaginary, in the same vein as Russia's Eurasian narrative. I first situate the discussion by briefly exploring the many uses of the Silk Road allegory by external actors and Russia's rival terminology of Eurasia. Then, I move to analyzing the birth and framing of the US Silk Road narrative, its administrative and policy locus. Finally, I investigate its elusive geopolitics, and its role as a vehicle for the US selective projection of what Central Asia is and should be.

Central Asia[1] appears to be a highly fertile region for producing inflated imaginaries aimed at both domestic and external actors. Myriad historical allegories and geographical metaphors have been used in attempts to define the region, and its global reemergence following the collapse of the Soviet Union makes it an important theater for soft power influence and muscle-flexing capability. The historical Silk Roads, linking China and Southeast Asia to Europe, the Mediterranean world, and the Indian Ocean, disappeared as a key world trade route in the sixteenth century. Since the Russian conquest and even more during Soviet time, Central Asia was almost entirely cut off from its southern and eastern neighbors and entirely turned toward interaction with the northern metropolis of St. Petersburg and Moscow. This "cul-de-sac" location terminated with the Soviet collapse and the revival of close interaction with China and, to a lesser extent, with Afghanistan, Iran, Pakistan, and India.

Since the 1990s and more openly even since 2011, official Washington has embraced the evocative and romantic concept of the Silk Road in formulating its policy for Central Asia. This usage should be evaluated not only in terms of its policy relevance or irrelevance, but also through the perspective of the social sciences. This article uses the critical geopolitics approach to understand US foreign policy assumptions and projections about post-Soviet Central Asia and its broader environment. Critical geopolitics considers geopolitics as a discourse, situated in a specific constellation of worldview perceptions and political culture, not an objective knowledge that would merely translate the "reality" of state interests (Ó Tuathail 1996). In this article, I argue that the US version of the Silk Road can be interpreted as a geopolitical imaginary, in the same vein as Russia's Eurasian narrative. Russia has been studied for years in its promotion of a Eurasianist reading of its relationship to the so-called Near Abroad. China has also entered the realm of Silk Road mythmaking since the early 1990s, and India's foreign policy narratives contain mythological features as well. In this article, I situate the discussion by briefly exploring the many uses of the Silk Road allegory by external actors and Russia's rival terminology of Eurasia. Then, I move to analyzing the birth and framing of the US Silk Road narrative and its administrative and policy locus. Finally, I investigate its elusive geopolitics and its role as a vehicle for the US selective projection of what Central Asia is and should be.[2]

Non-US visions of the Silk Road: competing imaginations and interests

All international actors, not just the United States and Europe, noticed the supposed "revival" of the Silk Roads that followed the collapse of the Soviet Union. Central Asian authorities obviously were among the first to seize on a term that allowed them to quickly and inexpensively establish a branding strategy as their new states made a late entrance onto the international scene. Many state-produced documents and official speeches by senior officials from the region begin by locating Central Asia "at the crossroads" of the Old Continent. Turkey, Iran, and many Asian countries also saw it as a useful instrument to promote their own interests in Central Asia. They incorporated the notion into their official discourse, including specific national characteristics (Ganguli 2011; Fedorenko 2013). A brief overview of this broader use of the Silk Road metaphor helps us encapsulate the polymorphous instrumentalization of the term.

Japan, for example, emphasized the historical role that Central Asia played in the spread of Buddhism from India toward East Asia. In so doing, it presented itself as the country best qualified to show the Central Asian states the path to economic modernization without losing their cultural identity. Tokyo has used the Silk Road metaphor on several occasions, including launching a "Silk Road diplomacy" in 1997 under Prime Minister Hashimoto Ryutaro, establishing a "Silk Road Energy Mission" in 2002 with the hope of becoming more involved in the energy market, and a simpler "Central Asia Plus Japan" strategic dialog in 2004 (Len, Tomohiko, and Tetsuya 2008). South Korea plays with the Silk Road allegory, too, and has argued that the Central Asian region – to which it adds Mongolia – is the possible cradle for the ethnic and linguistic origins of Koreans. This historical and cultural argument is used to strengthen economic ties with Central Asia, especially with Uzbekistan, which has developed a strong partnership with Seoul, complemented by a Soviet-era Korean diaspora that is well-integrated in the region and very active in promoting bilateral relations (Calder and Kim 2008; Evans 2014).

India has its own Silk Road narrative, part of an historical meta-discourse that claims ancient cultural links between the two regions dating to the time of the Avesta and Buddhism and lasting until the Babur Empire in the sixteenth century (Laruelle and Peyrouse 2011; see also Kavalski 2010). The Mughal Empire constitutes the real jewel in the crown of "Silk Road" arguments that the Central Asian states and India advance in order to exalt their age-old relations (Gayer 2010; Kumar 2007, 17; Foltz 2007, 156;. See also Mukhtarov 2003). As Alam and Subrahmanyam (2007, 229) have shown, this

Indo-Central Asian space, in which Iran and a large part of the Ottoman Empire must be included, formed "a single domain of circulation, an ecumene with powerful shared cultural values and symbols." This joint historical past is constantly promoted as the driver of current Indo-Central Asian relations; ironically leaving in the shadow the more recent history of friendly Indo-Soviet relations (Gopal 2005).

China is obviously the elephant – or dragon – in the Silk Road room. China began to invest massively in Central Asia in the early 2000s, with an exponential trade and investment strategy that made it, after the 2008 economic crisis, the region's leading trade partner. In 2012, China's trade with Central Asia reached $46 billion, almost double that of Russia. Chinese state-run enterprises invested in energy – first in Kazakh oil and the Sino-Kazakh pipeline, then in Turkmen gas and the Sino-Central Asia pipeline – but also in road and railway improvements, while private firms and the Xinjiang Production and Constructions Corps (*Xinjiang shengchan jianshe bingtuan* – XPCC) traded in the booming wholesale markets (Laruelle and Peyrouse 2012).

China has used the Silk Road metaphor since the early 1990s (Swanström 2011), but Beijing made it an official policy only in September 2013, repackaging several (until then) dissociated strategies into a highly integrated whole. Chinese President Xi Jinping publicized the launch of the "Silk Road Economic Belt" during his visit to Kazakhstan in September 2013 (Jiao and Yunbi 2013; Marantidou and Cossa 2014). The Belt will follow a lengthy axis going from Xi'an to Lanzhou, Urumqi, Kazakhstan's border, and then, according to the map published by the Xinhua News Agency, crossing Central Asia and Afghanistan and going west through Iran and Turkey to reach the Mediterranean and Europe (Tiezzi 2014). One year later, in fall 2014, Xi Jinping announced that China would contribute $40 billion to set up the Silk Road Fund, which provides investment and financing support to carry out infrastructure, resource, industrial, and financial cooperation. However, the Chinese Silk Road Economic Belt is duplicated by a "Twenty-First Century Maritime Silk Road," which connects China with the Southeast Asian countries, Africa, and Europe via the Indian Ocean.[3] While the US Silk Road strategy (below) focuses only on the revival of continental trade, China is developing a dual, continental and maritime, strategy. This maritime strategy is backed not only by trade but by strategic partnerships to form a "pearl necklace" (in the Chinese government's words) in South and Southeast Asia through the establishment of a series of permanent Chinese military bases to secure energy supplies, such as those in Chittagong in Bangladesh, the Coco Islands in Myanmar, Habantota in Sri Lanka, Marao in the Maldives, and Gwadar in Pakistan (Zajec 2008, 18–19).

As seen from Beijing's perspective, the Silk Road offers another opportunity, related to domestic issues, which few external observers have noticed. The Chinese narrative on the Silk Road downplays the severity of the Uyghur issue by incorporating Xinjiang into a deep-rooted Han history and obscures Islam by highlighting the pre-Islamic periods of the Han and Tang dynasties. It therefore contributes to China's rewriting of history and to its celebration of the alleged historical continuity between ancient and contemporary China. This would obscure the long centuries where the Turkestani world – which includes both Central Asia and Xinjiang – was developing independently from a remote China. James Millward (2009, 65–66) perceptively notes,

> What in the West are celebrated as Silk Road exchanges and interconnectivity are, in China, portrayed rather as evidence that the world is beating a path to China's (once again) open door. Rather than as a transnational bridge between civilizations, the Silk Road here is nationalized as China's doorstep.

Eventually, one must contrast these Silk Road allegories with Moscow's use of the Eurasia metaphor. Russia is the only big external player that does not operationalize the Silk Road narrative to justify its involvement in the region. With its Eurasian terminology, Russia possesses a unique brand that makes implicit sense both with domestic and Central Asian audiences, albeit in different ways depending on the country. Projecting Central Asia as being part of Eurasia moves the region from being one of the centers of the old continent – as in the Silk Road – to being the southern periphery of a geographical entity whose core is constituted by Russia and its Siberian landmass. The notion of Eurasia disconnects Central Asia from its Chinese neighbor and even more from its southern neighbors – Iran, Afghanistan, as well as the Indian subcontinent.

Russia's Eurasian projection on Central Asia is based on a sophisticated toolkit – conceptually, historically, and policy-wise. The term "Eurasia" has been in use since the end of the nineteenth century and was further conceptualized by the Eurasianist interwar movement (Bassin, Glebov, and Laruelle 2015). The Soviet construction repeated, in many aspects, this Eurasian tradition inherited from the Russian tsarist empire and its colonization of Central Asia. The term reached Soviet public opinion in the last year of perestroika though the figure of Lev Gumilev (1912–1992), a semi-dissident historian and ethnographer. Today, Gumilev is considered an untouchable figure in many Russian academic institutions, especially in the national republics, as well as in Kazakhstan. The emergence of a neo-Eurasianist movement in post-Soviet Russia, embodied by the prolific geopolitician Alexander Dugin, blurred the line between a revival of Russian imperial traditions blended with fascist ideology and Eurasianism in its strictest sense. However, many other neo-Eurasianist variants structured themselves among intellectual and political elites in the republics (Tatarstan, Yakutia-Sakha, Buriatia, Tuva, etc.) and preserved the terminology from its exclusive appropriation by the Russian "imperial" narrative (Laruelle 2008). The critical role of Kazakhstan in promoting its own brand of Eurasianism has helped to maintain a large discursive reality. Russian President Vladimir Putin's pet project of building a Eurasian Union[4] gave a new plasticity to the term, making it less ideological and philosophical, more pragmatic and economic-oriented, less purely imperial and Russian-centered, and more open to Central Asia (Dutkiewicz and Sakwa 2015).

Origins and Framing of the US "New Silk Road"

US Secretary of State Hillary Clinton announced the new US Silk Road vision during a speech in Chennai, India, in July 2011:

> Historically, the nations of South and Central Asia were connected to each other and the rest of the continent by a sprawling trading network called the Silk Road. Indian merchants used to trade spices, gems, and textiles, along with ideas and culture, everywhere from the Great Wall of China to the banks of the Bosporus. Let's work together to create a new Silk Road. Not a single thoroughfare like its namesake, but an international web and network of economic and transit connections. That means building more rail lines, highways, energy infrastructure, like the proposed pipeline to run from Turkmenistan, through Afghanistan, through Pakistan into India. It means upgrading the facilities at border crossings, such as India and Pakistan are now doing at Waga. And it certainly means removing the bureaucratic barriers and other impediments to the free flow of goods and people. It means casting aside the outdated trade policies that we all still are living with and adopting new rules for the twenty-first century.

However, US Silk Road terminology was not born from Secretary Clinton's 2011 speech; it has a long history of its own. Following the collapse of the Soviet Union in 1991, Western countries quickly latched onto the Silk Road metaphor and started to speak of an emerging East-West axis that would link the newly independent states to Europe. A symbol of this geopolitical construct was the European and US-backed Transport Corridor Europe-Caucasus-Asia (TRACECA) project, launched in 1993. It aimed to open up Central Asia and the South Caucasus through the creation of a vast transport and communications corridor.[5] TRACECA still functions today as a European assistance project, but it did not impact the main regional trade trends. The United States had its own Silk Road strategy, too: in 1999, and again in 2006, the Congress tried – without success – to target US assistance to the former Eurasian space, specifically to Central Asia and the South Caucasus, through the Silk Road Strategy Act of 1999.

This first US Silk Road conceptualization had several objectives: first to reduce the dependence of the South Caucasus and Central Asia on Russia; second, to prevent Iran from becoming a hub of regional commerce; and third to bolster Turkey's role as a strategic partner of the United States and Europe, as well as a key intermediary between them and the newly independent states. At that time, most think tanks interested in the Central Asian region were also expanding their horizons to the South Caucasus, with the two areas being seen as part of the same policy and research portfolio. The Central Asia-Caucasus Institute and the Jamestown Foundation, both with conservative and openly anti-Russian agendas (Abelson 2006; Tsygankov 2009) tracing back to the 1990s, continue to closely work on both regions even today.

This East-West Silk Road construct became largely outdated in the early 2000s; the much anticipated Central Asia-South Caucasus geopolitical and trade axis failed to materialize, except through the modest GUAM (Georgia-Ukraine-Armenia-Moldova) project, a short-lived anti-Russian alliance. Georgia, Ukraine, Moldova, and Azerbaijan still share an agenda of departure from Russia-led regional integration projects, but they do not invest in joint strategies or institutions. The South Caucasian states' inability to resolve their intraregional tensions led the West to lose patience waiting for a locally built solution. The terrorist attacks of September 11, 2001 shifted attention toward Kabul. International military involvement in Afghanistan at the end of 2001 opened up a window of opportunity for those who wanted to develop a Central Asia–South Asia axis, already envisioned in the 1990s when the now-defunct US firm Unocal took the lead for a gas pipeline project from Turkmenistan to South Asia through Afghanistan (TAPI). This southern reorientation, strongly promoted by Professor S. Frederick Starr, became institutionalized in 2005, with the State Department's creation of a Bureau for South and Central Asian Affairs, while the South Caucasus remained a part of the Bureau of European and Eurasian Affairs.[6] This bureaucratic division played a critical role in pushing the State Department to elaborate a narrative justifying why and how to unite Central Asia and South Asia into one strategic bloc.

Secretary Clinton's announcement of a new US Silk Road strategy in 2011, which explicitly included Central Asia and South Asia and was symbolically proclaimed in India, therefore appears as the continuation and repackaging of the 2005 reorganization of the State Department Bureau. The official credence the secretary of state lent to this second Silk Road concept can be explained through obvious strategic rationales: first, to prepare for the US withdrawal from Afghanistan by leaving a (hopefully) viable development project for Afghanistan's reintegration into the wider region and, second, to confirm the new status of its strategic ally, India, seen as a possible US gateway to the

Indian Ocean and Central Asia (Shukla 2014). Professor Starr again is one of the few advocates still to remind us of the first Silk Road project, envisionning a unified process with two phases: one Silk Road toward the South Caucasus and another toward South Asia (Starr 2011b, 9–10). However, this falls short of an explanation. Neither the State Department nor the think tank community has been able to develop a viable justification of the exact nature of the link between the two Silk Roads and their apparently contradictory logics. Has the first merely disappeared from the agenda or is it considered accomplished? What links the South Caucasus and South Asia, other than shared location adjacent to Central Asia?

Since 2011, the Silk Road has become the main metaphor that US senior officials use to frame their strategy for Central Asia and Afghanistan. In contrast, the metaphor is infrequently employed regarding India and Pakistan, with which Washington has long had deep bilateral ties and does not need any new rhetoric. During his trip to India in June 2013, new Secretary of State John Kerry only briefly mentioned the Silk Road in passing, without any other emphasis (Kerry 2013). In January 2015, President Barack Obama, a guest of honor at India's Republic Day, celebrated the US-India strategic partnership without mentioning the Silk Road strategy at all (Obama 2015).

No US presidential administration has had an official strategy or concept documents entitled "Silk Road," a glaring absence that indicates the lack of importance accorded to the idea at the higher level of the US government. But the Silk Road narrative lies at the core of US diplomatic relations with Central Asia and became an important tool for the State Department's Bureau of South and Central Asian Affairs. However, even the cabinet-level department does not officially consider it to be a full-fledged "strategy," which would require more robust administrative support, budgetary allocations, and personnel. Several high-level officials have tried to downplay the Silk Road narrative, repeating that it is more of a "vision" or "mindset" than a strategy per se (Pyatt et al. 2012). From 2011 to early 2014, the most common definition of the Silk Road to appear on the State Department website was as a "vision" – something like a hope for the future. Since spring 2014, preference seems to be given to defining it as an "initiative,"[7] an upgraded terminology likely related to the draw-down from Afghanistan, even if State Department officials have never clarified the difference between "vision" and "initiative." Only Deputy Assistant Secretary for Central Asia and South Asia Fatema Sumar hinted at the difference, in June 2014, stating that the US Silk Road is "not just a vision. I really want to use the word initiative here, it is a very carefully well-though-of project, a series of projects, very much developed" (Sumar 2014).

The US administration is not the only advocate of the Silk Road metaphor. Several Washington-based think tanks, mostly of conservative orientation but also some bipartisan ones, have embraced the concept, often enticed by grants from US federal agencies or embassies from the Central Asian region to deepen the conceptual apparatus of the Silk Road. Publications from the Central Asia-Caucasus Institute (CACI) at the Johns Hopkins University School of Advanced International Studies have led the think-tank output, with its founder and chairman, S. Frederick Starr, being the most prominent and most vocal herald of it since the 1990s. Other institutions active in this regard include the Jamestown Foundation, the Center for Strategic and International Studies, the Atlantic Council, and the Rethink Institute.

Lacking both an official status and a clearly stated policy commitment, the terminology related to the US Silk Road reflects its fuzziness. A co-authored 2010 paper from Andrew Kuchins and S. Frederick Starr tries to promote the concept of a "Modern Silk Road"

(MSR). Later, Starr distinguished the "New Silk Roads" (plural) from "the New Silk Road" (singular) strategy of the United States. The former consists of all projects underway related to emerging continental corridors in the region, which are overseen by countries, international organizations, and private donors or corporations. The latter, on the other hand, is narrowed to just US government policies designed to align with these corridors and to make them the driver of US activities in the region (Starr 2011b, 9). This distinction is welcome and significant, because the United States is only one actor among many on the New Silk Roads, and its New Silk Road sometimes compete with the Silk Roads advanced by regional actors.

The US Silk Road: an elusive geopolitical construct

The choice of the Silk Road narrative to give a certain depth to a relatively unstructured US policy toward Central Asia encapsulates the evolution of US strategic thinking since the collapse of the Soviet Union. The 1990s-era emphasis on "democracy promotion" and access to "free markets" was partly reversed in 2001: the focus of US attention shifted to a security-driven policy after 9/11, and the Defense Department became the main engine of US policy in the region (McCarthy 2007). Since 2010, in preparation for the withdrawal from Afghanistan planned for 2014, the US narrative shifted once again, this time to a trade-driven policy, with more initiative given to the State Department. This emphasis on trade is less controversial than the democracy promotion agenda of the 1990s and welcomed by the Central Asian authorities, guaranteeing functional channels of communication. It also allows for combining US bilateral policy toward each country with an assumption that regional cooperation is the unavoidable milestone of "stability and prosperity" for the region, as the State Department likes to repeat (Pyatt 2012). The US Silk Road initiative has thus been developed around four axes: building a regional energy market, facilitating trade and transport, improving customs and border procedures, and linking businesses and people (Biswal 2015). Nonetheless, the expert and scholarly community working on the region has widely criticized the initiative on several grounds.

Policy criticisms

The first group of critics are policy-oriented.[8] They claim that the US Silk Road advances a grand design that is unrealistic given the decline in US financial assistance and diplomatic involvement in Central Asia. Second, the value-added of the US Silk Road is very limited, because the projects that will be implemented – whatever their benefits – are already championed by international financial organizations such as the World Bank and the Asian Development Bank. What makes the US brand attractive to the region is not trade and infrastructure but US culture, lifestyle, political values, know-how, higher education, professional training, and science and technology. Third, the Silk Road concentrates on regional cooperation while the countries of the region are known for their lack of enthusiasm for externally imposed regional identities. It focuses on interconnectivity, transportation, and trade, but it has never been demonstrated that regional trade could be the main conduit of development in Central Asia. It does not offer any solution at how to provide jobs in agriculture and reverse the trend of migrants leaving impoverished rural regions for provincial cities and capitals, and then for foreign countries.

Conflating the NDN and the Silk Road

A second group of criticisms is related to the conflation of the Northern Distribution Network (NDN) with the US Silk Road. Long-term strategic thinking on opening Central Asia to South Asia is part of US strategy of consolidating Central Asian states' sovereignty away from Russia by strengthening their links to other regional powers (Mankoff 2013). The success of the opening to the south will depend on two elements: reintegrating Iran into the regional picture, and India consolidating its rise as a regional power.

But this long-term prospect has been conflated with a short-term agenda; namely, facilitating the US military presence in Afghanistan. The NDN was a bi-directional system of air, land, and sea supply routes devised by the Pentagon to provide troops and equipment to the International Security Assistance Force (ISAF) and to exit equipment once the mission terminates. As the route through Pakistan was overloaded, insecure, and dependent on the vicissitudes of US-Pakistani relations, the US came to rely on the Central Asian states, especially Uzbekistan and, to a lesser extent, Kazakhstan. By integrating the NDN into its Silk Road grand design, Washington took several risks. First, it opened the door to considerable criticism from advocacy groups that denounced a US policy ready to cooperate with dictatorial regimes in Central Asia, while the Pentagon's goal was purely transactional (Lee 2012). The result was a US policy that appeared ready to funnel considerable sums of money to local kleptocratic elites. It also contributed to a proliferation of narratives on how "the US-led Northern Distribution Network (NDN) is so powerfully making the case for the renewal of continental trade" (Starr and Kuchins 2010, 15). This mental leap ignores the fact that the NDN is not a trade route based on any confirmed commercial rationality, but a set of military logistics, motivated by security and strategic concerns. Contrary to what its more enthusiastic advocates have claimed, the NDN has not contributed to the development of regional cooperation in Central Asia; rather, it has highlighted the dysfunction that already existed, with regimes starting to compete over contracts and lucrative benefits (Lee 2012).

Linking the establishment of the NDN to a fundamental transformation in regional relations in Central and South Asia has failed. A transactional policy based on military rationales represents just a fleeting "moment" in history, while the Silk Road vision necessitates decades of societal transformation. The enthusiasm for the NDN as a window of opportunity for US policy to promote regional trade seems to have blinded even those who know the social fabric and local cultures well. For instance, Starr and Kuchins (2010, 1, 7) affirmed that putting into place regional transport mechanisms to reintegrate Afghanistan with countries on its northern and southern borders could help these countries begin "reaping these [trade] benefits within 18 to 24 months" as the international community concentrates on a simple and easy remedy to solve the problem, such as a "prompt removal of existing 'corks' preventing the quick transit of goods, especially bureaucratic impediments." After decades of failed development policies from the World Bank and International Monetary Fund, it was wishful thinking to believe that "bureaucratic impediments" could be resolved in less than two years. This naïveté harmed Central Asia and especially Afghanistan by validating unrealistic economic projections of a sustainable Afghan budget based on customs revenues (and mining royalties that will not be received for years, if not decades). Many Afghanistan specialists have denounced this fallacy and warned of the grave political consequences of these miscalculations (Cordesman 2011).

The combination of short-term military goals with long-term economic development policy has also created an error in geographical representation. A long-term strategic vision that links Central and South Asia relies on the conventional assumption that Asia's power is on the rise and will dominate the twenty-first century. Yet a US-centric vision of the world has often been laid out. Starr and Kuchins (2010, 14) argued, for instance, that "no country, international agency, or financial institution is better positioned than the United States to lead the removal of existing impediments to continental transport and trade in Afghanistan and the adjoining region," as well as that the "physical distance from the proposed transit corridors actually gives America a further comparative advantage" (Starr 2011a, 16). This emphasis on the legitimacy of the United States as the driving force behind Asia's reorganization seems contradictory to assertions of Asia's increased autonomy from the Western world and norms. It is also reminiscent of a unipolar world vision in which Washington would be able to impact domestic realities everywhere in the world.

Missing pieces of the puzzle

Third – and a core component of this elusive geopolitical approach – the US Silk Road strategy claims to offer a comprehensive vision of the shifts of the geopolitical "tectonic plates" in Eurasia and Asia. This is a forward-looking idea, founded on the long-term decline of Western power and trans-Atlantic trade in favor of the Asia-Pacific region, but there are too many pieces missing from the puzzle to make the US Silk Road vision an accurate and genuine assessment. The roles of Russia, China, Iran, and Europe are not articulated in Washington's call for linking Central Asia to South Asia. Stating that the Silk Road strategy "benefits all and is against no one" (Starr 2011a, 16) – a statement supported by the State Department, which insists on the supposed complementary of the US project with the ones advanced by other external actors – obscures the strategic motivations behind the reorganization of economic dependencies in this part of the world.

Russia plays a key economic role in Central Asia by offering jobs to millions of labor migrants, not to mention its political, strategic, and cultural influence. But the US Silk Road does not explain how it plans to exclude or accommodate Central Asia's relationship with Russia. The same goes for China, which has developed its own "Silk Road Economic Belt." The United States remains silent on how the Silk Road fits into the Pentagon's strategic turn toward the Asia-Pacific – that is, China – as the key battlefield for future "Great Games" of the twenty-first century (Kuchins 2013). The same goes for Iran. If one follows the economic logic of the US Silk Road in seeking to open Central Asia to southern seas and the Persian Gulf, then Iran is a logical – yet missing – piece of US strategy. Central Asia is being penalized in terms of regional development and trade by the absence not of Afghanistan, which has been economically marginal for centuries, but of Iran, the only legitimate historical regional power (Kucera 2011; Peyrouse 2014). Pakistan's role as a commercial outlet for Central Asian products makes sense only because Iran is currently excluded from the regional picture.

Finally, Europe is also missing from the US picture. While the first US Silk Road that linked Central Asia and the South Caucasus to Turkey and the Mediterranean closely corresponded to European logics, the second Silk Road of 2011 represented a "de-Europeanization" of Central Asia, at a moment when Brussels was working hard on a new Strategy for Central Asia and on its Eastern Partnership policy (Emerson and Boonstra 2010). Furthermore, few in Central Asia see South Asia as a successful model of development and would prefer to strengthen links with Europe. Interestingly, in the chorus of voices favoring the US Silk Road, the Jamestown Foundation is the only one to note, "The new Silk Road strategy should be a joint European-US endeavor" (Starr 2011b, 6), but it has failed to offer any constructive steps toward accommodating European perspectives on the region.

A flawed historical metaphor

The idea that Central Asia constitutes a vital hub of Asia-Europe commerce and a natural transit route for the twenty-first century goes against any statistical analysis or world trade. Today, three-quarters of the world's trade is carried out by sea, and continental trade is not likely to dethrone maritime trade just because, once upon a time, caravans traveled along these routes. Uzbekistan is and will remain one of only two countries in the world that is doubly landlocked (the other being Lichtenstein). As Richard Pomfret (2011, 49) states, "Dramatic images of new silk roads or continental crossroads present a misleading picture. Central Asia's location advantages are not those of the fourteenth century."

The US Silk Road approach emphasizes what appears to be the center of a map of Eurasia – not a network of contemporary trade flows. Former Assistant Secretary of State for Central and South Asian Affairs Robert O. Blake, Jr., proclaimed, "This [US] vision is deeply rooted in history"; it "represents both an expansion and, to an important degree, a reworking of the original model" (Blake 2011, 42). Although the comparison can be compelling, it is purely a product of the Western vision of the Orient. Blake's ambitious statement about reworking the original model of the Silk Roads is totally disconnected from the realities of sea-based world trade.

At stake here is a form of anachronism that goes by the name of "presentism," according to which past successes are destined to repeat themselves. Such analogies have strong symbolic power, but they are a misleading basis for policy. The Silk Roads lost their *raison d'être* with the great maritime discoveries of the sixteenth century, as well as with changes to the local context, such as Iran's passing under Savafid domination and the uprising of the Sikhs in Punjab in the eighteenth century. The disappearance of the Silk Roads was due to internal evolutions in the region as well as a changing global context. It preceded imperialism and can by no means be explained solely by the "Great Game" between Russia and Great Britain in the nineteenth century.

US Deputy Secretary of State William Burns stated, "The old Silk Road transported not only goods and people, but ideas, cultures, and technology. It helped create great civilizations and foster great innovations. Central Asia can have a similarly historic impact today" (Burns 2009). This begs the question as to how the United States, as the world's leading technological superpower, can promote the idea that great innovations will come from building roads and railways. Such thinking represents a nineteenth-century vision of "progress" and development.

The Silk Road metaphor thus confuses a geographical crossroads with a core. Contrary to what US officials have repeated (such as Clinton's "Afghanistan's bustling markets sat at the heart of this network"), Afghanistan was never the nerve center of the Silk Road; and contrary to what has been said by Professor Starr ("Over two millennia Afghanistan was the place where trade routes to India, China, the Middle East, and Europe all converged"; 2011a, 11), the routes did not "converge" in Afghanistan, rather they passed through it; such was the way of trade. Having been a transit area is the historical reality of the region, but this is not the same as being the heart of a network.

This reality is less romantic, but it is more in tune with what the region is today: a periphery of multiple other cultural and trade centers, and a "center" only in the geographical sense of the term

Geopolitizing US involvement

Finally, the Silk Road narrative tends to "geopoliticize" US involvement in the region. It gives ground to a politically oriented narrative about a US "mission" in the region and develops an ideological justification that obscures the limited character of actual US involvement on the ground.

The US Silk Road tends, for instance, to conflate the Silk Road with the "Heartland." First articulated by Sir Halford Mackinder (1861–1947) in 1904, the Heartland is a fundamental concept of geopolitics that deeply influenced German *Geopolitik*, especially Karl Haushofer (1869–1946). According to Mackinder (1904), the Heartland is the geographical pivot of history, the center of the world-island linking Europe, Asia, and Africa, and it largely incorporates the territory of the Russian Empire, Afghanistan, and China's West from Xinjiang and Tibet to the Yangtze River. The Heartland would be the territory most contested for geopolitical domination. Mackinder predicted a century-long competition between continental Germany and maritime Great Britain to control the heart of the Heartland – Eastern Europe. Mackinder famously summarized this notion: "Who rules East Europe commands the Heartland; who rules the Heartland commands the World-Island; who rules the World-Island controls the world" (Mackinder 1919, 194).

Since the cold war era, the notion of Heartland resurfaced in US geopolitical theories, albeit with a major geographical shift. While the original, European conception of Heartland was centered on Russia, from the Baltic Sea–Black Sea axis to the Siberian landmass, the US version has moved it southward to focus on Central Asia, the South Caucasus, and Afghanistan, with the potential addition of Tibet. This was visible in both George Kennan's (1947) containment strategy ("The Sources of Soviet Conduct") and, later, in Zbigniew Brzezinski's "Grand Chessboard" theory. The former security advisor to President Jimmy Carter, Brzezinski (1998) defined Central Asia and the Caucasus as the "Eurasian Balkans," and viewed them as a potential zone of confrontation among the great powers, claiming, "Eurasia remains [...] the chessboard on which the combat for global primacy will unfold." According to Professor Starr (2007, 6), "The emergence of transport and trade-oriented countries in Greater Central Asia is a natural and inevitable process driven by the forces of the modern global economy and not by mere geopolitics." Yet, this argument does not take into account that contemporary geopolitics expresses itself through the language of economics and trade.

Today's Silk Road appears to pursue this geopoliticized vision of a region critical to the destiny of the West and the leading great power status of the United States, with "connectivity" and trade routes replacing traditional power competition (Petersen 2011). In 1904, Mackinder could legitimately exalt how railways were revolutionizing interactions inside the Heartland:

> True, that the Trans-Siberian railway is still a single and precarious line of communication, but the century will not be old before all Asia is covered with railways. The spaces within the Russian Empire and Mongolia are so vast, and their potentialities in population, wheat, cotton, fuel, and metals so incalculably great, that it is inevitable that a vast economic world, more or less apart, will there develop inaccessible to oceanic commerce. As we consider this rapid review of the broader currents of history, does not a certain persistence of

geographical relationship become evident? Is not the pivot region of the world's politics that vast area of Euro-Asia which is inaccessible to ships, but in antiquity lay open to the horse-riding nomads, and is to-day about to be covered with a network of railways? (Mackinder 1904).

Contemporary proponents of US Silk Road share many elements of Mackinder's enthusiastic vision of continental trade, with one major exception: they exclude Russia from it, while Mackinder's Heartland was definitively articulated around Russia's vast territories.

Conclusions

The US government's policy toward Central Asia has always navigated between a larger geopolitical design, à la Brzezinski, and more modest practices and outcomes, given the high resilience of Central Asia's authorities to externally imposed agendas and the predominant Russian and Chinese influences.

The Silk Road motif does not differ from the succession of grand narratives previously selected by the United States to script its involvement in the region. As since 2001 it has been deeply articulated within the US strategy in Afghanistan, the US storyline stands as the counterpart of the "discourse of danger" famously captured by Heathershaw and Megoran (2011). From the much-vaunted "war on terror" to the hype on "spillovers risks," the United States has been promoting a binary analysis – stability and prosperity through the Silk Road or the war on terror – that downplays the complexity of local situations and locally built agendas. By emphasizing a historical metaphor, the Silk Road provides a kind of meta-narrative for US involvement in the region that is mostly interpreted, by the other external actors and by local players, as a kind of "Great Game" in disguise. This grand design is ambivalent toward Russia, China, and Iran and seems to underline a geopolitical rivalry for influence and legitimacy in the region. Transforming the "geopolitical Heartland" into a "trade Heartland" accentuates the geopoliticization of US policy in the region. By playing with fuzzy geographical boundaries, it also conflates scales – confusingly endorsing improvement in small-scale cross-border trade, the launch of a regional energy market, and development of transcontinental trade – and therefore intertwines the modest economic realities of Central Asian countries with a kind of exalted macro view of Asia-Europe trade.

Critical geopolitics explores geopolitical perceptions as context-specific, rooted in political culture and policy traditions. The US Silk Road narrative presents itself as a textbook case. By considering resources and development to be synonymous, and by not addressing the role of perceptions – historical distrust and identity anxieties largely drive the Central Asian and South Asian states' foreign policy – the Silk Road narrative uncovers many of the neo-liberal assumptions of the Washington Consensus institutions that stipulate that infrastructure creates trade, trade creates jobs, and jobs create political democracy. It also unveils a "mirror game" with Russia's Eurasian narrative, which predates the US one. Both the Eurasian and Silk Road readings of Central Asia give prevalence to the geographical location of the region, as if Central Asia's destiny was shaped by its location more than any other feature.

Despite multiple and contradictory variations, compared with its US Silk Road counterpart, the Russian Eurasian terminology benefits from the same advantages as the Chinese Silk Road concept. It updates century-long traditional spheres of influence, while the US narrative seems artificial and without historical roots – except if one believes that the United States directly inherited the British role in conquering the India subcontinent and counterbalancing Russia and China. The Eurasian narrative accompanies concrete and identifiable public policies backed by budget allocations, intense diplomatic activities, and a combination of hard power and soft power tools, whereas the US narrative rests mostly on a level of wishful thinking. Russia's investment in building a concrete, palpable Eurasian Economic Union, and China's intense diplomatic and financial investments in the region confirm the long-term commitment from both countries to influence Central Asia. Both the Russian Eurasia and the Chinese Silk Road are also motivated by what appears to be a very long-lived assumption: emphasis on century-long historical interaction and on territorial contiguity allows for discretely asserting that Central Asia is part of Russia's and China's natural "geopolitical bodies"; claims that the United States cannot advance. However, the future of the US Silk Road narrative will probably not be tested by Russia, but by the competing Chinese Silk Road narrative, already well developed in terms of cultural nation-branding (tourism and patrimonialization) and economic investments, and that now seems to have taken hold at the policy level.

Disclosure statement

No potential conflict of interest was reported by the author.

Notes

1. Central Asia is a flexible geographical and cultural notion. By Central Asia, here I define the five post-Soviet republics, but the term often associates Afghanistan, Mongolia, and Azerbaijan too, depending on the context.
2. I would like to thank Cory Welt for his comments on an earlier draft of the paper.
3. "China's Initiatives on Building Silk Road Economic Belt and twenty-first Century Maritime Silk Road." Accessed 17 March 2015. http://www.xinhuanet.com/english/special/silkroad/
4. The Eurasian Economic Union includes Russia, Belarus, Kazakhstan, Kyrgyzstan, and Armenia.
5. Originally, TRACECA involved the five countries of Central Asia and the three South Caucasian states, but it has been strengthened by the addition of Moldova, Ukraine, Turkey, Romania, and Bulgaria. The program is now increasingly connected with pan-European road projects planned in the framework of the EU Neighborhood Policy.
6. http://www.state.gov/p/sca/
7. The term "vision" received 116 mentions (34 in 2011, 47 in 2012, 29 in 2013, 6 in early 2014), while "initiative" is on the rise with 75 mentions (14 in 2011, 21 in 2012, 11 in 2013, 24 in 2014, and 5 in January 2015). I thank Cory Welt for providing me with these numbers.
8. These policy critics have not, to my knowledge, been presented into an analytical paper, but they are heard regularly at Washington-based events organized on the US Silk Road at CACI, Jamestown, CSIS, and at GW's Central Asia Program.

References

Abelson, Donald E. 2006. *A Capitol Idea: Think Tanks and US Foreign Policy.* Montreal: McGill Queens University Press.

Alam, Muzaffar, and Sanjay Subrahmanyam. 2007. *Indo-Persian Travels in the Age of Discoveries, 1400–1800.* New York, NY: Cambridge University Press.

Bassin, Mark, Sergey Glebov, and Marlene Laruelle, eds. 2015. *Between Europe and Asia: The Origins, Theories, and Legacies of Russian Eurasianism.* Pittsburgh, PA: Pittsburgh University Press.

Biswal, Nisha Desai. 2015. "The New Silk Road Post-2014: Challenges and Opportunities." Presentation at the Woodrow Wilson International Center for Scholars, Washington, DC, January

22. Accessed March 17. http://www.wilsoncenter.org/event/the-new-silk-road-initiative-post-2014-challenges-and-opportunities

Blake, Robert O. 2011. "Keynote Address." In *Afghanistan, and the New Silk Road: Political, Economic, and Security Challenges*, edited by Central Asia, 42–46. Washington, DC: Jamestown Foundation.

Brzezinski, Zbigniew. 1998. *The Grand Chessboard: American Primacy and Its Geostrategic Imperatives*. New York, NY: Basic Books.

Burns, William J. 2009. "Silk Road Trade and Investment: New Pathways for US-Central Asia Economic Ties." Remarks at the US Chamber of Commerce, Washington, DC, October 7. Accessed March 17, 2015. http://www.state.gov/p/us/rm/2009a/130389.htm

Calder, Kent E., and Viktoriya Kim. 2008. "Korea, the United States, and Central Asia: Far-Flung Partners in a Globalizing World." *Korea Economic Institute Academic Papers Series* 3: 1–13.

Clinton, Hillary Rodham. 2011. "Secretary of State Hillary Rodham Clinton Speaks on India and the United States: A Vision for the 21st Century." July 20. Accessed March 17, 2015. http://chennai.usconsulate.gov/secclintonspeechacl_110721.html

Cordesman, Anthony H. 2011. "The Afghan War 10 Years on." *Center for Strategic and International Studies*, October 11. Accessed May 31, 2015. http://csis.org/publication/afghan-war-ten-years

Dutkiewicz, Piotr, and Richard Sakwa, eds. 2015. *Eurasian Union: The View from within*. London: Routledge.

Emerson, Michael, and Jos Boonstra. 2010. *Into Eurasia: Monitoring the EU's Central Asia Strategy. Report of the EUCAM Monitoring*. Brussels: Centre for European Policy Studies.

Evans, Alison. 2014. "South Korea's 'New Silk Road' to Central Asia: Diplomacy and Business in the Context of Energy Security." *US-Korea 2012 Yearbook*. Washington, DC: Johns Hopkins University, School of Advanced International Studies, US-Korea Institute, 67–81. Accessed May 31, 2015. http://uskoreainstitute.org/wp-content/uploads/2014/04/Evans_YB2012.pdf

Fedorenko, Vladimir. 2013. *The New Silk Road Initiative in Central Asia*, Rethink Paper no. 10. Washington DC: Rethink Institute.

Foltz, Richard C. 2007. "Cultural Contacts between Central Asia and Mughal India." In *India and Central Asia: Commerce and Culture, 1500–1800*, edited by Scott Cameron Levi, 155–175. New Delhi: Oxford University Press.

Ganguli, Sreenati. 2011. "The Revival of the Silk Roads Topic: A Contemporary Analysis." In *Mapping Central Asia: Indian Perceptions and Strategies*, edited by Marlene Laruelle and Sebastien Peyrouse. Farnham, UK: Ashgate, 61–74.

Gayer, Laurent. 2010. "From the Oxus to the Indus: Looking Back at India-Central Asia Connections in the Early Modern Age." In *China and India in Central Asia: A New, "Great Game"?* edited by Marlene Laruelle, Jean-François Huchet, Sebastien Peyrouse and Bayram Balci, 197–214. London: Palgrave Macmillan.

Gopal, Surenda. 2005. *Dialogue and Understanding: Central Asia and India. the Soviet and Post-Soviet Era*. Kolkata: Shipra.

Heathershaw, John, and Nick Megoran. 2011. "Contesting Danger: A New Agenda for Policy and Scholarship in Central Asia." *International Affairs* 87: 591–594.

Jiao, Wu, and Zhang Yunbi. 2013. "Xi Proposes a "New Silk Road" with Central Asia." *China Daily*, September 8. Accessed March 17, 2015, http://usa.chinadaily.com.cn/china/2013-09/08/content_16952304.htm

Kavalski, Emilian. 2010. *India and Central Asia: The Mythmaking and International Relations of a Rising Power*. New York: I.B. Tauris.

Kennan, George. 1947. "The Sources of Soviet Conduct." *Foreign Affairs* 25: 566–582.

Kerry, John. 2013. "Remarks on the US-India Strategic Partnership." June 23. New Delhi, India. Accessed March 17, 2015. http://www.state.gov/secretary/remarks/2013/06/211013.htm

Kucera, Joshua. 2011. "Iran Left out of the New Silk Road." *The Atlantic*, November 22. Accessed March 17, 2015. http://www.theatlantic.com/international/archive/2011/11/us-plan-for-a-new-silk-road-faces-a-big-speed-bump-iran/249048/

Kuchins, Andrew. 2013. "Why Washington Needs to Integrate the New Silk Road with the Pivot to Asia." *Asia Policy* 16: 175–178.

Kumar, B. B. 2007. "India and Central Asia: Links and Interactions." In *India and Central Asia: Classical to Contemporary Periods*, edited by J. N. Roy and B. B. Kumar, 3–33. New Delhi: Astha Bharati.

Laruelle, Marlene. 2008. *Russian Eurasianism: An Ideology of Empire*. Washington, DC, and Baltimore, MD: Woodrow Wilson Center Press/Johns Hopkins University Press.

Laruelle, Marlene, and Sebastien Peyrouse, eds. 2011. *Mapping Central Asia: Indian Perceptions and Strategies*. Farnham, UK: Ashgate.

Laruelle, Marlene, and Sebastien Peyrouse. 2012. *The "Chinese Question" in Central Asia: Domestic Order, Social Changes, and the Chinese Factor*. New York: Columbia University Press.

Lee, Graham. 2012. *The New Silk Road and the Northern Distribution Network: A Golden Road to Central Asian Trade Reform?* Occasional Paper no. 8. Open Society Foundation. Accessed March 17, 2015. http://www.opensocietyfoundations.org/sites/default/files/OPS-No-8-20121019.pdf

Len, Christopher, Uyama Tomohiko, and Hirose Tetsuya, eds. 2008. *Japan's Silk Road Diplomacy: Paving the Road Ahead. Silk Road Papers*. Washington, DC: Central Asia-Caucasus Institute.

Mackinder, H. J. 1904. "The Geographical Pivot of History." *The Geographical Journal* 23: 421–437.

Mackinder, H. J. 1919. *Democratic Ideals and Reality*. London: Constable and Company.

Mankoff, Jeff. 2013. *The United States and Central Asia after 2014*. Washington, DC: Center for Strategic and International Studies, January. Accessed May 31, 2015. http://csis.org/files/publication/130122_Mankoff_USCentralAsia_Web.pdf

Marantidou, Virginia, and Ralph A. Cossa. 2014. "China and Russia's Great Game in Central Asia." *The National Interest*, October 1. Accessed May 31, 2015. http://nationalinterest.org/blog/thebuzz/china-russias-great-game-central-asia-11385

McCarthy, Michael J. 2007. *The Limits of Friendship: US Security Cooperation in Central Asia*. Maxwell Air Force Base, AL: Air University Press.

Millward, James A. 2009. "Positioning Xinjiang in Eurasian and Chinese History: Differing Visions of the 'Silk Road'." In *China, Xinjiang, and Central Asia*, edited by Colin Mackerras and Michael Clarke, 55–74. London: Routledge.

Mukhtarov, A. 2003. "Some Aspects of the History of Cultural Relations of Central Asia with India." In *India and Tajikistan: Revitalizing a Traditional Relationship*, edited by M.Singh, 49–56. Kolkata: Maulana Abul Kalam Azad Institute of Asian Studies.

Ó Tuathail, Gearoid. 1996. *Critical Geopolitics: The Politics of Writing Global Space*. London: Routledge.

Obama, Barack. 2015. "Remarks by President Obama at US-India Business Council Summit." Taj Palace Hotel, New Delhi, India, 26 January. Accessed March 17, 2015. http://www.whitehouse.gov/the-press-office/2015/01/26/remarks-president-obama-us-india-business-council-summit

Petersen, Alexandros. 2011. *The World Island: Eurasian Geopolitics and the Fate of the West*. New York: Praeger.

Peyrouse, Sebastien. 2014. "Iran's Growing Role in Central Asia? Geopolitical, Economic, and Political Profit and Loss Account." *Al-Jazeera*, April 6. Accessed March 17, 2015. http://studies.aljazeera.net/en/dossiers/2014/04/2014416940377354.html

Pomfret, Richard. 2011. "Trade and Transport in Central Asia." In *Central Asia and the Caucasus: At the Crossroads of Eurasia in the Twenty-First Century*, edited by Werner Hermann and Johannes Linn, 43–62. Thousand Oaks, CA: Sage.

Pyatt, Geoffrey. 2012. "Delivering on the New Silk Road." *US Department of State*. July. Accessed March 17, 2015. http://www.state.gov/p/sca/rls/rmks/2012/194735.htm

Pyatt, Geoffrey, Michaela Prokop, Gael Raballand, and Gulshan Sachdeva. 2012. "Discussing the 'New Silk Road' Strategy in Central Asia." *Central Asia Policy Forum* 2 (June): 1–7.

Shukla, P. P. 2014. *India-US Partnership: Asian Challenges and beyond*. New Delhi: Wisdom Tree.

Silk Road Strategy Act of 1999, H.R. 1152, 106th Cong. 1999.

Starr, S. Frederick, ed. 2007. *The New Silk Road: Transport and Trade in Greater Central Asia*. Washington, DC: Central Asia-Caucasus Institute & Silk Road Studies Program.

Starr, S. Frederick. 2011a. *Afghanistan beyond the Fog of Nation Building: Giving Economic Strategy a Chance. Silk Road Papers*. Washington, DC: Central Asia-Caucasus Institute.

Starr, S. Frederick. 2011b. "Opening Keynote: A Strategy for Central Asia after the US Military Withdrawal from Afghanistan." In *Afghanistan, and the New Silk Road: Political, Economic, and Security Challenges*, edited by Central Asia, 9–17. Washington DC: Jamestown Foundation.

Starr, S. Frederick, and Andrew Kuchins, eds. 2010. *The Key to Success in Afghanistan: A Modern Silk Road Strategy. Silk Road Papers*. Washington, DC: Central Asia-Caucasus Institute.

Sumar, Fatema. 2014. "The TUTAP Interconnection Concept and CASA-1000." *Center for Strategic and International Studies*. 6 June. Accessed May 31, 2015. http://csis.org/event/tutap-interconnection-concept-and-casa-1000

Swanström, Niklas. 2011. *China and Greater Central Asia: New Frontiers?* Washington, DC: Central Asia-Caucasus Institute & Silk Road Studies Program.

Tiezzi, Shannon. 2014. "China's 'New Silk Road' Vision Revealed." *The Diplomat*, May 9. Accessed March 17, 2015. http://thediplomat.com/2014/05/chinas-new-silk-road-vision-revealed/

Tsygankov, Andrei. 2009. *Russophobia*. New York: Palgrave Macmillan.

Zajec, O. 2008. "La Chine affirme ses ambitions navales [China affirms its naval ambitions]." *Le Monde Diplomatique* [The Diplomatic World] (September): 18–19.

Benevolent hegemon, neighborhood bully, or regional security provider? Russia's efforts to promote regional integration after the 2013–2014 Ukraine crisis

Andrej Krickovic and Maxim Bratersky

ABSTRACT

Russia has tried to use economic incentives and shared historical and cultural legacies to entice post-Soviet states to join its regional integration efforts. The Ukraine crisis exposed the weaknesses of this strategy, forcing Russia to fall back on coercive means to keep Kiev from moving closer to the West. Having realized the limits of its economic and soft power, will Russia now try to coerce post-Soviet states back into its sphere of influence? Fears of such an outcome overestimate Russia's ability to use coercion and underestimate post-Soviet states capacity to resist. Rather than emerging as a regional bully, Russia is trying to push Eurasian integration forward by becoming a regional security provider. The article relates these efforts to the larger literature on regional integration and security hierarchies – bridging the two bodies of theory by arguing that regional leaders can use the provision of security to promote economic integration. Despite initial signs of success, we believe that the new strategy will ultimately fail. Eurasian integration will continue to stagnate as long as Russia's economic and soft power remain weak because Russia will be unable to address the economic and social problems that are at the root of the region's security problems.

Introduction

Over the last few years, Russia has tried to establish a "soft" hegemony in the post-Soviet region, using economic incentives and Soviet legacies of shared history and culture to entice post-Soviet states to join its regional integration efforts. To further this goal, Moscow has even been willing to take responsibility for the provision of regional collective goods such as security, free trade, energy resources, and financial stability. However, Russia's ability to play the role of benign regional hegemon has been challenged by the crisis in Ukraine. Russia's efforts to entice Ukraine into its economic and political orbit have backfired, and Moscow now finds itself using military and covert means in order to destabilize the situation in that country and prevent it from moving closer to the West. Moscow's hope is that once the country's turn to the West fails, Ukraine will have no choice but to participate in Russian-led Eurasian integration.

This article will examine the ways in which Russia's approach to regional integration has changed in the wake of the Ukraine crisis. We argue that the Ukrainian crisis has forced Russia to recognize the limits of its ability to use its economic and soft power (as the concept is defined by Joseph Nye) to pursue the regional integration agenda it values so dearly (Nye 2011). However, this does not necessarily mean that Russia is now bent on using hard or coercive means to achieve its goals. It is unlikely that Moscow will adopt the same tactics it has used in Crimea and eastern Ukraine in other parts of the post-Soviet space. The leadership realizes that the costs of doing so are prohibitively high, as its behavior in Ukraine has already raised concerns and fears in other post-Soviet states, forcing the Russian government to reallocate precious economic resources toward calming these fears and reassuring their partners.

Rather than becoming a regional bully, Russia is trying to use the advantage it still holds in military and other hard-power resources to position itself as the region's security hegemon, helping the region's weak authoritarian regimes to deal with internal and external security threats, particularly those rising from the spread of Islamic radicalism represented by ISIS and instability in Afghanistan. In doing so, Russia is pursuing a strategy very much in line with Lake and Morgan (1997) and Lake's (2009) concept of regional security hierarchy. By providing for these states' security, Moscow hopes to gain the loyalty of the region's other states and their participation in its project of regional economic integration. We argue that Russia is thus trying to use its hard power to achieve what are primarily economic or (more precisely) geoeconomic goals.

The shift toward using hard power is neither the result of traditional Russian imperialism reasserting itself (Laqueur 2015; Socor 2014), nor is it primarily motivated by the regime's need to distract public attention away from domestic problems (Stoner and McFaul 2015). Rather, it reflects the dearth of soft and economic power resources available to Russia to pursue the larger goal of regional integration, which Russia's elites see as being a key to maintaining the country's status as a great power in world politics in the years to come.

Russia's more aggressive use of hard-power tactics and its willingness to employ military force in Ukraine, and most recently even outside the post-Soviet space in Syria, has given rise to fears in the West that Russia is a full blown revisionist power bent on overturning the Western-led global order (Giles et al. 2015; Socor 2014). If Russia indeed emerges as revisionist challenger to the Western-led liberal order, the West will also have to share some of the blame. While Russia bears a good deal of the responsibility for escalating the crisis, the West also made a critical mistake in pushing for Ukraine's Western integration without taking Russia's interests into account (Mearsheimer 2014). In doing so, Western leaders took advantage of Russia's perceived economic and soft-power weaknesses, not recognizing that this would provoke a hard-power response. As a result, Russia has become further estranged from the Western-led order and sees the use of coercive and hard power as the only tools available to it to defend and promote its interests.

Russia's Drive for Post-Soviet Regional Integration

Russia has invested heavily in the process of regional integration of the post-Soviet space. Russia's elites see it as a key to the country's economic development and its survival as a key geopolitical player and "great power" (Krickovic 2014). Despite robust economic growth through much of the 2000s, Russia's economy is still overly dependent on the export of natural resources and hydrocarbons, and this "resource curse" has hampered the development of domestic institutions and the growth of the high-tech industries and private enterprise (Kudrin and Gurvich 2015). Even before Western sanctions and the bottom dropped out of the global oil market in 2014, the Russian Ministry of Economic Development forecast that Russia's percentage of world GDP would decline from its current level of 4% to less than 3% by 2030 (Kuvshinova 2013). Russian leaders believe that Eurasian integration is of key importance for reversing their country's economic and geopolitical decline. From this perspective, Russian-led Eurasian economic integration will create a protected economic space where Russian firms and capital can develop and grow, helping it modernize its economy and become less dependent on Western markets and the exports of natural resources to them. (Chebanov 2010). A larger Euraisan economic space will help Russia compete with the larger economic blocs dominated by the other great powers, including those promoted by the West but also with China (Krickovic 2014). In this way, Russia will lay the necessary economic foundations that will allow it maintain its status as a great power and become one of the poles in a future "multipolar" world order" (Putin 2011).

This new push for integration began in the late 2000s, and it is a significant departure from earlier policies. The Commonwealth of Independent States (CIS)[1] was initially designated as the main vehicle for regional integration in the 1990s. Leaders and experts in Russia and other post-Soviet states had high hopes that the CIS would one day become the "EU of the East." Yet, integration made little progress under the auspices of the CIS. According to one study, less than 10% of the thousands of documents and resolutions adopted by CIS bodies have actually been ratified by member countries (Moskvin 2007). Regional disintegration and the dissolution of economic and political bonds that had been created in the Soviet and Tsarist periods continued. Intraregional trade as a percentage of total trade of the CIS region fell by almost 40% between 1994 and 2008 (Gurova and Efremova 2010). Russia was responsible for many of these failures. Despite its pro-integration rhetoric, Russia was not really willing to make the sacrifices and efforts needed to make regional integration work. This was apparent early on when Russia withdrew its support for maintaining the CIS as a ruble zone in 1992. Throughout the first two decades of the post-Soviet period, Russia's leaders were reluctant to take on new foreign policy burdens, fearing that this would detract precious economic resources from domestic reform (Kubicek 2009). As a result, Russian-led regional integration took on a "virtual" character; it was heavy on rhetoric but short on actual substance (Trenin 2011).

Russia had neither the will nor the resources to counter the upsurge of nationalism throughout the former Soviet Union. Instead of accepting regional leadership and pushing integration forward, it conducted a very specific foreign policy toward the former republics, pretending to do business as usual, avoiding conflict and de facto sponsoring their economies (specifically the economies of Ukraine, Belarus, and Moldova) through subsidized supplies of energy and natural resources (Bordachev and Skriba 2014). According to one authoritative estimate, Russia has been subsidizing the Ukrainian economy by 5 to 10 billion USD annually since 1991 (Gaddy and Ickes 2014). This policy served to reassure the Kremlin that since Russia subsidized these countries, it still maintained influence in the post-Soviet space. Nevertheless, this policy was self-deceiving; the elites of the neighboring countries gladly took advantage of these economic opportunities and often enriched themselves personally form these arrangements (Sakwa 2015). At the same time, however, they steadfastly defended their sovereignty and resisted Russia's efforts to tie them into Russian-led regional institutions. This was particularly true of Ukraine. In 1991, Ukraine chose to be only a participant, but not a member of the CIS. Ukraine failed to ratify the CIS treaty and instead only accepted associate member status in the organization. It turned down the invitation to join the Russian-led Customs Union (CU) and only accepted observer status in Eurasian Economic Community in 2000.

Russia encountered less resistance to integration in Central Asia. Kazakhstan has always supported Eurasian economic integration and made it a priority of its foreign policy (Sultanov 2015). Other Central Asian states were also interested in integrating with the richer Russian economy. However, Russia did not reciprocate their interest in integration because of these countries' economic weakness (Naumkin and Ivanov 2013). Apart from very active dialog on integration with Kazakhstan, Russia confined itself to promoting political stability and regional security through regional security schemes such as the Shanghai Cooperation Organization (SCO) and the Collective Security Treaty Organization (CSTO). It only played a limited role in regional economic affairs by providing economic aid and sponsoring regional agreements to address issues such as water scarcity or energy security (2013).

Given its residual military capabilities, Russia found it much easier to establish itself as a regional peacekeeper and conflict manager. At first, Russia reluctantly embraced this role in response to the growing ethnic and civil conflict throughout the region in the 1990s (Shashenkov 1994). Russia carried out peacekeeping operations in Moldova's Trans-Dniester region and in Georgia's region of South Ossetia in 1992 and then later in Georgia's Abkhazia region and in Tajikistan in 1994. Initially, the primary goal of these operations was to stop violence and prevent anti-Russian political forces from taking power. These goals were later expanded to include preventing military peacekeeping by outside powers and keeping countries such as Georgia or Moldova from joining NATO (because of their unresolved domestic conflicts) (Mankoff 2009). Nevertheless, as was the case with economic integration, Russia adopted a cautious approach to security leadership, and any impulses Russia may have had to use military means to advance regional hegemony were dampened by its concerns about the costs of such a policy and by the adverse effects it could have on the course of internal reform (Lynch 2000).

Russia began to show a renewed commitment toward regional integration after the financial crisis of 2008. The crisis demonstrated the limits of the previous model of economic development based on hydrocarbon exports and shattered hopes that Russia could regain its lost international status by becoming an "energy superpower" (Tsygankov 2013). Since then, Russia has intensified its efforts to promote regional integration. Moscow moved away from the previous strategy, which aimed to bring all of the former Soviet states (excluding the Baltic countries) under the same tight institutional umbrella toward a flexible, multi-layered approach that includes bilateral relations with post-Soviet states as well as smaller multilateral groupings like the Customs Union, the Eurasian Economic Union, and the CSTO (Bratersky 2010). These relationships exclude states such as Georgia and Azerbaijan, which are more interested in integration with powers in the Western bloc, and instead focus on building relations with states like Belarus, Kazakhstan, Armenia, and the Central Asian states, which are more amenable to integration.

The primary institutional vehicle for achieving Russia's goals has been the CU, which was established by Russia, Belarus, and Kazakhstan in 2009 with the goal of eliminating tariffs and customs controls between their countries and creating a genuine common market and economic space. The CU was transformed into the EEU January 2015, as additional measures were implemented to harmonize legislation among the three markets and set up an arbitration mechanism to settle disputes. With Russia's financial backing the EEU has also established a $10 billion crisis fund. Belarus drew $3 billion from the fund in 2012, helping it to meet its international debt obligations and avoid having to go to the International Monetary Fund for assistance. Armenia and Kyrgyzstan joined the EEU in 2015, and Tajikistan has been officially invited to join and is undergoing the ascension process for membership. Armenia was due to sign an association agreement with the EU along with a free trade deal in late 2013, but several months ahead of the agreement's conclusion it instead decided to join the EEU.

In promoting this Eurasian vision, Moscow recognized the declining utility of hard and coercive power in advancing regional integration and instead sought to advance its project through a mixture of economic incentives and soft power (Tsygankov 2013). Access to Russia's growing domestic markets and Russian finance, as well as discounted energy prices were all used to entice the post-Soviet countries to join. Russia also has considerable soft power in the post-Soviet space. The post-Soviet states all share common cultural legacies as former members of "Soviet Civilization" (Sinyavsky 2015). On the elite level this includes shared political and economic networks rooted in Moscow, where many of the region's elite continue to go to for business and education. Yet, there is also a popular component, exemplified by the ubiquitous presence of Russian pop music, movies, and television series throughout the region. Russia has tried to cultivate this cultural influence through its project of establishing a Russkiy Mir (Russian World). The project, which enjoys strong support from the Russian government, is designed to promote the development of a common cultural and linguistic space that will unite the post-Soviet countries based on the shared legacy of Russian language, history, and culture. Modeled on established institutions that other countries have used to promote their soft power, such as Great Britain's British Council and France's Alliance Francaise, the project invests heavily in the promotion of educational and cultural exchanges and the establishment of Russian language education throughout the region (Kudors 2010). In effect, Russia has tried to position itself as a benevolent regional hegemon that can provide regional collective goods, be they in the form of security, access to markets, acting as a lender of last resort in times of crisis, or the preservation of shared cultural legacies.

Ukraine: A Failure of Russian Economic and Soft Power

In the prelude to the 2013 Ukraine crisis, Russia made sustained efforts to use its economic and soft power to bring Ukraine into its Eurasian integration project. These efforts had to compete with those of the European Union to begin the process of Ukraine's European integration via the signing of a Deep and Comprehensive Free Trade Agreement (DCFTA) and an "association agreement" that would begin the country's political and economic transformation according to EU standards. For its part, Moscow offered Kiev preferential access to its market through the Customs Union. Economists close to the Kiev government claimed that membership in the CU could boost Ukraine's GDP by as much as 15% by 2030 (Ivanter et al. 2012). They argued that Eurasian integration promised an immediate improvement in Ukraine's balance of trade and stability of its balance of payments, while a DCFTA with the EU would require major trade liberalization and entail worsening of trade conditions and direct economic losses for Ukraine (Glazev 2013). Russia's economic advocacy also had a coercive element to it. Citing the need to prevent its own markets from the threat of re-export of European goods from Ukraine, Moscow warned that it would cancel preferential trade agreements it had with Ukraine if the country signed the DCFTA. Russia introduced limited trade restrictions in the summer of 2013 to remind Ukraine's political and business leaders just how dependent the country was on Russian markets (exports to Russia constitute over 30% of Ukrainian exports). According to some estimates, these trade restrictions cost Ukraine as much as $2.5 billion in lost trade (The Economist 2013).

Moscow was also willing to provide Kiev with much-needed financial aid. Ukraine's financial position was dire, with the country owing nearly USD 60 billion in loans due by July 2015. Few were willing to lend to Ukraine. The EU offered only USD 618 million in aid as part of the association agreement. Russia's decision to buy USD 15 billion in Ukrainian bonds was thus a real lifeline to Ukraine's struggling economy. Even the head of the IMF, Christine Lagarde, admitted in April 2014 that Ukraine's economy would have collapsed if it were it not for Russia's purchase of these bonds (Adamczyk 2014). Another economic lever was Ukraine's energy dependence on Russia, with more than half of Ukraine's gas coming from Russia. As part of the financial rescue package deal that President Viktor Yanukovych brokered with the Russians and which ultimately led him to reverse course on the government's plans to sign the association agreement with the EU, Russia agreed to lower gas prices by 30%, which would have saved Kiev almost $5 billion in 2015 (Bloomberg 2013).

Russia tried to take advantage of its still-considerable soft power in Ukraine, manifest in the two countries' strong historical and cultural ties. These efforts were part of the broader, "Russkiy Mir" project. Russian popular culture had a wide following in Ukraine as did the Russian media, which often disseminated the Kremlin's views on current events – including the issue of Eurasian integration. Moscow also invested heavily in pro-Russian media and cultural organizations and financed NGOs and civil society groups in Ukraine that would support Russia's interests, particularly in Crimea and the eastern portion of the country. These groups were mobilized from time to time to promote the Russian position on controversial issues such as the status of the Russian language and military cooperation with the US and NATO (Bogomolov and Lytvynenko 2012).

Russia enjoyed more subtle and less conspicuous soft-power influence. Ukraine's political and business elites were tied to the post-Soviet "old boy" power networks centered in Moscow. This gave Moscow considerable leverage in terms of the elite's material interests, which were often closely allied with Moscow's. Ukraine's elites also shared a common worldview and political and business culture with their counterparts in Moscow. They often looked to Moscow for guidance. "For many of them Moscow remains the preferred, although not necessarily the only destination for business and leisure, a source of inspiration for new ideas and practices." (Bogomolov and Lytvynenko 2012, 13). Yanukovych wanted to bring state institutions under his centralized control, and he saw Russia's centralized semi-authoritarian system as a model to be emulated. Experts were expected to enter a section on relevant "Russian experience" when writing government policy papers (Bogomolov and Lytvynenko 2012, 3).

Russia's use of soft-power resources and economic incentives seemed to score a significant success when in November 2013 Kiev indefinitely postponed the signing of the association agreement with the EU – stunning many Western observers who believed that the deal was a foregone conclusion. This prompted one unnamed Russian official to gloat, "It's like stealing the bride right before the wedding…this is another victory for President Putin in the international arena." (Vedemosti 2013) This "geopolitical victory" proved to be short-lived. Months of massive pro-EU demonstrations followed as many Ukrainians saw Yanukovych's decision as selling out the larger national interest to Russia. A sizable number of Ukrainians continued to support close ties with Russia and to reject NATO membership. Yet, Moscow proved unable to attract counter demonstrations in support of Eurasian integration or to stem the tide of public opinion, which was beginning to decisively shift against Yanukovych (Sakwa 2015).

The events that led to Yanukovych's ouster are still poorly understood and highly contested, with some claiming that it was a popular revolution (Higgins and Kramer 2015) while others see it as an unconstitutional coup organized by a small minority of right-wing militants with the possible connivance of the West (Cohen 2014). Nevertheless, the results, in terms of Russian foreign policy are clear. In the end, all of Russia's efforts to use economic incentive and soft power to entice Ukraine to join the Eurasian integration project failed. Moscow found itself facing its nightmare scenario – the coming to power of a pro-Western government that was determined to turn Ukraine toward the path of Western integration and possibly even NATO membership.

This turn of events exposed Russia's lack of soft power as well as its limited ability to use economic inducement to achieve its foreign policy goals – even in its own immediate neighborhood. In fact, Russia's attempts at using soft power had the opposite effect, mobilizing Ukrainians to resist what they saw as "Russian imperialism." This failure forced Russia to turn to other, more coercive, hard-power means to achieve its goals. Moscow looked to protect its core strategic and military interest in the country by orchestrating the annexation of Crimea. The Kremlin was also determined to isolate the new government in Kiev and bring it to heel by stirring up ethnic conflict in the east. Toward this goals, it helped to incite domestic malcontents in the Donbas and in Crimea who were bolstered by radical elements from Russia itself. These forces were given generous military aid and support, including the deployment of regular Russian troops, to save them from probable defeat at the hands of the Ukrainian authorities in August 2014.

Eurasian Integration after Ukraine – Russia's Declining Ability to Use Instruments of Soft Power

Russia's instruments of economic and soft power are beginning to decline even further in the wake of the Ukraine crisis. Russia' economy has been battered by Western sanctions and the steep decline in the price of oil. Russian GDP growth was beginning to stagnate even before the crisis; in 2014, GDP grew by less than 1% and the ruble lost almost 60% of its value against the US dollar. GDP declined by 3.7% in 2015 and most projections expect it to continue to fall through 2016 (Matlack 2016). This economic downturn has reduced Russia's ability to use economic incentives and inducements to promote its Eurasian project. Trade among EEU member states fell by nearly 13% in the first quarter of 2014 compared to the same period in 2013 (Coalson 2014). As a result, Russia has become less attractive as a market for other post-Soviet states. Russian businesses are also less willing to invest in neighboring countries as they face a credit crunch at home. Western financial sanctions make it difficult for them to refinance and reschedule their own debts, much less make new investments abroad (Makhovsky, Solovyov, and Antidze 2015).

Over the late few decades, Russia has become a major destination for migrant labor throughout the post-Soviet space. Russia is the country with the second largest number of immigrants in the world, after the United States. Remittances by guest workers represent a critical source of income for the region's most impoverished countries, accounting for 30% of Kyrgyzstan's and a staggering 52% of Tajikistan's GDP (Hille 2015). Dependence on this source of income has given Russia tremendous political and economic leverage. For example, the government of Tajikistan agreed to extend Russia's right to base troops on its territory to 2042 in exchange for Russia raising the quota for the number of guest workers from Tajikistan that would be allowed to work in Russia. With the economic downturn in Russia and the steep decline in the value of the Russian ruble, many of these migrants are now going home. According to figures from Russia's Federal State Statistics Service, net migration to the country dropped by as much as 10% from January to October 2014 (Kolesnikov and Gabuev 2015). Migrant laborers are also sending back less money to their home countries. In the first quarter of 2015, transfers to Uzbekistan, Tajikistan, and Kirgizstan declined by more than 40% (Mirzayan and Pak 2015). Russia is becoming less attractive as a destination for migrant labor, and Russia may be losing a valuable source of soft-power leverage over these states.

The crisis in Ukraine has also exposed the limited appeal of Russia's soft power – even within the post-Soviet space. Russia's approach to soft power has relied heavily on common cultural legacies. These continue to have a strong presence in the post-Soviet space, although Russian language use has been declining throughout the region since the breakup of the USSR. However, as Nye makes clear in his work on soft power, cultural attractiveness alone is not enough to ensure political influence over another state or to mobilize members of another society to their cause (Nye 2011). A country must also have an attractive economic and political model that others admire and seek to emulate. While some domestic interest groups in post-Soviet states may have close connections to the Russian political and economic establishment, Russia's authoritarian political regime and corrupt oligarchic capitalism have little popular appeal beyond its borders.

Most post-Soviet states have authoritarian political regimes, and this has not significantly hindered Russia's ability to project its influence. The situation was different in Ukraine, however, where politics is contested and there is a strong tradition of popular activism and protest. Russia was unable to mobilize popular support for its Eurasian project amongst a wider swath of the Ukrainian public (Silayev 2014). Though the EU was unable to offer much in the way of material incentives, the European model of political and economic governance, which Ukraine would be obliged to accept if it signed the agreements with the EU, offered a way out of the morass of cronyism, corruption, and political manipulation that have plagued Ukraine ever since its independence. For those who took to Maidan square to protest the Yanukovych government's decision to back away from the EU, a turn toward the Russian-led Eurasian Union promised to maintain the status quo or – even worse – to increase authoritarians and open the country to economic predation by Russian oligarchs.

Russian Ethnic Nationalism Rears its Ugly Head

With its ability to use soft power and economic incentives declining, Russia may shift to using hard-power resources – military force and political pressure – in order to push integration forward. Moscow is the dominant military power in the region. While the outright use of military force in the form of invasion may be prohibitively costly in today's world, it has enhanced Russia's ability to engage in unconventional warfare; i.e. covert warfare through local proxies combined with political and economic destabilization (Gvosdev 2014). These hard-power tools were on full display during Moscow's sophisticated operation to annex Crimea and (less effectively) its support of separatists in Ukraine's Donetsk and Lugansk regions. Many post-Soviet states are relatively weak in terms of the domestic legitimacy of their regimes and the stability of national borders and are therefore vulnerable to the kinds of internal security threats that Russia has the ability to provoke and manipulate (Bremmer 2009).

Moscow justified the annexation of Crimea by appealing to Russia's right to defend its ethnic Russian kin throughout the post-Soviet space. Putin made this the central theme of his 18 March 2014 Crimea speech, in which he announced Crimea's "return" to Russia and justified his policies in Ukraine to the nation. The speech can be read as a call to irredentism and a repudiation of the legitimacy of the post-Soviet division of borders. Putin laments that in 1991 "overnight the Russian people became one of the biggest, if not the biggest, ethnic group in the world divided by borders" (Putin 2014). The speech argues that in annexing Ukraine, Russia is correcting a historical wrong and suggests that the Crimean precedent can be repeated in other post-Soviet countries where Russian ethnic minorities face the threat of persecution, such as Kazakhstan and Moldova. According to Vladimir Socor (2014), "This view resembles Serbian leader Slobodan Milosevic's thesis about the Serbian nation as a 'divided nation,' entitled by virtue of 'historical injustice' to reclaiming territories from Yugoslavia's former constituent republics."

These policies have caused much concern throughout the post-Soviet space, awakening fears that Russia may try to play out the Crimean scenario in their countries, or at the very least use the threat of doing so as a tool of political blackmail. Official reaction to Russia's Ukraine policy has ranged from outright condemnation and firm support for Ukraine's sovereignty (in the Baltic States, Georgia, and Azerbaijan) to muted criticism and more general appeals for a peaceful political settlement (Uzbekistan and Armenia), to plain silence on the issues (Tajikistan, Kyrgyzstan, and Turkmenistan). Even states whose close economic and political ties with Russia have forced them to tacitly accept Moscow's policies (Kazakhstan and Belarus) have found subtle ways to voice their displeasure.

For example, during his annual marathon press conference for representatives of the foreign press in January 2015, Belarusian President Aleksandr Lukashenko rejected the idea that Belarus was part of Moscow's "Russkiy Mir" project and assured his audience that the Belarusian armed forces were now prepared to protect the country's borders "from Brest [on the border with Poland] to Vitebsk [on the border with Russia]."(BELTA 2015) Belarus is the country with the closest ties with Russia. The two are officially part of a "union state" project. The Belarusian population closely identifies with Russia and has been largely sympathetic to Russia's policies in Ukraine. However, results from a national survey conducted by the Independent Institute for Socio-Economic and Political Studies (IISEPS) in December 2014 seem to indicate that Russia's actions in Ukraine have heightened anxieties about Belarusian national sovereignty and eroded support for union with Russia. In this survey, 58.4% of Belarusians they would say "no" to a referendum on unification, up from 31.6% in 2007 and 47% in March 2014 (Korovenkova 2015).

There is also subtle evidence that the threat of Russian irredentism has already begun to affect the bargaining calculus between Russia and Kazakhstan. Russia has been unhappy with the Kazakh government in that it has not offered it the expected level of support on Ukraine. Astana has refrained from directly criticizing Crimea's annexation and Russia's support for separatists in the east and has approached the subject very carefully, stressing the complex historical context under which the crisis has occurred and stressing the need for an end to the violence and a negotiated settlement. Its support has stopped short of formal recognition, and it failed to support the retaliatory sanctions against agricultural products that Russia put in place in response to EU sanctions, a move that has significantly weakened the cohesiveness of the EEU's trade and customs policies. Yet, more troubling for Moscow, it has also sought to hedge against growing Russian influence in Kazakhstan by improving its ties with China and the EU.

When Russia's firebrand ultra-nationalist politician Vladimir Zhirinovsky called for Kazakhstan and other Central Asian republics to be incorporated into Russia at a public rally in Crimea (significantly attended by Putin himself), Kazakh President Nursultan Nazarbayev reacted decisively, demanding that Russia censure Zhirinovsky and asserting that Kazakhstan was ready to leave the EEU at any time if it believed its sovereignty was at risk (Tengri News 2014a). Putin answered with a statement of his own a few days later at an annual "town hall" style meeting with pro-Kremlin university students. Responding to a question by a young student whether a "Ukraine scenario" was possible in Kazakhstan, Putin praised Nazarbayev for his political wisdom and genius in leading a "territory" such as Kazakhstan that had no "history of statehood" (Tengri News 2014b). Though ostensibly complimentary, the statement can be interpreted as an indirect threat to Kazakh sovereignty and a warning of what may follow if Kazakhstan does indeed choose to leave the EEU (Suslov 2015).[2]

The use of these kinds of coercive tactics comes at a steep price. If Moscow continues to go down this path it will erode the legitimacy of the Eurasian project and foster resistance on the part of subordinate states. There are already some indications that this has begun to happen. Though Russia, Belarus, and Kazakhstan went forward with plans to establish the EEU in 2015, Russia's efforts to imbue the new union with political and foreign policy structures was scuttled by Nazarbayev and Lukashenko, who openly rejected further political integration and stressed that integration would only continue on a "purely economic" basis (Lillis 2014). Both countries have also asserted their independence during the Ukraine crisis. Rather than unequivocally backing Russia (their ostensible ally) in its fight against the Ukrainian government, both have hedged against Russian domination by opening up channels of communication and cooperation with Kiev. Kazakhstan recently announced plans to export coal to Ukraine, where most coal mines are now under the control of pro-Russian separatists in the Donbas (Tengri News 2014c).

Moreover, EEU members and candidate countries have taken advantage of the Ukraine crisis to extract costly concessions from Russia in exchange for continued support for integration. On the eve of the EEU treaty signing Belarus successfully pressured Russia to provide it with a USD 2 billion loan and USD 1.5 billion rebate for customs duties it previously paid to Russia for the re-export of Russian oil (Falyakhov 2014). EEU member state Kyrgyzstan is asking for a payment of USD 1 billion to compensate for the losses that higher EEU import tariffs will incur to its re-export of Chinese goods to Russia and other EEU member states. Kazakhstan is putting pressure on Russia to allow it to export its gas to Europe using Russia's pipeline infrastructure (but without paying Russian export duties) and demanding that it be allowed to restrict exports from Russia to protect its domestic industries – a clear violation of the free trade zone that the EEU is supposed to create (Kommersant 2015). In the end, the use of force and coercion in Ukraine has made it more costly for Russia to continue its regional integration project at a time when its economic resources are diminished. According to Suslov (2015), "The Ukraine crisis has ruined any prospects of real political and economic integration and reduced the EEU project to Russia buying off the loyalty of its allies."

Positioning Itself as a Regional Security Hegemon and Guarantor of Stability

In the aftermath of the Ukraine crisis, it is tempting to conclude that Moscow is moving toward a more coercive strategy for regional integration in which it will now use the threat of separatism and ethnic conflict and its considerable "asymmetric warfare" capabilities to pressure the smaller post-Soviet states to fall in line. However, such a conclusion fails to capture the true nature of the shift in Russia's regional leadership strategy. It is true that there have been some voices in Moscow that advocate for a more coercive turn in Russia's policy toward the former Soviet states (Dugin 2014; Prokhanov 2014). But these voices have been in the minority. The leadership in Moscow has backed away from coercive strategies even in Ukraine. It has distanced itself from the irredentist Novorossiya project and used its influence to clean out the most fervent proponents of this project from the rebel leadership in Donetsk and Lugansk (Dergachev and Krilov 2015). The initial euphoria over Crimea has given way to a more sober assessment of the costs of a more coercive approach – both in terms of the stiff resistance it is likely to meet from the smaller states in the post-Soviet space as well as isolation from the West. Moreover, as has been examined in the above section, the post-Soviet states have actually been able to use Russia's post-Ukraine international isolation to extract economic concession from Moscow in return for their continued support for integration.

At the same time, there is a growing appreciation in Moscow that Russia's ability to continue to use economic incentives to ensure these states' loyalty is now limited by Moscow's own economic woes. As a result, Russia has begun to shift its policy of regional integration away from economic issues and toward security and hard-power issues, areas where it continues to enjoy a distinct advantage. Russia is not simply trying to use its hard and coercive power to reestablish empire. Rather, Russia's approach is to establish its influence in these states by taking the lead on regional security issues and by positioning itself as the main guarantor of the security of the region's regimes against internal and transnationals security threats.

The Arab Spring and the rise of ISIS are of grave concern to the leadership in Central Asia and the Caucusus, both indirectly as a result of demonstration effects that may destabilize their own regimes (which are rife with corruption and weak in legitimacy), as well as directly, as jihadists that are currently fighting in the Middle East eventually come home. These problems take on increased importance as the United States winds down its presence in Afghanistan, opening up the prospect of further instability and the Taliban's return to power. These threats are of growing importance as two of the most significant Central Asian nations, Kazakhstan, and Uzbekistan, face the prospects of leadership successions due to the advanced age of their authoritarian presidents. Thus far, the region's leaders have been able to keep the lid on public discontent. But recent events in Tajikistan are a reminder of just how fragile many of these regimes are. On 4 September 2015, the former deputy defense minister led a group of armed Islamic militants in a failed coup against the country's authoritarian and secular government that left 22 people dead and scores injured (Pamfilova 2015). In May of the same year, Colonel Gulmurod Khalimov, commander of the country's OMON forces (a special police unit used in paramilitary and anti-protest actions), publicly defected to ISIS vowing in an online video that he would return to Tajikistan to establish Sharia law and warned the country's leadership that "we are coming to slaughter you" (Reuters 2015).

In recent months, Russia has stepped up its security commitments throughout the region. It has ramped up troop deployments and military aid to Central Asian states and increased military exercises and training within the framework of the CSTO, the Russian-led regional security body (Ritm Evrazii 2015). The topic of instability in the Middle East and the threat from ISIS were at the forefront of the CSTO's September 2015 meeting in Dushanbe, Tajikistan. At Russia's initiative member states agreed to a number of reforms designed to address the ISIS threat. These included increasing troop levels of the CSTO's rapid reaction force (currently at 4000 troops under Russian command) and providing it with more modern weaponry and equipment, as well as updating the organization's crisis management mechanisms and reforming its charter and other legal documents in order to make them more responsive (Mir24 2015). Despite a direct request from the Kyrgyz government, the CSTO failed to send forces to Kyrgyzstan in May–June 2010 when riots between ethnic Uzbeks and Kyrgyz broke out in the southern part of that country, largely because the legal framework for such an intervention was not in place. These latest changes may be intended to remedy this situation. Russia has also used the CSTO to police cyber space throughout the region. The coordinated efforts of regional security bodies have led to the closure of over 57,000 websites that are deemed to pose a threat to regional stability, including many that are accused of actively recruiting fighters for ISIS (Sputnik News 2015). According to President of Kyrgyzstan, Almazbek Atambayev,

> The recruitment of our citizens to participate in the armed conflict on the side of ISIS is particularly troubling, as many have returned to continue their terrorist activities and recruit others to their cause in the countries of the region. (Mir24 2015)

Russia's intervention in Syria can also be understood in this light. By supporting the Assad regime, Moscow is demonstrating its commitment to its Central Asian allies. The leadership of these countries can identify with the Assad regime. Most of them head secular authoritarian regimes that have narrow bases of social support based on clan or tribal affiliations and which face the prospect of active Islamist insurgencies. Many of these regimes do not fully trust the US and fear that any support they may get from Washington will require concessions on their part on democracy and human rights. According to prominent Russian security expert Konstantin Eggert (2015),

> Putin wants to demonstrate to the whole world that if you are an ally of the US, as was Egyptian president Hosni Mubarak, you will be told at the most critical moment "solve your own problems." But if you are an ally of Russia we will send you warplanes and tanks.

There are signs that Russia's readiness to respond to the region's growing security problems is helping push the integration agenda forward. After prolonged foot-dragging, Kyrgyzstan finally joined the EEU in August 2015. The fact that it did so at a time when Russia's economy is in crisis suggests that growing security concerns played a decisive role in the decision to finally accept membership (Mikheev 2015). EEU member countries have intensified cooperation with Russia (both bilaterally and under the auspices of the CSTO) to thwart the threat from terrorism, holding military exercises as well as increasing intelligence sharing (Korostikov 2015).

Russia has taken advantage of the renewed fighting in Nagorno-Karabakh to boost its influence and position itself as the lead outside mediator in the conflict between Armenia and Azerbaijan (Bryza 2016). Though the two countries are wary that Moscow is using the conflict to promote its regional hegemony, they are also heavily dependent on Russia. Armenia is a member of the CSTO and counts on Russia as a final guarantor if its security. Azerbaijan relies on Russia for a large proportion of its arms sales. Both sides are thus forced to grudgingly accept Russia's regional security leadership, despite their misgivings (Lukyanov 2016).

Russia has been able to play on regional security concerns to push forward security proposals that reflect its more narrow security concerns. CSTO member states have begun serious discussions about forming a regional air defense system (Tass 2015). Such a system would, of course, be of little use against Islamist insurgents, but would in reality serve to counter threats from NATO and the US. Up until now, Azerbaijan, Uzbekistan, and Turkmenistan have jealously guarded their sovereignty and independence and stayed out of Russian-led integration efforts. However, as the Islamist threat grows many experts in these countries are now re-evaluating their stance toward Eurasian integration, with the hopes that Russia can protect them against this new threat (Temnikov 2015).

Realizing the limits of its economic resources, Russia has now invited China to participate in its projects of Eurasian economic integration. Putin and Chinese President Xi Jinping have signed a memorandum of cooperation between the Eurasian Economic Union and China's Silk Road initiative whereby the two powers would work together to promote the region's economic development. Though many of the details still need to be ironed out, the Russian and Chinese expert communities are hard at work developing concrete proposals for economic cooperation that go beyond traditional areas such as energy and infrastructure and now include high tech, manufacturing, and the development of cross-regional production networks (Bordachev, Likhacheva, and Zhang 2014). Moscow's acceptance of China's growing economic presence in Central Asia represents a dramatic reversal in policy. For years, Moscow resisted Chinese proposals for joint economic cooperation through regional structures such as the Shanghai Cooperation Organization (SCO), and instead preferred to push for economic integration through structures such as the CU and EEU, which are controlled by Moscow (Tsygankov 2013). Russia is now resigned to establish a division of labor with China when it comes to

Eurasian integration, whereby Russia takes on the responsibility for security and China for economic development.

Russia has gone from a strategy of trying to use economic incentives and soft power to push forward regional integration to one in which it is looking to use its hard-power resources to achieve the same goals by positioning itself as the main security provider for the region's states. This shift in strategy reflects a realization of the limits of its economic and soft-power resources as well as a newfound confidence in its hard-power capabilities. However, while the means and resources employed to push integration forward may have shifted more to the military and security realm, the fundamental motivations behind the push for Eurasian integration remain geoeconomic: to create an integrated economic space where Russia will be the dominant economic power and thereby advance the larger goal of Russia's economic development and modernization.

Broader significance for the study of regions and regionalism

While the focus has been on contemporary Russia, this study makes several contributions to our broader understanding of regionalism, and in particularly, the role that regional powers play in pushing regional integration forward. This issue is one of growing interest in international relations, as emerging powers such as China, Brazil, and Russia have shown a keen interest in pushing forward various regional integration schemes (Acharya 2007; Hurrell 2012). The study of regionalism has traditionally been split into two tracks: the study of regional economic integration and the study of regional security complexes and hierarchies. There is a growing interest in linking and bridging both dimensions of "regioness" and in analyzing the possible patterns of interaction between them (Nolte 2010). This study illustrates the way in which regional powers can use hard power to pursue what are predominantly economic goals. Studies of regional economic integration primarily focus on the ways in which regional powers use their economic dominance or soft power to push forward the regional integration agenda (Flemes 2012). This study shows that even when a regional power's ultimate goal is economic integration, it does not necessarily have to only rely on its economic or soft-power to push forward the integration agenda. Hard-power tools or the provision of security (for example, through the deployment of peacekeeping forces) can also be used as a tool for regional powers that lack economic or soft-power resources.

Russia's new strategy for regional integration is in line with Lake and Morgan (1997) and Lake's (2009) work on regional security hierarchies. According to these studies, dominant states provide order and security, and in turn, make demands on subordinate states, which benefit from the order and therefore come to regard the leadership of the dominant state as legitimate and necessary for the maintenance of order. Dominant states form a kind of "contractual relationship" with subordinates in which protection is exchanged for loyalty.

> Key is that both the dominant and subordinate states understand that the dominant state has the right to make certain demands, rooted in its 'special responsibilities' for social order, and the subordinate state has an obligation to comply with those commands if made. (Lake 2009, 38)

As the dominant military power in the region, Russia is attempting to assert its authority by establishing hierarchical relations with the region's smaller states that commit it to providing for their internal security in exchange for their participation in Russian-led regional integration projects. Lake and Morgan (1997) and Lake (2009) focus on the security dimensions of the emerging hierarchical relationship. However, Russia's goals are shaped as much by economic interests as they are by security. Russia is trying to leverage its security leadership to achieve its broader regional economic goals and advance regional economic integration.

The analyses of Lake and Morgan (1997) and Lake (2009) primarily focus on the external security threats that subordinate states face from other states; that is, in providing regional order, the dominant state keeps subordinate regional states from fighting one another. This article also examines the leadership role that powerful regional states can play in protecting states from internal security threats, particularly those that have transnational or global dimension, such as Islamic radicalism. These kinds of threats are arguably more important to developing states that are still undergoing the process of nation and state building and where the legitimacy of domestic political institutions is weak (Ayoob 1995). As such, this is a heretofore ignored dimension of security leadership that may be of particular significance to regional powers that are trying to exercise their authority over developing states. It is these kinds of internal threats, rather than threats from other states, that constitute the most acute threat to many of the post-Soviet countries, whose experiences of state- and nation-building are weak and whose domestic political regimes are poorly institutionalized and highly authoritarian. Here, Russia has a comparative advantage over other possible security sponsors (such as the United States) in that it is much more tolerant of these regime's violations of human rights and anti-democratic practices.[3]

Conclusion

The crisis in Ukraine exposed Russia's limited ability to use material incentives and soft power to integrate the post-Soviet space under its leadership. As a result, Russian leaders have chosen to rely on the country's still considerable hard power and military capabilities to push integration forward. Russian leaders are fully aware of the pitfall of this strategy, but they chose to pursue it because they see reliance on hard power as the only option now open to them (Bordachev 2015). The alternative would mean giving up the pursuit of regional dominance and the Eurasian integration project, and is not considered to be a viable option. Continued regional dominance and Eurasian integration are regarded as being critical to

Russia's economic revival and its ability to maintain its status as a great power in international politics, and thus worth the risk (Karaganov 2014).

Russia's decision to pursue regional integration by the most effective means that it possesses – the use of hard and military power – has heightened conflict between Russia and the West to the point where many politicians and experts believe that the two sides are in a "new cold war." It would be easy to place the blame squarely on Russia for this turn of events. But the West – and particularly the architects of the EU's Eastern Policies – also made a critical mistake in pushing for Ukraine's Western integration without taking Russia's interests into account. In doing so, Western leaders took advantage of Russia's soft-power weaknesses, not appreciating that this would provoke a hard-power response. A more farsighted strategy would have also given Russia some stake in Ukraine's future, instead of letting the question of Ukraine's future devolve into a zero-sum contest between East and West. Russia may have been willing to accept Ukraine's European integration if it had also promised some tangible benefits for Russia. German Chancellor Angela Merkel's proposal that the EU and EEU can begin negotiations on a free trade agreement after the Ukraine crisis is settled has piqued the interest of Moscow (Trenin 2014). If such an agreement had been offered from the very beginning, we may have been able to avoid the current crisis entirely. As of now, it may be too little too late.

A Ukrainian strategy that included Russia would also have allowed Russia to develop its economic and soft power so that it would not have to rely on its hard and coercive power to push regional integration forward. Such a Russia could be a factor of regional and global stability. Instead, we now have a Russia that has become estranged from the Western-led world order and which sees the use of coercive and hard power as the only tools available to it to defend and promote its interests.

In the absence of economic and soft power it may be logical for Russia to move to a strategy where it relies on its hard power to push forward its foreign policy goals. However, such a strategy is deeply problematic. Russia may be overestimating the usefulness of its hard-power resources in achieving goals, such as Eurasian integration, which will also require it to effectively exercise economic and soft power if they are to truly be successful. Reliance on hard power can detract from its ability to develop its soft power and economic resources. Investment in the military and the embracing of military commitments on its periphery and beyond places a burden on the Russian economy and creates resentment and fear of Russia among many of the smaller states in the region. Even Russia's ability to successfully play the role of regional security provider will be limited as it does not have the tools to address the economic and social problems that are at the root of the region's internal security problems (Cooley 2012).

According to Andrei Kortunov (2015),

To quote Mark Twain: 'When the only tool you have is a hammer you tend to see the world as being made up of nails.' Because hard power is the only effective tool we have we see it as the solution to all problems.

While Russia and the rest of the world certainly face many new and pressing security challenges, Moscow's preoccupation with the use of hard power may give rise to a skewed world view which exaggerates the degree to which the world is becoming more dangerous and disorderly and overemphasizes the effectiveness of traditional military means in addressing the actual threats that are emerging in an increasingly globalized and interconnected world.

Notes

1. The current full members of the CIS are Armenia, Azerbaijan, Belarus, Kazakhstan, Kyrgyzstan, Moldova, Russia, Tajikistan, and Uzbekistan. Ukraine and Turkmenistan are only associate members, as they never officially ratified the CIS founding treaty. Georgia withdrew from the organization in 2009 in the aftermath of the 2008 Russo-Georgian war. In 2014, the Ukrainian government announced that it will cut all its ties with the organization and submitted a bill to the Ukrainian parliament to begin the process. However, as of writing, the Ukrainian authorities have not yet made a final decision to leave the CIS.
2. We are indebted to Dmitry Suslov for pointing out the significance of this episode in Russian–Kazakh relations. Suslov (2015).
3. The US also supports authoritarian regimes when it deems it to be in its larger national interests to do so. However, the US criticizes the kind of gross human rights violations that often occur when authoritarian governments repress internal opposition. For example, the US was very vocal in its criticism of Uzbekistan's government after it massacred protester in Andijan in 2005. This prompted Uzbekistan to back away from the security ties it was developing with the US (forcing the closure of a US airbase) and to increase security cooperation with Russia and China.

Disclosure statement

No potential conflict of interest was reported by the authors.

ORCID

Maxim Bratersky http://orcid.org/0000-0003-2966-0056

References

Acharya, Amitav. 2007. "The Emerging Regional Architecture of World Politics." *World Politics* 59: 629–652.

Adamczyk, Edward. 2014. "Ukraine's Economy Would Have Collapsed without Russian Aid – IMF Chief." *United Press International*, April 4. http://www.upi.com/Top_News/World-News/2014/04/04/Ukraine-economy-would-have-collapsed-without-Russian-aid-IMF-chief-says/2821396632040/

Ayoob, Mohammed. 1995. *The Third World Security Predicament: State Making, Regional Conflict, and the International System*. New York: Lynne Reiner.

BELTA. 2015. "Lukashenko za pyat' chasov press-konferentsii otvetil bolee chem na 60 vopros' zhurnalistov." [During a Five Hour Press Conference Lukashenko Answered Over 60 Questions Posed By Journalists.] January 16. http://www.belta.by/president/view/lukashenko-za-pjat-chasov-press-konferentsii-otvetil-bolee-chem-na-60-voprosov-zhurnalistov-65431-2013

Bloomberg. 2013. "Russia Offers Ukraine Cheaper Gas to Join Moscow-led Group." December 2. http://www.bloomberg.com/news/articles/2013-12-01/russia-lures-ukraine-with-cheaper-gas-to-join-moscow-led-pact

Bogomolov, Alexander, and Oleksandr Lytvynenko. 2012. "A Ghost in the Mirror: Russian Soft Power in Ukraine." *Chatham House Briefing Papers*. London. http://www.chathamhouse.org/sites/files/chathamhouse/public/Research/Russia%20and%20Eurasia/0112bp_bogomolov_lytvynenko.pdf

Bordachev, Timofei. 2015. *Interview with authors*. Moscow, January 15.

Bordachev, Timofei, and Andrei Skriba. 2014. "Russia's Eurasian Integration Policies." In *The Geopolitics of Eurasian Economic Integration*, edited by David Cadier, 16–22. London: LSE IDEAS Special Report.

Bordachev, Timofei, Anastasia Likhacheva, Xin Zhang. 2014. "What Asia Wants, or the "Four C's": Consumption, Connectivity, Capital & Creativity." *Valdai International Discussion Club Paper Series*, no. 1. http://vid-1.rian.ru/ig/valdai/Paper_Asia_eng.pdf

Bratersky, Maxim. 2010. "Regional'nyye ekonomicheskiye ob'yedineniya skvoz prizmu mirovoy politicheskoy ekonomii." [Regional Economic Integration through the Prism of the World Political Economy.] *SShA-Kanada* 8: 19–33.

Bremmer, Ian. 2009. "The Post-Soviet Nations after Independence." In *After Independence: Making and Protecting the Nation in Postcolonial and Post-Soviet States*, edited by Lowell Barrington, 141–162. Ann Arbor: University of Michigan Press.

Bryza, Mathew. 2016. "Putin Fills Another U.S. Leadership Void in Nagorno-Karabakh." *Washington Post*. April 11. https://www.washingtonpost.com/opinions/nagorno-karabakh-conflict-is-too-dangerous-to-ignore/2016/04/11/1e32fc44-ff23-11e5-9d36-33d198ea26c5_story.html

Chebanov, Sergei. 2010. "Strategecheski interesy rossii na postsovetskom prostranstve." [Russia's Strategic Interests in the Post-Soviet Space.] *Mirovaia ekonomika i mezhdunarodnoye otnosheniia* 8: 23–40.

Coalson, Robert. 2014. "Despite Ukraine Crisis, Russia Pursues Eurasian Integration Dream." *Radio Free Europe/Radio Liberty*, May 28. http://www.rferl.mobi/a/russias-eurasian-integration-dream-steams-ahead-despite-ukraine-crisis/25401714.html

Cohen, Steve. 2014. "Cold War Again: Who's Responsible?" *The Nation*, April 1. https://www.thenation.com/article/cold-war-again-whos-responsible/

Cooley, Alexander. 2012. *Great Games, Local Rules*. Oxford: Oxford University Press.

Dergachev, Vladimir, and Dmitry Krilov. 2015. "Proekt Novorossiya zakryt." [The Novorossiya project is closed.] *Gazeta.ru*, May 5. http://m.gazeta.ru/politics/2015/05/19_a_6694441.shtml

Dugin, Alexander. 2014. "Proyekt Novorossiya." [Project Novorossiya.] *RIA Novosti Ukraini*, April 21. http://med.org.ru/article/4790

Eggert, Konstantin. 2015. "Siriiskaya ruletka Vladimira Putina." [Vladimir Putin's Syrian roulette.] *Radio Deutsche Welle Russia*, October 1. http://www.dw.com/ru/a-18752815.

Falyakhov, Rustem. 2014. "Ot Evrazii odni ubitki." [From Eurasia only losses.] *Gazeta.ru*, October 21. http://www.gazeta.ru/business/2014/10/17/6264305.shtml

Flemes, Daniel, ed. 2012. *Regional Leadership in Global Perspective*. Hamburg: GIGA.

Gaddy, Clifford, and Barry W. Ickes. 2014. "Ukraine: A Prize Neither Russia Nor the West Can Afford to Win." *Brookings Foundation*, May 22. http://www.brookings.edu/research/articles/2014/05/21-ukraine-prize-russia-westukraine-gaddy-ickes

Giles, Keir, Philip Hanson, Roderic Lyne, James Nixey, James Sherr, and Andrew Wood. 2015. *Chatham House Report; The Russian Challenge*. London: The Royal Institute of International Affairs.

Glazev, Sergey. 2013. "Takyie raznie integratsiii: Chemu uchit''vostochnoe partnerstvo." [Diverse approaches to integration: What the 'Eastern Partnership' teaches us.] *Rossiya v globalnoi politike* 6, http://www.globalaffairs.ru/number/Takie-raznye-integratcii-16252

Gurova, Irina, and Mariya Efremova. 2010. "Potentsial regionalnoi torgovli SNG." [The Potential for Regional Trade in the CIS.] *Voprosyekonomiki* 7: 108–122.

Gvosdev, Nikolas. 2014. "The Bear Awakens: Russia's Military Is Back." *The National Interest*, November 12. http://nationalinterest.org/commentary/russias-military-back-9181

Higgins, Andrew, and Andrew Kramer. 2015. "Leader Was Defeated Even Before He Was Ousted." *New York Times*, January 3. http://www.nytimes.com/2015/01/04/world/europe/ukraine-leader-was-defeated-even-before-he-was-ousted.html?_r=0

Hille, Katherine. 2015. "Russia: Dangers of Isolation." *Financial Times*, January 9. http://www.ft.com/cms/s/0/657967da-9725-11e4-845a-00144feabdc0.html#axzz3wB7TmU2t

Hurrell, Andrew. 2012. "Regional Powers and the Global System from a Historical Perspective." In *Regional Leadership in Global Perspective*, edited by Daniel Flemes, 15–30. Hamburg: GIGA.

Ivanter, Viktor Valery, Vladimir Yasinskiy Geets, Alexander Shirov, and Andrey Anisimov. 2012. "The Economic Effects of the Creation of the Single Economic Space and Potential Accession of Ukraine." In *Eurasian Integration Yearbook 2012*, edited by Vladimir Vinokurov, 19–41. St. Petersburg: Eurasian Development Bank.

Karaganov, Sergei. 2014. "Russia Needs to Defend its Interest with an Iron Fist." *Financial Times*, March 5. http://www.ft.com/intl/cms/s/0/1b964326-a479-11e3-9cb0-00144feab7de.html

Kolesnikov, Andrei, and Aleksander Gabuev. 2015. "Prospects for the Eurasian Economic Union." *Moscow Carnegie Center Eurasia Outlook*, January 13. http://carnegie.ru/commentary/?fa=57699

Kommersant. 2015. "Embargo po-kazaxskii." [Embargo Kazakh-style.] February 6. http://www.kommersant.ru/doc/2661304

Korostikov, Mikhail. 2015. "Islamskoe gosudarstvo probivaet put' v Tsntralnuyu Aziyu." [The Islamic State Barges into Central Asia.] *Kommersant*, December 7. http://www.kommersant.ru/doc/2871941

Korovenkova, Tatiana. 2015. "Lukashenko budet ubezhdat' elektorat chto dlya krizis vinovata Rossiya (Lukashenko will Convince the Electorate that Russia is Responsible For the Crisis)." *Naiyny.by*, January 17. http://naviny.by/rubrics/politic/2015/01/17/ic_articles_112_188014/

Kortunov, Andrei. 2015. *Interview with authors*, Moscow, October 14.

Krickovic, Andrej. 2014. "Imperial Nostalgia or Prudent Geopolitics? Russia's Efforts to Reintegrate the Post-Soviet Space in Geopolitical Perspective." *Post-Soviet Affairs* 30: 503–528.

Kubicek, Paul. 2009. "The Commonwealth of Independent States: An Example of Failed Regionalism?" *Review of International Studies* 35: 237–256.

Kudors, Andis. 2010. "Russian World" – Russia's Soft Power Approach to Compatriots Policy." *Russian Analytical Digest* 81: 2–6.

Kudrin, Alexander, and Evsey Gurvich. 2015. "A New Growth Model for the Russian Economy." *Russian Journal of Economics* 1: 30–54.

Kuvshinova, Olga. 2013. "Rossiya gotovitsya k desyati toschim godam." [Russia Prepares for Ten Thin Years.] *Vedomosti*. November 7. http://www.vedomosti.ru/newspaper/articles/2013/11/07/rossiya-gotovitsya-k-desyati-toschim-godam

Lake, David. 2009. "Regional Hierarchy: Authority and Local International Order." *Review of International Studies* 35: Supplement S1: 35–58.

Lake, David, and Patrick M. Morgan. 1997. *Regional Orders: Building Security in a New World*. University Park, PA: Pennsylvania State University Press.

Laqueur, Walter. 2015. *Putinism: Russia and Its Future with the West*. London: St. Martin's Press.

Lillis, Joanna. 2014. "Did Putin's Eurasian Dream Just Suffer Geostrategic Setback?" *Eurasianet*, April 24. http://www.eurasianet.org/node/68322

Lukyanov, Feodor. 2016. "Trebuetsya chestnyi broker." [In Need of an Honest Broker.] *Gazeta.ru*, April 7. http://www.gazeta.ru/comments/column/lukyanov/8164907.shtml

Lynch, Dov. 2000. *Russian Peacekeeping Strategies in the CIS*. London: Palgrave Macmillan.

Makhovsky, Andrei, Dmitry Solovyov, and Margarita Antidze. 2015. "Russia's Financial Crisis May Bury Putin's Eurasian Dream." *Reuters*, January 14. http://www.reuters.com/article/us-russia-crisis-cis-idUSKBN0KN16720150114

Mankoff, Jeffrey. 2009. *Russian Foreign Policy: The Return of Great Power Politics*. Plymouth: Rowmann & Littlefield.

Matlack, Carol. 2016. "Russia's Great Downward Shift." *Bloomberg*. January 28. http://www.bloomberg.com/news/articles/2016-01-28/russia-s-economy-faces-long-term-decline

Mearsheimer, John. 2014. "Why the Ukraine Crisis is the West's Fault."*Foreign Affairs* 93 (Sep/Oct): 77–89.

Mikheev, Sergei. 2015. "Vstuplenie v EAES – strategicheski vygodnyi shag dlya KR." [Joining the EEU – A Strategically Beneficial Step for the Republic of Kyrgizstan.] *Vechenii Bishkek*, November 24. http://www.vb.kg/doc/329472_sergey_miheev:_vstyplenie_v_eaes_strategicheski_vygodnyy_shag_dlia_kr.html

Mir24. 2015. "Lideri stran ODKB podveli itogi vstrechi v Dushanbe." [CSTO Leaders Sum Up Results of Dushanbe Meeting.] September 15. http://mir24.tv/news/community/13254225

Mirzayan, Gevorg, and Mikhail Pak. 2015. "Ne grozi yuzhnomu frontu." [Do not disturb the Sothern Front.] *Lenta.ru*, October 20. http://lenta.ru/articles/2015/10/20/khorasan/

Moskvin, Lev. 2007. *SNG: Raspadilivozrozhdenie? vzglyad 15 let spustya* [The CIS: Failure or Rebirth? The View 15 years Later]. Moscow: Institut Sotsiologii RAN.

Naumkin, Vladimir, and Igor Ivanov. 2013. *Russia's Interests in Central Asia: Contents, Perspectives, Limitations*. Moscow: Spetskniga.

Nolte, Detlef. 2010. "How to Compare Regional Powers: Analytical Concepts and Research Topics." *Review of International Studies* 36: 881–901.

Nye, Joseph. 2011. *The Future of Power*. New York: Public Affairs.

Pamfilova, Victoria. 2015. "False Start of a Coup in Tajikistan"*Vestnik Kavkaza*, September 8. http://vestnikkavkaza.net/analysis/False-start-of-a-coup-in-Tajikistan.html

Prokhanov, Alexander. 2014. "Vkluchat Novorossiyu v sostav Rossii eshe ne vremya i ne rezon." [To Include Novorissiya in Russia in Premature and Unreasonable Yet.] *Sobesednik*, July 1. http://sobesednik.ru/politika/20140701-aleksandr-prohanov-vklyuchat-novorossiyu-v-sostav-rossii-esh

Putin, Vladimir. 2011. "Novyi integratsionii proekt dlya Evrazii – budushee kotoroe razvivaetsya segodnya." [The New Integration Project for Eurasia – The Future Which is Happening Now.] *Izvestia*, October 3. http://ivestia.ru/news/502761

Putin, Vladimir. 2014. *Transcript of Address by President of the Russian Federation*. March 18. http://eng.kremlin.ru/news/6889

Reuters. 2015. "Commander of elite Tajik police force defects to Islamic State." May 28. http://www.reuters.com/article/us-mideast-crisis-tajikistan-idUSKBN0OD1AP20150528

Ritm Evrazii. 2015. "V Tadjikistane i vo vsei Tsentralnoi Azii idet planovoe usilenie rossiskogo voennogo prisustviya." [In Tajikistan and throughout Central Asia Russia is Systematically Strengthening its Military Presence.] October 8.

Sakwa, Richard. 2015. *Frontline Ukraine: Crisis in the Borderlands*. London: I. B. Tauris.

Shashenkov, Maxim. 1994. "Russian Peacekeeping in the 'Near Abroad'." *Survival* 36: 46–69.

Silayev, Nikolai. 2014. "How to Sell 'Russia'? Why Russian Soft Power Does Not Work." *Russia in Global Affairs* 1 (Jan/Mar). http://eng.globalaffairs.ru/number/How-to-Sell-Russia-16506

Sinyavsky, Andrei. 2015. *Soviet Civilization: A Cultural History*. New York: Arcade.

Socor, Vladimir. 2014. "Putin's Crimea Speech: A Manifesto of Greater-Russia Irredentism." *Eurasia Daily Monitor* 11 (56): March 25. http://www.jamestown.org/regions/centralasia/single/?tx_ttnews%5Btt_news%5D=42144&tx_ttnews%5BbackPid%5D=53&cHash=377ed0c543440239db55d887478f2c53

Sputnik News. 2015. "CSTO Thwarts 50,000 ISIL Recruitment Websites in Central Asia." September 15. http://sputniknews.com/world/20150915/1027014422.html#ixzz3qAE9q02B

Stoner, Kathryn, and Michael McFaul. 2015. "Who Lost Russia (This Time)? Vladimir Putin." *The Washington Quarterly* 38: 167–187.

Sultanov, Bulat. 2015. "Kazakhstan and Eurasain Integration" In *Eurasian Integration – The View from Within*, edited by Richard Sakwa and Piotr Dutkiewicz, 97–110. London: Routledge.

Suslov, Dmitri. 2015. *Interview with Authors*. Moscow. February 15.

Tass. 2015. "Minoborony: peredacha yzla Balxash pozvolit razvivat' edinuyu regionalnuyu sistemu PVO-PRO." [Ministry of Defense: The Command Center At Balkhash Will Help Develop A Common Regional Air And Missile Defense System.] December 5. http://tass.ru/armiya-i-opk/2501888

Temnikov, Roman. 2015. "EAES kak garant bezopasnosti." [The EEU as a Guarantee of Security.] *Gazeta Kaspiy*, October 29. http://kaspiy.az/news.php?id=32835#.Vnp7rvl97IV/

Tengri News. 2014a. "Kazakhstan may leave EEU if its interests are infringed: Nazarbayev." August 27. http://en.tengrinews.kz/politics_sub/Kazakhstan-may-leave-EEU-if-its-interests-are-infringed-Nazarbayev-255722/

Tengri News. 2014b. "President Vladimir Putin of Russia on Kazakhstan and Its Future." August 30. http://en.tengrinews.kz/politics_sub/President-Vladimir-Putin-of-Russia-on-Kazakhstan-and-its-future-255793/

Tengri News. 2014c. "Kazakhstan Recently Announced Plans to Export Coal to Ukraine." June 2. https://en.tengrinews.kz/markets/Kazakhstan-wants-to-export-coal-to-Ukraine-China-and-Europe-253890/

The Economist. 2013. "Trading Insults." August 24. http://www.economist.com/news/europe/21583998-trade-war-sputters-tussle-over-ukraines-future-intensifies-trading-insults

Trenin, Dmitri. 2011. *Post-Imperium: A Eurasian Story*. New York: Carnegie.

Trenin, Dmitry. 2014. "Konets soglasiya: chego xochet' Evropa ot Rossii?" [The end of Agreement: What does Europe want from Russia?] *RBk Daily*, December 1. http://www.rbc.ru/opinions/politics/01/12/2014/547c4489cbb20f0a1ef8b7b1

Tsygankov, Andrei. 2013. *Russia's Foreign Policy: Change and Continuity in National Identity*. Plymouth: Rowman & Littlefield.

Vedemosti. 2013. "Ukraina ostanovila podgotovku k podpisanyu soglasheniya s ES." [Ukraine Halts Preparations for Agreement with the EU.] October 22. http://www.vedomosti.ru/politics/articles/2013/11/21/ukraina-ostanovila-podgotovku-k-podpisaniyu-soglasheniya-s.

(Dis-)integrating Ukraine? Domestic oligarchs, Russia, the EU, and the politics of economic integration

Julia Langbein

How do the politics of economic integration pursued by the European Union (EU) and Russia in their shared neighborhood affect domestic change in these countries? Do the two external powers further economic integration with one or the other, and how do their strategies shape the survival of rent-seeking domestic elites? Examining the case of Ukraine's car industry, the paper reveals a considerable degree of disengagement by both the EU and Russia. Both external actors offer domestic elites surprisingly few opportunities for economic integration but rather pursue a "policy" of de facto exclusiveness that caters to the domestic interests of the EU and Russia. So far, the EU has strongly promoted trade liberalization to facilitate market access for European car producers but has not created opportunities for foreign-led restructuring of Ukraine's car industry, thereby leaving the sector without a chance to benefit from liberalization. Russia, in turn, compromised existing trade linkages with Ukraine to protect its own domestic car industry. What is more, the strategies of both the EU and Russia even provided opportunities for Ukrainian oligarchs with stakes in the domestic car industry, who were not interested in transparent forms of economic interaction in the first place, to pursue rent-seeking strategies that undermined any chance for sustainable development of the industry. While Russia's disengagement has caused trade disintegration that may have contributed to Ukraine's reluctance to join Russia-led integration regimes, the EU is well advised to create opportunities for sustainable integration if it does not want to become a factor of further destabilization in Ukraine.

Introduction

Long before the Ukrainian crisis and the war in eastern Ukraine, scholars had emphasized that the European Union (EU) and Russia pursue competing interests in the countries belonging to their shared neighborhood (Dimitrova and

Dragneva 2009; Casier 2012; Delcour and Wolczuk 2013).[1] Since the end of 1990s, the EU has sought to transfer its rules and norms to these countries, first via the conclusion of Partnership and Cooperation Agreements (PCA) and more recently through the European Neighborhood Policy and the Eastern Partnership initiative (Langbein 2015). Russia in turn has attempted to integrate these countries in Russia-led integration regimes, such as the Eurasian Customs Union. The Russian leadership fiercely opposes the EU's involvement in the region, especially since the latter started to negotiate Association Agreements and Deep and Comprehensive Free Trade Agreements (DCFTAs) with Ukraine, Georgia, Moldova, and Armenia in 2009 (Delcour and Wolczuk 2013). This special issue uses the term "contested neighborhood" to account for the competing interests of the EU and Russia in the region (Ademmer, Delcour, and Wolczuk 2016).

That said, none of the countries located in the contested neighborhood should be considered as "pawns" in a geopolitical competition between the EU and Russia or victims of geography. While recognizing that their location is, in, and of itself, problematic in geopolitical terms, previous research has emphasized that the preferences of rent-seeking domestic elites mitigate the effect of the EU and Russia on domestic change. In some countries like Belarus or Azerbaijan, rent-seeking incumbent elites consist predominantly of formal veto players, such as the president and his entourage. In other countries like Ukraine, these elites consist predominantly of informal veto players, such as competing oligarchic groups in Ukraine who work through the Ukrainian government to enhance their rent-seeking (Franke, Gawrich, and Alakbarov 2009; Balmaceda 2013; Dimitrova and Dragneva 2013). Notwithstanding the particular character of domestic elites, the EU and Russia are more likely to achieve their policy goals in the region – if their goals fit the preferences of incumbent elites (see, for example, Ademmer and Börzel 2013; Dimitrova and Dragneva 2013; Delcour and Wolczuk 2013; Langbein 2015; Ademmer and Delcour 2016).

This article takes a closer look at the dynamics underlying the three-way interaction between the EU, Russia, and rent-seeking domestic elites and its consequences for domestic change in the contested neighborhood. First, the article studies the politics of economic integration pursued by the EU and Russia vis-á-vis the contested neighborhood and their effects on domestic opportunity structures. Do the EU and Russia, indeed, offer domestic elites opportunities to foster closer economic integration with one or the other external power? Or rather, do the EU and Russia offer opportunities for domestic elites to undermine integration? Second, and related to this, the article investigates how the politics of economic integration pursued by the EU and Russia shape the stability of the local regime: Do domestic elites use the opportunities offered by the EU and Russia to advance their rent-seeking, thereby undermining sustainable economic development (cf. Buzogány 2016; Wolczuk 2016)?

This article examines these questions through an in-depth empirical analysis of the Ukrainian automotive industry since the breakup of the Soviet Union. The

Ukrainian automotive sector is a strategic case for such a study because it was highly integrated with the Russian automotive industry during Soviet times. We would therefore expect Russia to keep the sector within its sphere of influence and use interdependencies in order to exert political and economic leverage (Ademmer, Delcour, and Wolczuk 2016). At the same time, we would expect the EU to have a general interest in integrating Ukraine into the larger European market and in economically stabilizing the country, especially in an industry as pivotal to the Ukrainian economy as the car industry.

Instead, the paper reveals a considerable *disengagement* of both the EU and Russia from the Ukrainian car industry. Both external actors offer domestic elites surprisingly few opportunities for economic integration but rather pursue a "policy" of de facto exclusiveness that caters to the domestic interests of both the EU and Russia. Put differently, the case study shows how unilateral decisions being taken by external actors to defend their purely domestic interests produce externalities that can crucially affect domestic change in third countries.

While the EU strongly promoted trade liberalization to facilitate market access for European car producers, Ukrainian producers were unable to use trade liberalization as an opportunity for development. Ukrainian brands could not compete with EU brands on the liberalized market on an equal footing. In contrast to the EU's engagement in prospective candidate countries from Central and Eastern Europe prior to their accession in 2004/2007, the EU did not create opportunities for foreign-led restructuring of Ukraine's automobile industry through financial incentives or institution building (Bruszt et al. 2015). Unlike the EU, Russia was eager to promote protectionism and compromised existing trade linkages with Ukraine to protect its own domestic car industry, thereby contributing to the declining competitiveness of Ukrainian brands.

Last but not least, the analysis of Ukraine's automotive sector provides further evidence for the argument that strategies of external actors can stabilize or even strengthen rent-seeking domestic groups (see also Börzel and Pamuk 2012; Balmaceda 2013; Wolczuk 2016). In fact, the strategies of both the EU and Russia had an even more devastating effect on the Ukrainian car sector since so-called "oligarchs," who were not interested in market economic reforms in the first place and hence did not try to adjust to competitive pressures in a profit-maximizing way, had captured the Ukrainian state. Both the EU and Russia did not offer opportunities for the emergence of state capacities to promote market integration and upgrading. Instead, both the EU and Russia provided opportunities for Ukrainian oligarchs with stakes in the domestic car industry to pursue rent-seeking strategies that put sustainable development of the industry at great risk. The oligarchs even formed short-term alliances with Brussels whenever EU preferences (trade liberalization) helped them to pursue their own interests (see also Ademmer and Börzel 2013), no matter whether their actions had a deteriorating effect on the Ukrainian car industry or on the Ukrainian economy writ large.

To sustain these arguments, the article will proceed as follows: the next section traces developments in the Ukrainian automotive industry since the breakup of the Soviet Union and identifies three time periods; collapse (1991–1999), upward trend (2000–2008), and second collapse (2009–2015). For each period, the paper presents in-depth analysis of domestic strategies that shaped the development of the Ukrainian automotive industry and scrutinizes whether and how the interaction with external actors encouraged domestic elites to foster sustainable development or facilitated rent-seeking. The conclusion summarizes key findings and discusses their implications for the politics of economic integration pursued by Russia and the EU in their contested neighborhood.

Developments in post-Soviet Ukraine's automotive industry

Ukraine was the second largest automotive producer within the Soviet Union. The Zaporizhia Automobile Building Plant (ZAZ) located in Zaporozhye was the most important car factory in Ukraine and produced relatively cheap small-sized cars. The Zaporozhets, and to some extent the Tavria, became very popular model lines in the former Soviet Union (Yegorov 2004, 207). However, with the end of state socialism, Ukraine's automotive sector collapsed. Passenger car production declined from 155,600 units in 1991 to only 10,136 units in 1999. Interestingly, beginning in 2000, Ukraine's car production witnessed a dramatic upward trend, with production numbers increasing to slightly more than 400,000 units in 2008. Yet in 2009, Ukraine's car industry once more witnessed a sharp decline, with production numbers dropping to around 65000 units. Production numbers declined even more drastically after 2011 to less than 26,000 units in 2014 (Figure 1).

The following sections examine how the EU and Russia shaped these outcomes by structuring opportunities for domestic elites to support or undermine the development of Ukraine's car industry.

Collapse of Ukraine's automotive industry (1991–1999)

Domestic strategies
Key domestic actors representing Ukraine's automotive industry during 1990s were car producers such as Ukraine's largest company, AvtoZAZ, which was state-owned until 1998, on the one hand, and an emerging group of private-owned car dealers, Bogdan Corporation and UkrAVTO, on the other hand. Bogdan was founded in 1992 and started its business by selling Russian cars on the Ukrainian market. In 1996, Bogdan became the official dealer for the South Korean automobile firm KIA in Ukraine. Until 2009, Bogdan belonged to Ukrprominvest, an investment company led by Petro Poroshenko (Ukraine's current president) and his father, Oleksiy (Olszanski and Wierzbowska-Miazga 2014).[2] UkrAVTO emerged out of the privatization of the Ukrainian Motor Vehicles Maintenance Department that consisted of diverse car manufacturing, parts, sales and maintenance agencies during

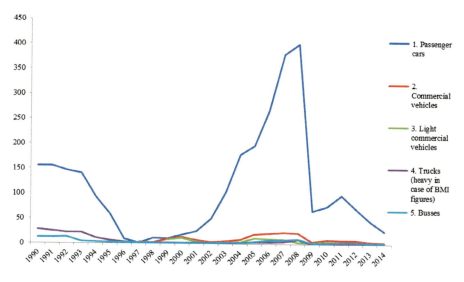

Figure 1. Expansion of the sector – Volume of motor vehicle production in Ukraine – by type (in thousands).
Source: VDA International auto statistics 1995–2001 for 1990–1996; OICA Production Statistics for 1997–2014.

Soviet times. Tariel Vasadze established the company in 1995. UkrAVTO became the main dealer of Toyota in Ukraine and later also for Nissan, Daimler Chrysler, and other foreign brands.[3] Both Bogdan and UkrAVTO also dominated the car dealing business with used cars from the West. The owners of the two companies, Petro Poroshenko and Tariel Vasadze, became leading Ukrainian oligarchs by the end of 1990s.

Throughout 1990s, car producers and car dealers pursued diverging interests. Ukraine's car producers were keen to keep import tariffs for used and new cars on a high level in order to limit the inflow of technologically more competitive foreign brands. Ukrainian car dealers wanted to keep import tariffs on final and used cars at a low level in order to meet increasing domestic demand for foreign brands after the breakup of the Soviet Union. With the introduction of the new customs tariff in 1992,[4] the Ukrainian government under Prime Minister Leonid Kuchma pursued an ambivalent policy to cater to diverse domestic interests in the sector: on the one hand, the government liberalized the import of used cars from the West, which put domestic producers under extreme competitive pressures. The decision accommodated the interests of car dealers and the broader public whose demand for Western second-hand cars increased significantly during the early 1990s (Yegorov 2004). On the other hand, the Ukrainian government introduced a 20 percent import tariff on new cars. In addition, the government granted domestic car producers value-added tax (VAT) exemptions to increase their competitiveness on the Ukrainian market and cushion the negative effects of increasing imports of used cars from the West.[5] Yet the majority of locally produced cars were not sold but were exchanged for material through a chain of intermediaries. Cash-flow problems brought the production of Tavria at Ukraine's main car producing plant AvtoZAZ to a practical standstill by the end of 1997, although its capacity was at 170,000 cars per year (Yegorov 2004).

Notwithstanding AvtoZAZ's deep crisis in late 1996, the South Korean firm Daewoo expressed its interest in investing US$150 million in a joint venture with AvtoZAZ (Yegorov 2004). Daewoo's decision must be seen in the context of an internationalization strategy pursued by its headquarters to increase the competitiveness of the company in the global automotive market. In 1994 and 1996, Daewoo had already invested in the automotive sectors of Central and East European countries (CEEC), such as Poland and Romania, to gain access to the EU market. Daewoo's investments in Ukraine aimed at using the country as a hub to get access to the Commonwealth of Independent States (CIS) market. Daewoo's plan in Ukraine foresaw the production of 70,000 cars in the first year, both Tavria models and Daewoo's own brands, and 250,000–300,000 cars per year until the fifth year, which would have not only saved the jobs of AvtoZAZ's 28,000 employees but would have created new jobs. During negotiations with the Ukrainian government, the company made commitments to promote technology transfer by attracting Ukrainian sub-contractors. Daewoo's investment plan foresaw having 70 percent the parts and components produced in Ukraine within 10 years (Yegorov 2004, 210).

In turn, the Ukrainian Cabinet of Ministers promoted the adoption of a law "On the Stimulation of Automobile Production" in 1997 that was based on several government decrees aiming at facilitating the joint venture between AvtoZAZ and Daewoo (OECD (Organization for Economic Cooperation and Development) 2001).[6] The law privileged automobile manufacturers with at least US$150 million investment (the exact scope of Daewoo's planned investment) by granting them various tax exemptions until 2008 and by doubling import duties on other car brands to 60 percent of the price (Yegorov 2004, 210). Furthermore, the Ukrainian government agreed to impose an import ban on used cars older than five years and established a US$5000 minimum customs value on imported used cars (Schneider 2001).[7] Ukrainian car dealers heavily opposed the law. However, in 1997, when the law privileging Daewoo was debated, leading Ukrainian car dealers like Bogdan's Petro Poroshenko and UkrAvto's Tariel Vasadze were just about to enter Ukrainian politics and turn into powerful oligarchs. They still lacked political influence to prevent the adoption of the law. The Kuchma government was keen to preserve AvtoZAZ from going bankrupt, and the law was passed (Yegorov 2004, 210).

Despite these protectionist measures, the Korean investor did not fulfill the expectations it had raised during negotiations with the Ukrainian government. The new Tavria and Daewoo models produced at the AvtoZAZ plant were too expensive to compete with imported used cars, the major bulk of which had still been imported before the introduction of special customs duties following the legislative changes discussed earlier. Further, Ukrainian second-hand car dealers started to use the practice of "temporary import" of cars, which allowed them to circumvent high import tariffs (Yegorov 2004, 212).[8] As a result, AvtoZAZ-Daewoo managed to produce less than 25,000 cars and sales totaled only 10,700 vehicles in 1998. Results for 1999–2001 did not look much better. Further, Daewoo did not keep its promise to build a network of domestic suppliers of parts and components. For Daewoo models, they were imported from elsewhere; e.g. from Daewoo's plants in the CEEC. For Tavria models, 60 percent of parts and components were still imported from Russia in 1998/1999 (Yegorov 2004, 211). By 1999, 90 percent of all cars sold in Ukraine were used cars from the West (Yegorov 2004, 207).

To what extent did the Russian and EU policy toward Ukraine contribute to or at least not help in preventing the collapse of the Ukrainian car industry?

Russian impact

The competitiveness of Ukrainian car brands both in terms of quality and price was affected by three decisions taken by the Russian government: first, in the early 1990s, the Russian government began to divert exports of fuel and gas from the former Soviet Union to Western markets to gain higher revenues for its natural resources (Yegorov 2002). As a consequence, Ukrainian car producers had to pay higher fuel and gas prices that led in turn to an increase in the costs of production. Thus, the price of Ukrainian Tavria models increased from US$2000 in 1994 to US$3000–$3500 in 1996, but their quality did not improve commensurately, which made it more difficult for AvtoZAZ to compete with used cars imported from the West or cheaper Russian cars.

Secondly, after the breakup of the Soviet Union, Ukrainian firms continued to purchase goods from Russian firms on the basis of credits. Yet in 1992, the Russian government prohibited issuing credits to Ukrainian firms, as Russia's central bank was afraid to import even higher inflation from neighboring Ukraine (Lloyd and Freeland 1992). For Ukraine's car manufacturers, Russia's decision accelerated the collapse of traditional trade links, as Ukrainian manufacturers had used 80 percent of parts and components from Russia during Soviet times. As a result, the quality of Ukrainian brands decreased even more dramatically (Yegorov 2002). Only when Daewoo invested in Ukraine did that cash flow help to re-establish trade links with Russia.

Thirdly, after the breakup of the Soviet Union, the Russian market stopped being an important export destination for Ukrainian cars. Despite a free trade agreement (FTA) between Russia and Ukraine concluded in 1993, at best 10–15 percent of the actual trade flows between both countries were covered through the FTA (Le Le Gall 1994, 77). The rest was regulated through bilateral agreements, sometimes at the level of firms. This development, however, became increasingly difficult due to Ukraine's liquidity problems. Further, throughout 1990s, the Russian government heavily protected its domestic automotive industry, which had been in predominantly private hands since 1993, by tariffs ranging between 40 and 46 percent on imports of final cars, including those from Ukraine. In 2000, out of more than

1 million new cars sold in Russia, only 48,000 were imported (Pavlínek 2002, 60). Ukraine's share in this respect was marginal.[9]

When Daewoo invested in Ukrainian AvtoZAZ, Russian car-makers were not concerned about the introduction of the various protective measures explained above. Russian brands were still competitive on the Ukrainian market. The Russian financial crisis of 1998, resulting in the depreciation of the ruble, made Russian Lada models much cheaper than Korean models despite high import tariffs that were supposed to protect AvtoZAZ-Daewoo.

All in all, during 1990s, Russia's trade policy toward Ukraine contributed to the declining competitiveness of Ukrainian car brands and to increasing domestic demand for Western second-hand cars or Russian brands. In fact, Russia's policies, such as increasing prices for Russian fuel and gas as well as a strongly protected Russian automotive sector, constrained rather than facilitated a sustainable development of the Ukrainian automotive industry. By doing so, Russia compromised existing trade linkages with Ukraine for protecting its own national car industry.

EU impact

Obviously, the EU did not object to the rather liberal trade policy pursued by the Ukrainian government in the automotive sector in the early 1990s, in particular, in the second-hand car market, which caused the collapse of the Ukrainian automotive industry: Passenger car production declined from 155,500 units in 1990 to only 2000 units in 1997 (Figure 1). Still, trade liberalization was not a result of EU leverage. Ukraine's trade policy in the car sector was rather an attempt to meet domestic demand at a time of economic scarcity.[10] At the same time, the EU did little to avoid the collapse. True, the EU offered de facto asymmetric liberalization. EU import tariffs on automotive products were 10 percent in 1995 and subsequently fell to 5 percent in 1996, while Ukraine's import tariffs stayed at 20 percent up to 1997 (European Commission 1992, 1994). However, regulatory harmonization with EU technical standards would have been essential to allow Ukraine's automotive industry to benefit from asymmetric liberalization that the EU had offered. Due to the country's Soviet past, Ukraine's car production still applied Soviet *Gosstandards*, which were based on different safety and environmental requirements than EU standards. Even though the PCA between the EU and Ukraine, which was enforced in 1998, foresaw Ukraine's regulatory approximation to the *acquis*,[11] the EU did not offer comprehensive financial and technical to support the process (Langbein 2015).

When the Ukrainian government tried to save the national car industry by concluding a deal with Daewoo that strongly discriminated against other foreign car importers, the EU sharply criticized the deal, arguing that it was contradicting several articles of the PCA, including articles on non-discrimination of foreign companies and state aid commitments, as well as rules of the World Trade Organization (WTO). The Commission threatened to no longer support Ukraine's bid to join the WTO, to withdraw aid, and to start a dispute settlement under Article 96 of the

PCA (European Commission 1998). But the Ukrainian government was keen to attract Daewoo's investment in order to save the country's national car industry and did therefore not enforce the regulations for the car sector that had been agreed upon in the PCA.

It is hard to say whether the collapse of Ukraine's automotive sector could have been avoided if the EU had pursued a different strategy of economic integration vis-á-vis Ukraine's automotive sector. Clearly, the EU did not force Ukraine to adopt such a liberal trade policy in the early 1990s, which basically destroyed the sector. That said, the EU did also not object to it. Notably, even vis-à-vis prospective candidate countries, the EU's trade policy was rather liberal as a consequence of the so-called Europe Agreements, which the EU negotiated with countries like Poland and Romania in the early 1990s, Poland introduced a 35 percent import tariff on final cars in 1992 and agreed to decrease it by 5 percent points every two years. In a similar vein, Romania gradually decreased its tariffs on car imports.

However, in Poland and Romania, the EU helped to attract some sustainable foreign investments through massive support for institution building beginning in the mid-1990s, as well as by providing financial assistance to cushion social costs of adjustment and by granting prospective candidates flexibility with implementing EU state aid regulations (Bruszt et al. 2015). The Romanian case, whose political economy was largely defined by rent-seeking and corruption in 1990s (Gallagher 2006), is particularly insightful for Ukraine, as it shows that the EU can strengthen a state's capacity to promote market integration and upgrading even in an unfavorable domestic environment.[12] The Ukrainian automotive industry did not only lack a state that was able to anticipate the negative consequences of trade liberalization and take the necessary steps to alleviate them, but also a committed external actor.

Recovery without restructuring (2000–2008)

Domestic strategies

After Daewoo's bankruptcy in 1999, the Ukrainian automotive industry witnessed an important change in ownership structures: leading Ukrainian car dealers acquired stakes in local car manufacturing plants. Daewoo was taken over by General Motors (GM), which was not keen to take the risk of starting full-scale production in Ukraine. Instead, GM signed a General Distribution Agreement with car dealer UkrAVTO. In 2002, UkrAVTO acquired 82 percent of AvtoZAZ's shares, and UkrAVTO's director Tariel Vasadze became AvtoZAZ's new president. Two years earlier, Ukrainian car dealer Bogdan had entered the car production business with the purchase of the LuAZ plant in Lutsk. Further, in 2000, AvtoInvestStroy (AIS Group) became the main founder and the general dealer of a Ukrainian–Russian joint venture named Kremenchuk-AutoGAZ,[13] which was renamed KraSZ in 2001.

The owners of the various plants, Tariel Vasadze (UkrAVTO), Petro Poroshenko (Bogdan), and Dmytro Svyatash and Vasyl Poliakov (AIS Group), became part of

Ukraine's "big business." These oligarchs were not only active in the car industry but also acquired stakes in other sectors such as media, food, or finance. In the early 2000s, all of them began to occupy political posts in various political parties. Due to the unpredictability of the state in a rent-seeking society, the key interest of oligarchs is not necessarily to achieve long-term development through entrepreneurial activities. Rather, oligarchs are interested in accumulating personal wealth and in stabilizing the system of patronage by blocking reforms aimed at breaking up the logic of rent-seeking and rent-giving. This is not a phenomenon peculiar to the car sector. It is also prevalent in sectors like energy and in other trade-related areas (Balmaceda 2013; Langbein 2015; Wolczuk 2016). In fact, key oligarchs, in particular, those affiliated with the former Party of Regions, such as Svyatash and Poliakov, were the worst public debtors in Ukraine in the past (Dimitrova and Dragneva 2013).

During 1990s, the dichotomy between Ukrainian car dealers and producers had triggered a somewhat ambivalent government policy oscillating between liberalization and protectionism. The personal union of dealers and producers, which emerged in the early 2000s, and the increasing political and economic power of these oligarchs led the Ukrainian government to pursue a more coherent approach toward Ukraine's automotive industry. At the same time, the presence of rent-seeking elites in the automotive industry foreclosed opportunities for developing the sector. Not only did the Ukrainian state refrain from creating the necessary institutional framework that would allow entrepreneurs to re-combine factor inputs and develop new and more competitive products. After Daewoo's commercial failure, the Ukrainian government's policy even allowed the new owners of car producing plants (i.e. the oligarchs) to gain short-term rents through asset stripping.

In the early 2000, the Ukrainian government agreed to relax the protection against competing car imports in response to increasing pressure of Ukrainian oligarchs with stakes in both the car dealing business and in car production. The import tariff for cars was reduced from 60 to 25 percent and the import tariffs for parts and components were reduced to 5 percent. Further, the government abolished the minimum customs value on cars and limited the import ban on used cars to cars older than eight years (Schneider 2001, 73). At the same time, the Ukrainian government continued to grant certain tax privileges to AvtoZAZ, including VAT exemption, which also applied to the sales of foreign brands assembled at AvtoZAZ's production plant.

These changes were supposed to attract semi-knock-down (SKD)[14] assembling at local production plants. Foreign investors were expected to start SKD assembling, which would allow them to circumvent the far higher import duty on cars and exploit tax exemptions. While SKD assembling would create some jobs related to the welding, painting, and assembling of cars, the key impetus for the government's decision came from Ukrainian oligarchs.

In this context, it is important to remember that in the early 2000s, Ukraine's economy started to grow. The purchasing power of Ukrainian consumers increased significantly due to simplified access to private credit as a result of an increasing share of foreign bank ownership (Lukianenko and Suchok 2011). Until 2006, Ukrainian consumers preferred budget cars priced up to US$10,000. But afterward, consumers tended to buy vehicles in the price range US$10,000–$15,000, and they became more interested in buying new rather than second-hand cars. By 2004, the Ukrainian market had become number 10 in Europe in terms of car sales, ahead of Switzerland, and Poland (Negyesi 2006). Foreign car producers could no longer afford to ignore the Ukrainian market. Hence, Ukrainian oligarchs with stakes in the car business saw their chance to make short-term profits at relatively low costs.

During 1990s UkrAVTO, Bogdan, and AIS had all gained experience as dealers of several foreign brands, which included after-sales services such as repair and quality controls. Hence, they lent themselves as ideal partners for SKD assembling. Ukraine's automotive industry witnessed a sharp increase in terms of production numbers from 2000 onward: Between 1999 and 2008, annual production increased by more than 390,000 units (see Figure 1). This production increase resulted from partnerships between domestic producers, such as AvtoZAZ, and foreign producers designed to facilitate SKD assembling of foreign car models at Ukrainian plants. UkrAVTO started to invest in SKD assembling of Opel models in 2002. AIS started assembling Russian GAZ Gazelle and Volga models, while Bogdan began assembling cars based on Russian VAZ kits.

In the following years, SKD assembling in Ukraine witnessed a boom: AvtoZAZ/Ukravto signed deals with Citroën, Fiat, and others, and Bogdan began SKD assembling of Hyundai and KIA cars in 2005. AIS began SKD assembling of Chinese Geely cars and South Korean SsangYong cars in 2007. Furthermore, in 2002, Eurocar (Evrocar) was established with Ukrainian capital in Solonomovo in the Transcarpathian Special Economic Zone (SEZ) located near the Ukrainian border with Slovakia and Hungary with the goal to attract SKD assembling of Skoda cars. Thanks to its location in an SEZ, Eurocar benefitted from numerous tax breaks including exemption from import and excise duties, import VATs and five-year exemptions from land tax and other taxes (OECD (Organization for Economic Cooperation and Development) 2001). In 2002, Eurocar started to produce duty-free Skoda models at a price 4 percent less than imported Skoda cars. However, Skoda and its parent company Volkswagen did not invest capital in Eurocar and did not sign a joint venture agreement, since they wanted to minimize the risk in view of Ukraine's bad business environment.[15]

From a developmental perspective, SKD assembling must certainly be preferred to mere reliance on car imports because SKD at least creates jobs for assembling and after-services. While SKD is often seen as the first step toward full-scale production, SKD assembling on its own cannot help a country to develop a strong and competitive car sector. In Ukraine, SKD did not help developing a suppliers network since the Ukrainian government did not tie benefits for SKD assembling, such as tax exemptions, to the condition that localization must be achieved within a given time period. Hence, local content rules for SKD assembling did not apply in Ukraine. Further, Ukrainian car producers did not invest their profits to improve the sector's position in transitional value chains; for example, by investing in new technology to facilitate complete-knock-down (CKD) assembling.[16] AvtoZAZ/Ukravto is the only plant engaged in CKD assembling, which began in 2004 with the in-house production of ZAZ Lanos using Daewoo licenses.

Russian and EU impact

Russia and the EU have hardly had an influence on the choices made by Ukraine's domestic elites in the early 2000s. The European Commission certainly welcomed the decision of the Ukrainian government in 2000 to lower import tariffs on parts and components to 5 percent and on final cars to 25 percent. However, it is unlikely that the Ukrainian government adopted a more liberal trade policy in response to EU leverage. Ukraine did not enjoy EU candidate status and had far more leeway to pursue its own national economic policy compared to countries like Poland or Romania. Further, President Leonid Kuchma's "multi-vectoral" foreign policy aimed at using the EU or Russia as opportunity structures to stabilize his power position internally vis-à-vis competing oligarchic groups (Bukkvol 2002). Hence, the Kuchma government would only accept EU demands if this step helped to gain support from certain oligarchs.

As mentioned earlier, since 1998, the European Commission had strongly criticized the preferential treatment of AvtoZAZ and threatened to postpone Ukraine's WTO accession should Ukraine not stop the discriminatory treatment against Western car imports (European Commission 1998). But only in 2000 did the Ukrainian government agree to liberalize automotive trade in response to the interests of Ukrainian oligarchs who had acquired stakes in car manufacturing. From this perspective, it is also hardly surprising that the EU's criticism of Ukraine's state aid policy in relation to the establishment of Special Economic Zones did not fall on fertile ground. In fact, the European Commission frequently lamented about tax exemptions offered to Eurocar and others, arguing that Ukraine would not comply with WTO and EU regulations.[17] But the EU was not able to limit the Ukrainian government's abuse of state aid as a means to protect domestic economic elites.

Further, the presence of the EU market had only have a marginal effect on the start of SKD assembling in Ukraine. Car producers from the EU and the United States did not intend to export cars made in Ukraine to the EU. For example, Skoda's goal was to get a foothold in the Ukrainian market and to expand to Russia and Belarus in anticipation of intensified regional economic integration and removal of trade barriers between these countries (Pavlínek 2008). Only Asian car producers wanted to use Ukraine as a springboard to gain access to the EU and CIS markets, hence the increasing SKD assembling of Chinese, Korean, and Japanese cars. Russian car producers like VAZ and GAZ, however, simply started to assemble their models in Ukraine to circumvent high import tariffs and remain competitive vis-à-vis other brands on the Ukrainian market.

While exports of final road vehicles to the EU increased slightly in 2000, exports to the EU declined again in the following years despite SKD and CKD assembling of foreign brands in Ukraine. The major export destinations of motor vehicles produced in Ukraine, of which cars make up the biggest share, were Russia and the other CIS countries (Figure 2), while the large majority was sold on the domestic market.

How has Russia's policy shaped the development of the Ukrainian automotive industry since the early 2000s? Similar to 1990s, the priority of Russian authorities and car producers was to strengthen the Russian car industry with a focus on establishing it as a producer of low-cost cars for the domestic market (Richet and Bourassa 2000). Russian authorities imposed import bans on Ukrainian car imports, which competed directly with low-cost Russian brands, and vice versa. Trade cooperation between both countries was hence hardly rule-based despite the existence of an FTA since 1993. As a result of ongoing trade wars, the Ukrainian government increased the import tariff on Russian-made cars to 31.7 percent in 2002, while all other car importers continued to pay a 25 percent surcharge. As a result, the share of Lada cars in the Ukrainian car market for new cars dropped from 54 percent in June to 37 percent in November 2002, which is why the Russian firm VAZ decided to expand SKD assembly of its Lada cars at Bodgan's production plant in early 2003 (Yegorov 2004). However, VAZ or other Russian car producers did not place long-term investments on the Ukrainian car market to facilitate integration of Ukrainian firms in transnational value chains with Russian car producers. Russian car producers preferred placing investments on the local market and building a suppliers network with Russian firms rather than with Ukrainian firms. Moreover, since the early 2000s, Russian car producers have been exposed to increasing competition with Western car imports, even in the low-cost segment. Therefore,

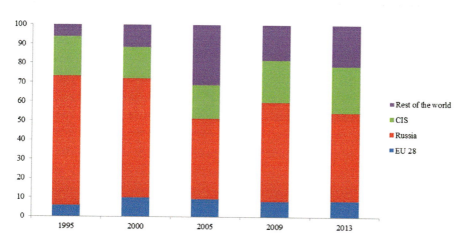

Figure 2. Main destinations of Ukraine's motor vehicle exports (in percent of total), 1995–2013.
Source: UNCTAD (Various years).

they were more interested in setting up joint ventures with Western producers like GM and Fiat to ensure better quality of their parts and components (Richet 2007).

Summing up, foreign investors, including European and Russian producers, played a limited role in restructuring Ukraine's automotive sector during 2000s. The discretionary, ad hoc actions of the Ukrainian government as shown in the Daewoo case, paired with weak institutions guaranteeing investors' rights, did certainly not increase foreign investors' trust in predictable policy-making. It is true that the Ukrainian government's decision to lower import tariffs on parts and components and rising purchasing power succeeded in attracting SKD assembling of foreign brands. However, foreign investors, from the EU in particular, were reluctant to place investments in Ukrainian car production in the form of CKD assembling or full-scale production. Ukraine's investment climate was still considered to be low and unpredictable. While a bad investment climate could hardly scare away Russian investors, throughout 2000s, Russian authorities and producers were keen to protect and strengthen Russian car production by imposing bans on Ukrainian car imports if needed or by focusing on the insertion of the Russian car industry in value chains with Western car producers.

Downturn 2.0 (since 2008)

Domestic strategies

The upward trend of the Ukrainian car industry lasted until 2008. Car production in Ukraine went down from more than 400,000 units in 2008 to 65,700 units in 2009 (Figure 1). Two events brought the car boom to an abrupt end: during WTO accession negotiations, the Ukrainian government agreed to reduce import tariffs for cars from 25 to 10 percent without a transition period upon the country's accession in 2008 (WTO 2008). As a result, foreign investors had weaker incentives to circumvent high import tariffs through SKD assembling and MNCs, especially from the EU (VW, Opel, Daimler Chrysler, Renault), began to scale back SKD assembling.[18]

Further, the Ukrainian car industry was dramatically affected by the financial crisis that hit Ukraine in October 2008 (Movchan 2009). The crash resulted in the devaluation of the Ukrainian Hryvna and cut domestic producers short off credits to buy consumer goods, including new cars. Data on personal disposable income dynamics reveal a decline of 15 percent in the year following the crisis. The numbers of car registrations went down by 73 percent from 2008 to 2009 (InvestUkraine 2012). At the same time, reduced import tariffs on cars, which the Ukrainian government had to introduce to fulfill Ukraine's obligations as a WTO member, led the share of imported cars in domestic car sales up from 52 to 80 percent between 2008 and 2012. In 2012, the majority of cars were imported from Russia followed by Germany, Japan, China, and Korea (GTAI (Germany Trade and Invest) 2013).

Car manufacturers like AvtoZAZ/UkrAVTO's Tariel Vasadze had heavily criticized the decision to lower import tariffs on cars upon accession to the WTO but Vasadze had lost political influence during the final phase of negotiations. In 2005,

Vasadze had joined Yulia Timoshenko's block as Member of Parliament. After the Orange Revolution in late 2004, Timoshenko became Prime Minister and Viktor Yushchenko became President of Ukraine. As Härtel (2012) details, Timoshenko and Yushchenko followed a joint political agenda in the beginning and strongly supported Ukraine's Western integration path, including membership in the WTO and the EU. Yushchenko, in particular, clearly linked Ukraine's prospects for democratization and economic liberalization to the country's clear commitment to a pro-Western/pro-EU policy. Timoshenko, in turn, gradually turned away from a clear pro-Western/pro-EU foreign policy orientation due to increasing electoral support beyond Western Ukraine.

Timoshenko had to give up her position as Prime Minister after her party lost the parliamentary elections of 2006 that brought her opponent Viktor Yanukovich from the Party of Regions to power. As the final phase of WTO accession, negotiations took place without Timoshenko in power; car producers like Tariel Vasadze had lost a strong political ally within the Ukrainian government and became increasingly marginalized. Petro Poroshenko, a close ally of President Yushchenko, had sold his share in Bogdan in the late 2008 as a consequence of the financial crisis (Olszanski and Wierzbowska-Miazga 2014) and therefore did not back up Vasadze's claims vis-à-vis the Ukrainian President.

Yushchenko was keen to finalize Ukraine's WTO accession and ready to accept EU demands for greater trade liberalization regarding the car sector. Oligarchs with stakes in the industry did not back him, so Yushchenko had nothing to lose in this respect. The EU had made the beginning of negotiations of an Association Agreement including a DCFTA, which Yushchenko often portrayed as a first step toward EU membership vis-à-vis his constituency, conditional upon Ukraine's membership in the WTO. By accelerating Ukraine's accession to the WTO, Yushchenko tried to weaken critical voices within his own party who accused him of Ukraine's sluggish reform process and stalled integration with Western alliances (Härtel 2012, 273).

Soon after Ukraine's WTO accession, the Ukrainian government under Yulia Timoshenko, who was re-appointed Prime Minister in late 2007 and took up her office in April 2008, pursued a more protectionist policy: About a year after Ukraine's WTO accession, in March 2009, a law introducing a temporary 13 percent import surcharge on all products excluding 'critical' imports came into force.[19] The list of products that were classified as 'non-critical' and thus subject to import surcharge included, among others, motor vehicles. Timoshenko's political ally and car producer Tariel Vasadze had lobbied heavily for this law because he was keen to increase incentives for foreign investors to invest in SKD or CKD assembling. Following harsh criticism raised by the WTO and the European Commission, the Cabinet of Ministers narrowed the list down to just two categories – cars and refrigerators (Movchan 2009). Nevertheless, the EU threatened to suspend negotiations of a DCFTA with Ukraine, which had begun in 2008. Further, in August 2009, the General WTO Council threatened the Ukrainian government to impose penalties if the import surcharge on these two categories would not be abolished. A month later, the EU's policy conditionality proved effective and the Ukrainian government abolished the import surcharge (Härtel 2012, 411). However, Ukraine's battle with the WTO and the EU would continue.

Viktor Yanukovich from the Party of Regions, who became Ukraine's President in 2010, pursued a more protectionist policy. Interestingly, in September 2011, UkrAVTO's Tariel Vasadze, who used to sit in the Ukrainian parliament for the Bloc Yulia Timoshenko, joined the Party of the Regions faction to maintain his political influence. Yanukovich increasingly underlined the potential negative effects of free trade with the EU although the EU made some concessions: At the start of DCFTA negotiations between the EU and Ukraine in 2008, the EU initially had favored an elimination of Ukraine's import tariffs for passenger cars imported from the EU. However, in 2012, the EU and Ukraine ultimately agreed upon a gradual reduction of import tariffs for passenger cars (which were at 10 percent after WTO accession) within a 10-year transition period and the right to adopt safeguard measures if car imports increase rapidly (Dabrowski and Taran 2012, 20). In view of increasing domestic criticism against the DCFTA, which was further fueled by a negative public relations campaign launched by the Russian government (Delcour and Wolczuk 2013), the EU made these concessions to secure the Ukrainian government's support for the DCFTA. Notwithstanding, powerful oligarchs like Tariel Vasadze with close relations to the Party of Regions lobbied the government not to sign the DCFTA, as they felt they were already sufficiently suffering as a result of Ukraine's WTO accession and the financial crisis.[20]

Ukraine's critical stance against WTO commitments and DCFTA negotiations changed only after a new government came to power following the Euromaidan protests against Yanukovich and his entourage. In June 2014, the new pro-Western Ukrainian government under President Petro Poroshenko and Prime Minister Arseniy Yatsenyuk signed the DCFTA with the EU. Despite the fact that Ukraine's automotive industry has hardly recovered given the general bad economic situation in the country, the ongoing war in eastern Ukraine, and tensions with Russia resulting in a shutdown of trade between the two countries in sectors like the car industry, the signed DCFTA kept the arrangements for the car sector that had been negotiated in 2012 with the Yanukovich government. After having decreased to 28,751 vehicles in 2014, automobile production in Ukraine has currently almost come to a complete halt, as motor vehicle production plunged in the first quarter of 2015 to 1485 vehicles (Unian 2015).

EU and Russian impact

The previous discussion suggests that EU leverage on Ukraine was increasing after the Orange Revolution. The new leadership under President Yushchenko and Prime Minister Timoshenko was at least in the beginning united in its desire to achieve closer relations with the West. Hence, WTO membership, which the EU defined as a precondition for the beginning of negotiations on the Association Agreement, was high on the agenda and somewhat compromised for a strategic assessment of its negative consequences and of potential measures to mitigate them. According to Valeriy Pyatnitskii, Ukraine's chief negotiator during WTO accession negotiations, the EU had exerted considerable pressure on Ukrainian negotiators to approve lower import tariffs on cars and lift tax exemptions for companies based in Special Economic Zones. The EU had promised investments in Ukraine's car industry in return, so the latter would be able to survive increasing competition from imported cars.[21] Other observers, however, argue that the EU, or more precisely the European Commission, was at no time in the position to promise investments, as this is first of all a decision of private business.[22] Instead, they argue, the Ukrainian government had simply not made a strategic assessment of the tentative effects of WTO accession on various sectors of the Ukrainian economy, and Brussels did not care about negative externalities of trade liberalization on the Ukrainian economy.

A similar approach shaped the EU position during DCFTA negotiations. Since membership was not in the cards, Ukraine's eventually low compliance with EU single market rules could not endanger the integrity of the internal market (Langbein 2014; Bruszt et al. 2015). Unlike during 1990s and the early 2000s, when the interests of incumbent elites under the leadership of President Kuchma did not always overlap with EU demands, the EU had increasing leverage on Ukraine's pro-Western governments after the Orange Revolution up until 2010 and after 2014. Brussels used this leverage to constrain Ukraine's opportunities for protecting the automotive sector and other industries against increasing competition. In turn, however, the EU hardly offered any opportunities for restructuring the sector. As a representative of DG Trade made it very clear still in the late 2014:

> We serve interests of EU producers. Increasing the competitiveness of the Ukrainian car sector would mean creating competitors to EU producers. Our goal is to improve the business climate in Ukraine, and reduce barriers to trade which will then also improve conditions for attracting foreign investment.[23]

The Russian government wanted to prevent Ukraine's WTO accession and also tried to discard the DCFTA by emphasizing its negative consequences on the Ukrainian economy, including trade disruptions with Russia. At the same time, Russia's policy regarding automotive trade with Ukraine remained rather protectionist after 2008. In 2012, Russian car producers successfully lobbied for a recycling tax on car imports, which also included Ukrainian imports. Although the share of exports to Russia made up only a minimal number of car sales for Ukrainian producers, Russia's protectionist measures certainly did not help to increase economic interdependencies between Russia and Ukraine, which Moscow could have used to exert political pressure on Ukraine later on. In a similar vein, Russia did not try to increase its leverage by establishing joint ventures with Ukrainian car producers. Although the Kremlin heavily criticized Ukraine for negotiating the DCFTA, it did not offer a viable alternative for the Ukrainian automotive industry either. Hence, the opposing position of Ukrainian car producers against the DCFTA was not so much rooted in the fear of trade sanctions Russia would impose. Rather, Ukrainian car producers were afraid of even further competition from Western car imports.

Conclusion

This article investigated how the politics of economic integration pursued by the EU and Russia in their contested neighborhood affected domestic change in these countries. By studying the Ukrainian car industry since the breakup of the Soviet Union, the article described how the EU and Russia shaped domestic choices for trade (dis-)integration. What is striking is the great level of disengagement of both Russia and the EU in a sector as pivotal to the Ukrainian economy as the automotive industry, which had a devastating impact on the sector's development resulting in disintegration rather than trade integration with one or the other market. But equally striking is the extent to which rent-seeking domestic actors have actively shaped the outcome and, in fact, often managed to stabilize their power position because of Russia's and the EU's disengagement.

Starting from the assumption that trade interdependencies are likely to increase external leverage on domestic politics (Ademmer, Delcour, and Wolczuk 2016), it is particularly surprising that Russia did not make use of existing linkages between the Russian and Ukrainian automotive industry after the breakup of the Soviet Union. Rather, Russia's trade policy toward the Ukrainian automotive sector was driven by the motivation to strengthen and protect Russia's own car producers. Russia imposed trade sanctions against Ukrainian car imports in violation of bilateral agreements when necessary. Russian government policies also compromised the existence of a suppliers network between Ukraine and Russia when its maintenance turned out not to be economically viable for Russia's national champions. Further, Russian car makers have made little investment in Ukraine's car industry. Unlike Russian oil companies or banks, whose investments not only increase Russia's economic but also political leverage on Ukraine, less strategic and successful sectors like the Russian car industry focused on ensuring their survival on the Russian market in the light of increasing competition from Western imports. While the Russian strategy may not be surprising from a business point of view, the case of the automotive industry shows that the Russian leadership has not used every chance to increase interdependencies with its "near abroad" and has not pursued a rule-based trade policy toward Ukraine.

That said, certain sectors of the Ukrainian industry, such as the military-industrial complex or machinery, are strongly interdependent with their Russian counterparts both in terms of imports and exports. They are certainly mores strongly affected by the complete shutdown of trade with Russia following the beginning of the Russo-Ukrainian war in 2014 than those industries that have been less dependent on Russia. In fact, trade barriers Russia imposed on Ukrainian car imports in 2014 can only be considered as the tip of the iceberg rather than the root cause for the second collapse of the Ukrainian car industry. The second downturn of the industry had already begun with the financial crisis in 2008. The fact that the Ukrainian car industry was hit so hard can be explained by its shallow integration in transnational production chains and generally low competeness. Unlike its counterparts from Central and Eastern Europe, Ukraine's car industry was more vulnerable to this external shock (Bruszt et al. 2015). Insights from other sectors, such as the food industry, indicate that the automotive sector is by far not the only sector where Russia has compromised interdependencies with Ukraine to protect Russian industry (Langbein 2015; Delcour 2016). This may also explain why Ukraine has never whole-heartedly embraced economic integration with Russia (see also Delcour and Wolczuk 2013; Langbein 2015).

As regards the role of the EU, Brussels did not offer the Ukrainian automotive industry more opportunities for development than Russia. The EU was keen to promote trade liberalization to facilitate market access for EU car producers. At the same time, the EU hardly offered any opportunities that would have helped the Ukrainian automotive industry to survive as an integrated part of European value chains. Unlike in EU candidate countries such as Romania, which was dominated by equally rent-seeking elites as Ukraine throughout 1990s, the EU left it mostly to domestic actors in Ukraine to establish institutions that would ensure an investor-friendly and competitive environment. Unsurprisingly, with rent-seeking domestic elites being in the majority, the Ukrainian government resorted to a domestic survival strategy that focused on producing cheap local brands for the Ukrainian market rather than creating the necessary institutional preconditions for foreign capital-led restructuring. As a consequence, Ukraine did not manage to attract foreign investment with long-term strategic interest. Notably, rent-seeking domestic elites, above all powerful oligarchs and their respective political allies, even used the EU as an opportunity structure to pursue their interests whenever EU policies fit their preferences (e.g. EU demands for trade liberalization), or they simply ignored EU demands (see also Ademmer and Börzel 2013). The EU allowed rent-seeking elites to maximize their benefits and to minimize constraints by simply not caring about the consequences of its policy for Ukraine's economic development (Wolczuk 2016).

All in all, the analysis of the developmental pathway of Ukraine's automotive industry showed that the EU and Russia have prioritized their own economic interests over fostering trade interdependencies with Ukraine. Both external powers wasted opportunities for increasing their leverage on Ukrainian politics. Considering Ukraine's comparatively high strategic importance for both the EU and Russia when compared to other countries in the contested neighborhood, similar dynamics should prevail in other neighborhood countries. From the perspective of the EU, the findings imply that Brussels needs to re-think its strategy toward countries like Ukraine that are expected to implement the DCFTA and a massive number of EU internal market rules in the years to come. The full potential of the DCFTA, including the benefits of trade liberalization, can only be exploited if the EU is ready to create more opportunities for domestic institution building and economic development and invest in building economic state capacities that allow the state to promote market integration and upgrading, even in sectors where immediate benefits to EU business are limited. To counter Russian propaganda that the DCFTA will negatively affect the Ukrainian economy, it is even more important to take such measures in sectors that have lost Russia as an export destination but are likely to develop a competitive advantage on the EU and other international markets. At the same time, the EU is well advised to take the preferences of ruling elites, including the oligarchs backing them, seriously. These actors cannot be ignored but could be offered a guarantee that their property rights will be protected in return for their support of domestic reforms that aim to undermine rent-seeking practices. Should the EU fail to promote sustainable economic development, it risks becoming a factor of further political and economic destabilization in Ukraine.

Notes

1. These include Armenia, Azerbaijan, Belarus, Georgia, Moldova, and Ukraine. From the perspective of the EU, these countries are grouped under the term "Eastern neighbors." Russia considers these countries as being part of its "near abroad."
2. For more information on the company's history see http://bogdan.ua/en/corporation.html.
3. For more information on the company's history see http://www.ukravto.ua/en/about_us/history.
4. Law of Ukraine "On the Unified Customs Tariff," No. 2097-XII, Feb. 5, 1992.
5. Interview with Ukrainian expert, Kyiv, Nov. 12, 2014.
6. Law of Ukraine "On the Stimulation of Automobile Production," No. 535/97, Sep. 19, 1997.
7. Cabinet of Ministers Degree "On Making Amendments and Additions to some Resolutions of the Cabinet of Ministers of Ukraine," No. 146, Feb. 16, 1998.
8. Import duties did not apply to the temporary import of cars for a period that varied throughout 1990s. Cars could be taken out of the country within that period but could again be temporarily imported right after. The practice was and still is used by Ukrainian car dealers and owners to circumvent import duties. This trend remained constant even though the "Temporary Clause on the Regime for Temporary Import of Goods, Property and Transportation Means," issued on Dec. 30, 1991, was amended in 1997 with additional limitations on the temporary entry of cars to Ukraine. Corporations and individuals were only allowed to bring one car per company/individual into Ukraine for a period of three years, exempt from customs fees.
9. According to UNCTAD International Trade Statistics, the value of Ukraine's car exports to Russia was 1.613.000 USD. As mentioned earlier, the price for Ukrainian Tavria models was US$35,000 in 1996 (Yegorov 2004). Assuming that the price did not change, not more than 461 Tavria models were sold in Russia in the year 2000.
10. Interview with Oleh Nazarenko, Ukrainian Association of Car Dealers and Importers, Kyiv, Nov. 12, 2014; Interview with Valeriy Pyatnizkii, former deputy Minister of Economy of Ukraine and chief negotiator for Ukraine's WTO accession, Kyiv, Nov. 13, 2014.
11. The acquis communautaire is the accumulated body of EU law and obligations. It comprises all the EU's treaties and laws, declarations and resolutions, international agreements, and the judgments of the Court of Justice. It also includes all EU measures

POLITICAL GEOGRAPHIES OF THE POST-SOVIET UNION

relating to Justice and Home Affairs as well as to the Common Foreign and Security Policy.

12. As Bruszt et al. (2015) show, the Romanian car industry experienced a similar downward trend during 1990s but has seen an impressive recovery since 2000 onward.
13. The joint venture was established in 1995 by Gorky Car Works of Russia and Kremenchuk Experimental Mechanical Plant.
14. SKD means assembling welded and painted cars.
15. Interview with Oleh Nazarenko, Kyiv, Nov. 12, 2014; Interview with Valeriy Pyatnizkii, Kyiv, Nov. 13, 2014.
16. CKD means assembling a product using a complete kit which contains all the necessary parts. Usually, the parts are manufactured in one country and then exported to another country for final assembly.
17. Interview with a representative of DG Trade, Kyiv, Nov. 10, 2014.
18. Interview with a former employee of AIS, Nov. 12, 2014.
19. "Law on Amending Certain Legislative Acts of Ukraine for Improving the Balance of Payments in Connection with Global Financial Crisis," Feb. 20, 2009, in effect from Mar. 6, 2009.
20. Interviews with Ukrainian experts, Kyiv, Nov. 13, 2014 and with Oleg Nazarenko, Kyiv, Nov. 12, 2014.
21. Interview with Valeriy Pyatnizkii, Kyiv, Nov. 13, 2014.
22. Interview with a Ukrainian expert who was involved in preparing Ukraine's accession to WTO, Kyiv, Nov. 13, 2014.
23. Interview with a representative of DG Trade, Kyiv, Nov. 10, 2014.

Acknowledgments

I would like to thank two anonymous reviewers and all contributors to this special issue of EGE for most helpful comments as well as Laura Milchmeyer, Nele Reich, and Felix Rüdiger for their invaluable research assistance. The usual disclaimer applies.

Disclosure statement

No potential conflict of interest was reported by the author.

Funding

This article results from research conducted in the framework of the project "Maximizing the Integration Capacity of the European Union (MAXCAP)" which has received funding from the European Union's Seventh Framework Programme for research, technological development, and demonstration under [grant agreement no 320115]. For further information, please consult www.maxcap-project.eu.

References

Ademmer, Esther, and Tanja A. Börzel. 2013. "Migration, Energy and Good Governance in the EU's Eastern Neighbourhood." *Europe Asia Studies* 65: 581–608.

Ademmer, Esther, and Delcour Laure.. 2016. "With a little help from Russia? The European Union and visa liberalization with post-Soviet states." *Eurasian Geography and Economics*. This symposium.

Ademmer, Esther, Laure Delcour, and Kataryna Wolczuk. 2016. "Beyond Geopolitics. An Introduction to the Impact of the EU and Russia in the 'Contested Neighbourhood.'" *Eurasian Geography and Economics*. This symposium.

Balmaceda, Margarita. 2013. *The Politics of Energy Dependency: Ukraine, Belarus and Lithuania between Domestic Oligarchs and Russian Pressure*. Toronto: Toronto University Press.

Börzel, Tanja A., and Yasemin Pamuk. 2012. "Pathologies of Europeanisation: Fighting Corruption in the Southern Caucasus." *West European Politics* 35: 79–97.

Bruszt, Laszlo, Julia Langbein, Visnja Vukov, Emre Bayram, and Olga Markiewicz. 2015. *The Developmental Impact of the EU Integration Regime: Insights from the Automotive Industry in Europe's Peripheries*. MAXCAP Working Paper No. 16. Berlin: Freie Universität Berlin.

Bukkvol, Tor. 2002. "Defining a Ukrainian Foreign Policy Identity: Business Interests and Geopolitics in the Formulation of Ukrainian Foreign Policy 1994–1999." In *Ukrainian Foreign and Security Policy. Theoretical and Comparative Perspectives*, edited by J. D. P. Moroney, T. Kuzio, and M. Molchanov, 131–154. Westport, CT: Praeger.

Buzogány, Aron. 2016. "Europeanisation and Private Governance Initiatives in Post-Soviet Countries: The Case of Forestry and Chemical Security in Ukraine." *Eurasian Geography and Economics*. This symposium.

Casier, Tom. 2012. "Are the Policies of Russia and the EU in their Shared Neighbourhood Doomed to Clash?" In *Competing for Influence: The EU and Russia in Post-Soviet Eurasia*, edited by R. E. Kanet and M. R. Freire, 31–55. Dordrecht: Republic of Letters Publishing BV.

Dabrowski, Marek, and Svitlana Taran. 2012. *The Free Trade Agreement between the EU and Ukraine: Conceptual Background, Economic Context and Potential Impact*, CASE Network E-Briefs No. 11, CASE Center for Social and Economic Research, Warsaw.

Delcour, Laure. 2016. "Multiple External Influences, Policy Conditionality and Domestic Change: The Case of Food Safety". *Eurasian Geography and Economics*. This symposium.

Delcour, Laure, and Kataryna Wolczuk. 2013. "Eurasian Economic Integration: Implications for the EU Eastern Policy." In *Eurasian Economic Integration: Law, Policy and Politics*, edited by R. Dragneva and K. Wolczuk, 179–203. Cheltenham: Edward Elgar Publishing.

Dimitrova, Antoaneta, and Rilka Dragneva. 2009. "Constraining External Governance: Interdependence with Russia and the CIS as Limits to the EU's Rule Transfer In The Ukraine." *Journal of European Public Policy* 16: 853–872.

Dimitrova, Antoaneta, and Rilka Dragneva. 2013. "Shaping Convergence with the EU in Foreign Policy and State Aid in Post-orange Ukraine: Weak External Incentives, Powerful Veto Players." *Europe-Asia Studies* 65: 658–681.

European Commission. 1992. *Council Regulation (EEC) No 3917/92 of December 21, 1992*. Brussels European Council. Accessed August 25, 2015. http://eur-lex.europa.eu/legal-content/EN/TXT/?qid=1458252038315&uri=CELEX:31992R3917

European Commission. 1994. *Council Regulation (EC) No 3281/94 of December 19, 1994*. Brussels European Council. Accessed August 25, 2015. http://eur-lex.europa.eu/legal-content/EN/TXT/?qid=1458252616826&uri=CELEX:31994R3281

European Commission. 1998. "Press Release, Commission May Start Dispute Settlement Procedure against Ukraine over Daewoo and Second Hand Car Discrimination." February 20, 1998. Brussels. Accessed August 15, 2015. http://europa.eu/rapid/press-release_IP-98-173_en.htm

Franke, Anja, Andrea Gawrich, and Gourban Alakbarov. 2009. "Kazakhstan and Azerbaijan as Post-Soviet Rentier States: Resource Incomes and Autocracy as a Double Curse in Post-Soviet Regimes." *Europe-Asia-Studies* 61: 109–140.

Gallagher, Tom. 2006. *Modern Romania: The End of Communism, the Failure of Democratic Reform, and the Theft of a Nation*. New York: New York University Press.

GTAI (Germany Trade and Invest). 2013. "Schwungvolle Entwicklung des ukrainischen Automobilmarkts(Sweeping Development of the Ukrainian Automotive Market)." February 21. Accessed August 15, 2015. http://www.gtai.de/GTAI/Navigation/DE/Trade/Maerkte/suche,t=schwungvolle-entwicklung-des-ukrainischen-automobilmarktes,did=764912.html

Härtel, André. 2012. *Westintegration oder Grauzonen-Szenario? Die EU- und WTO-Politik der Ukraine vor dem Hintergrund der inneren Transformation (1998–2009) (Integration with the West or EU and WTO Policies of Ukraine against the backdrop of domestic transition) (1998–2009)*. Münster: Lit-Verlag.

InvestUkraine. 2012. "Automotive." Industry Overview. Accessed August 24, 2015. http://www.investin.if.ua/doc/pub/Overview_Automotive_www.pdf

Langbein, Julia. 2014. "European Union Governance towards the Eastern Neigbourhood: Transcending or Redrawing Europe's East–West Divide?" *Journal of Common Market Studies* 52: 157–174.

Langbein, Julia. 2015. *Transnationalization and Regulatory Change in the EU's Eastern Neighbourhood. Ukraine between Brussels and Moscow*. London: Routledge.

Le Gall, Francoise. 1994. "A Trade and Exchange System Still Seeking Direction." In *Trade in the New Independent States*, edited by C. Michalopoulos and D. G. Tarr, 65–82. Washington, DC: World Bank Group.

Lloyd, John and Chrystia Freeland. 1992. "Russia Freezes Trade with Ukraine." *Financial Times*, September 23, p. 4.

Lukianenko, Iryna G., and Iaroslava A. Suchok. 2011. "The Influence of Foreign Bank Presence on Lending Levels in Ukraine." *Kyiv Mohyla Journal* 2: 1–10.

Movchan, Weronika. 2009. "Ukraine and the WTO: One Year On." *World Finance Review 2009*. Accessed August 20, 2015. http://www.worldfinancereview.com/pdf/Ukraine%20and%20the%20WTO.pdf

Negyesi, Pal. 2006. "Koreans Invest in Booming Ukraine Market." *Automotive News Europe*. February 20. Accessed August 14, 2015. http://europe.autonews.com/article/20060220/ANE/60217032/koreans-invest-in-booming-ukraine-market

OECD (Organization for Economic Cooperation and Development). 2001. *Investment Policy Review: Ukraine*. Paris: OECD.

OICA. (Various years). *Production Statistics for 1997–2014*. Accessed August 17, 2015. http://www.oica.net/category/production-statistics/

Olszanski, Tadeusz, and Agata Wierzbowska-Miazga. 2014. "Poroshenko, President of Ukraine." *OSW Analyses*. Accessed August 15, 2015. http://www.osw.waw.pl/en/publikacje/analyses/2014-05-28/poroshenko-president-ukraine

Pavlínek, Petr. 2002. "Restructuring the Central and East European Automobile Industry: Legacies, Trends, and Effects of Foreign Direct Investment." *Post-Soviet Geography and Economics* 43: 41–77.

Pavlínek, Petr. 2008. *A Successful Transformation? Restructuring of the Czech Automobile Industry*. Heidelberg: Physica.

Richet, Xavier. 2007. "The Emergence and Entry of Industrial Groups in Russia: the Case of the Car Industry." In *Globalisation in China, India and Russia: Emergence of National Groups and Global Strategies of Firms*, edited by J. F. Huchet, X. Richet, and J. Ruet, 325–350. New Delhi: Academic Foundation.

Richet, Xavier, and Frederic Bourassa. 2000. "The Reemergence of the Automotive Industry in Eastern Europe." In *The Globalization of Industry and Innovation in Eastern Europe: From Post-socialist Restructuring to International Competitiveness*, edited by C. von Hirschhausen and J. Bitzer, 59–94. Cheltenham: Edward Elgar.

Schneider, Klaus. 2001. "The Partnership and Co-operation Agreement (PCA) between Ukraine and the EU-idea and Reality." In *Ukraine on the Road to Europe*, edited by L. Hoffmann and F. Möllers, 66–78. Heidelberg: Physica.

UNCTAD. (Various years). *International Trade Statistics*. Accessed August, 14 2015. http://unctadstat.unctad.org

Unian (Ukrainian Independent Information Agency). 2015. "Automobile Production in Ukraine Almost Grinds to Halt." May 13. Accessed August 14, 2015. http://www.unian.info/economics/1077505-automobile-production-in-ukraine-almost-grinds-to-halt.html

VDA (Verband der Automobilindustrie e.V.). 1995–2001. *International Auto Statistics*. Frankfurt, M.: VDA.

Wolczuk, Kataryna. 2016. "Managing Opposing Flows of Gas and Rules: Ukraine between the EU and Russia. *Eurasian Geography and Economics*. This symposium.

WTO (World Trade Organization). 2008. *Report of the Working Party on the Accession of Ukraine. Addendum 1: Schedule of Concessions and Commitments on Goods*. Geneva: World Trade Organization.

Yegorov, Igor. 2002. "Transformation of the Ukrainian Economy in the Light of EU Enlargement and Development of Regional Co-operation." In *Regiok Europaja*, edited by B. Bezsteri and I. Levia, 25–33. Budapest: Regia Rex Nyomda.

Yegorov, Igor. 2004. "Much Fuss about Nothing: Restructuring Stalemate in the Ukrainian Car Industry." In *International Industrial Networks and Industrial Restructuring in Central and Eastern Europe*, edited by S. Radosevic and B. M. Sadowski, 207–221. Boston, MA: Kluwer.

Building identities in post-Soviet "de facto states": cultural and political icons in Nagorno-Karabakh, South Ossetia, Transdniestria, and Abkhazia

John O'Loughlin and Vladimir Kolosov

ABSTRACT

For states that have recently declared their independence but remained unrecognized "de facto states," building a national identity is critical in the face of international rejection of their political status. Key elements of this new or re-animated national identity are political and cultural icons symbolizing the new political entity but with historical antecedents. Following Anthony Smith's ethno-symbolism approach to the study of nationalism and motivated by Jean Gottmann's research on iconographies in political geography, the article reports the results of nationally representative samples in four post-Soviet de facto states, Nagorno-Karabakh, South Ossetia, Transdniestria, and Abkhazia. Respondents were asked to name up to five political and cultural figures that they admired. The collated results show a great array of local and Russian names in the four republics. Categorizing the names by historical era and by provenance allows a clarification of the extent to which nation building can rely on local heroes. Among the four republics, Nagorno-Karabakh stands out for its ethno-symbolic local character, while Transdniestrian respondents identified few iconic figures. South Ossetia shows a mix of local and Russian names while the respondents in Abkhazia were divided by nationality in their choices.

The hero and heroine embodied the innate goodness and 'true essence' of the nation, and it was his or her 'exemplum virtutis' that could help to restore a sense of dignity to downtrodden peoples and inspire and mobilize them to resist oppression and fight for self-rule. (Smith 2009, 69).

In this article, we report the empirical results of a study motivated by the quotation above from Anthony Smith (2009). Smith's promotion of symbolic elements in research in nation-building included the special role of images of heroes and idols in the formation and certification of group identity, both in-group solidarity and out-group exclusion. Who is an icon to a group and what kinds of backgrounds do these icons share? In the choice of icons, such as political leaders and cultural heroes of the past or the present, it is possible to gage the identity of an individual and the strength of ties to the nation. Examination of these symbols through large-scale surveys has not received sufficient attention from researchers of nationalism, partly because of the obvious difficulty and cost of collecting representative national responses. While there is a good deal of research on national memorials and museums that includes their emphases as well as their omissions, few have examined the iconographies of the mind. The collective pantheon of heroes that individuals who declare their belonging to a nation have chosen as a result of their education, political socialization, community influences, and ideological preferences allows researchers of national identity-building to understand the cohesiveness of the nation, to clarify its cleavages, and to indicate the range of spatial and temporal influences that solidify that identity. The role of such memories in the making of the post-socialist societies is vividly elaborated in Verdery 1999.

The numbers and reputations of local heroes have a strong influence on the formation of a territorial or national identity. A native or someone who has spent a significant part of life in a particular region or polity and who promoted its independence and subsequent construction is usually identified as a "founding father." Most nations have multiples of such individuals stretching back through the generations and certifying the long ties of the group to the land that they claim exclusively. A region often becomes personified through the specific assertion of a long history of settlement by the group – along the lines of an argument that "we were here first" (Murphy 2002), making a claim to a slice of the earth's surface by a self-imagined nation. Territorial claims are strengthened by evidence, real or imagined, of historical antecedence through archeological or archival sources.

In this article, we examine the special role of memories of human heroes and idols in the formation of national identity in the context of new (or renewed) identity formations in four "de facto" states that emerged from the breakup of the Soviet Union and subsequent wars in the 1990s. Though all four of our study sites (Nagorno-Karabakh, South Ossetia, Transdniestria. and Abkhazia) had some recognized territorial definitions through the Soviet project of autonomy, their successful separation from the newly independent Azerbaijan, Georgia, Moldova, and Georgia (respectively) after the collapse of the Soviet Union required a reinvigorated focus on defining the group, and its elements and exclusions, as it started the process of state-making and nation-building. We report on a set of open-ended questions that go straight to the theme of icon-making. Allowing up to five names, we asked "Who is an idol?" We repeat this question for both the political and cultural realms since we believe nation-building encompasses multiple realms of identity. Grouping named individuals by their provenance and historical era of prominence, we can assess the relationship between a respondent's perception of his/her own kind and the associations that are suggested by a particular name. By then correlating these groupings with the demographic characteristics of respondents, we can gain insights into the reasons why these icons are central to the group identity and to the success of the physical and psychic-boundary-making processes.

In this paper, we are building on the earlier work edited by Kolstø (2005) on the symbolic elements ("Myths and Boundaries") of nation-building in the Balkans. In turn, that work continued the emphasis of Barth ([1969] 1998) on how differences between ethnic groups are often "mythical" as well as "factual." Barth ([1969] 1998, 6) claimed that the "cultural features of greatest import (in separating groups) are boundary-connected." While cultural and political icons are not myths (after all, they are persons alive or dead who shaped the nation and its identity), by lauding "their people," members of the nation are making the same kinds of boundary definitions. In his well-cited article summarizing the status, properties, and futures of de facto states (now the preferred term, though he called them "quasi-states" at the time he wrote), Kolstø (2006) stressed the importance of building the new or re-invigorated nation by mobilizing the internal support of the local population through propaganda from the new regimes. Among these methods beyond new flags, anthems, constitutions, state holidays, and other symbols is the cultivation of the memory of the recent war and its victims, promotion of the image of the external enemy and continued danger of attack. In some cases, like South Ossetia and Nagorno-Karabakh, the appeal is abridged for a homogenized population through ethnic cleansing so that those who might have been sympathetic to the enemy are now gone (Kolstø 2006, 730). All four territories had defined boundaries in the ethno-territorial and physiographical cartography of the Soviet Union so the new republics can hark back to earlier expressions of delimited control or claims.

This article specifically examines the relationship between the recognition of the authority of certain political and cultural figures and the development of various forms of group identity among the populations of the four post-Soviet unrecognized states; Nagorno-Karabakh, South Ossetia, Transdniestria, and Abkhazia. The data that support our conclusions were embedded in an extensive questionnaire – over 120 separate items – and among these were two open-ended questions that allowed the possibility of choice for the respondent of up to five historical or contemporary figures for each of the political and cultural categories. A consequence of asking only about cultural and political icons was a restriction on the resulting examination of identity – in particular, we exclude the roles played by famous athletes and scientists. Authors in both the Barth ([1969] 1998) and Kolstø (2005) books underlined the role of cultural and political figures from history to the exclusion of other categories. We did not predefine what is "political" or what is "cultural." (For the wider project and the sampling design, see O'Loughlin, Kolossov, and Toal [2014]). This limiting assumption is quite acceptable, as previous qualitative research has emphasized the role of historical, cultural, and political figures in the examination of material objects (such as stamps or coins), flags, landscape elements, and texts (for example, Matjunin 2000; Penrose 2011; Penrose and Cumming 2011; Pointon 1998; Raento 2006; Raento et al. 2004).

Icons and national identities: the contribution of Jean Gottmann

In the burgeoning literature on nations and nationalism that shows no sign of abating, greater attention has been given in the last couple of decades to the ethno-symbolic elements of identity formation. Reacting to what was seen as an overemphasis on the role of states and elites in constructing nations and to the domination of the economic (modernist) dimensions of nation formation, scholars such as Anthony Smith drew attention to the social–psychological elements that they believe have received insufficient research consideration. Smith emphasized that he was not developing a new paradigm or a new theory of nationalism; instead "ethno-symbolism proposes … a radical but nuanced critique of the dominant modernist orthodoxy" (Smith 2009, 13). Smith considered nations as real sociological communities, always filled with dynamism and purpose. In arguing for greater scholarly attention to the pool of "symbolic resources" available to a nation or a reservoir of "invented tradition" (Hobsbawn and Ranger 1983) that could be mobilized for identity promotion, Smith advocated an examination of the iconographies which the geographer, Jean Gottmann (born in Kharkiv, then in the Russian Empire), first elaborated over 60 years ago. Though Gottmann's ideas are now poorly circulated in the sociological study of nationalism or indeed in modern political geography (Johnston 1996), his propositions about the role of iconographies in separating and defining a national territory (Muscarà 2005a, 2005b) are long overdue a close inspection. Few works in geography have tried to consider either personal or landscape iconographies in national imaginaries, though exceptions include Ma and Fung (2007) and Terlouw (2014).

In his memorial essay, Champion (1995) recalled Jean Gottmann's life project of conceptualizing what and how a political geography of the world would look like. Drawing on the French regional geography tradition (in particular, Paul Vidal de la Blache and his concept of "genre de vie" (type of life), and Albert Demangeon (Muscarà 1998)) in which he was trained and with which he remained imbued, Gottmann saw political units as more than administrative structures or more than their physical and human landscape characteristics.

> To be distinct from its surroundings, a region needs much more than a mountain or a valley, a given language or certain skills; it needs essentially a strong belief based on some religious creed, some social viewpoint, or some pattern of political memories, and often a combination of all three. Thus regionalism has some iconography as its foundation. (Gottmann 1951, 163)

The core of his views is the concept of the geographical space's partitioning (cloisonnement). In elaborating on the views of Vidal de la Blache, Gottmann believed that the key factor of its organization and change is circulation of people, ideas, goods, capital, technologies, etc. which collides with resistance, mostly of ideational rather than material nature, that he called iconographies. "Iconography works towards preservation of the established order and the strengthening of standing divisions,"(Gottmann 1994, 14). In *La Politique des Etats et leur Geographie*

(1952), he further elaborated his theories of "iconography," which he saw as a force for stability "representing the need which people have to group together around shared sets of beliefs, values and identities" (Champion 1995, 199). Perhaps the most forceful expression of Gottmann's notion of distinctiveness that is wrapped by a group's iconographies are these sentences:

> To be different from others and proud of one's own special features is an essential trait of every human group. No group greatly resents its example being followed, but none likes to follow another's lead. This basic character, inherent to human psychology and sociology, makes every unit of space inhabited by man essentially a human unit. The most stubborn facts are those of the spirit, not those of the physical world. (Gottmann 1951, 164)

Reflecting on these words almost 30 years later, he wrote,

> The stronger cosmopolitan or ecumenical trends are, the more regionalism develops to balance them. Some thirty years ago, with a youthful enthusiasm for daring generalization, I wrote that the most stubborn facts were those of the spirit, and that the most lasting partitions were those in the minds of men. These statements have long been resented by many geographers. At first, they seem even to have been interpreted by many social scientists (including some distinguished political scientists) as confirming the dominance of the spiritual over the temporal, and therefore denigrating the role of concrete geographical space. Such views were mistakenly based on the traditional opposition of physical and spiritual as two independent and even conflicting realms. (Gottmann 1980, 439)

In fact, Gottmann's views are closely aligned with those of Anthony Smith in his last works (2008, 2010) that reconsidered the modern nationalism projects that show no signs of ebbing even in a globalized world.

Gottmann did not elaborate much on his concept of iconographies, nor did he provide empirical validations or detailed case studies of how iconography works in practice. According to Bruneau (2000), he actually abandoned the concept in his later (1970s) publications. It has been left to others, in both the fields of cultural geography and in the sociological study of nationalism, to take up Gottmann's mantle and develop the "school of micro-nationalist studies ... that focuses on popular, everyday expressions of nationalist practices and institutions" (Smith 2009, 19). Smith argued for more attention to the "myth-symbols complex" for more than a quarter-century as an antidote to the heavy attention to the role of ethnic entrepreneurs, economic disparities, and state actions in the study of nationalism from the constructionist perspective. In his clearest amplification of the reasons for studying ethno-symbolism, Smith (2009) wrote "By revealing the workings of national ideas and sentiment among non-elites and the underlying, if intermittently expressed, importance to them of national ideas, the study of 'everyday nationhood' has undoubtedly enlarged our understanding of the field" (78–79). But he also recognized the significant gaps in the existing research

> Yet, it also suffers from a number of limitations. For one thing, it often fails to differentiate the various strata, regions and ethnicities of 'the people', each of whom or which may have different ideas and contain a variety of sentiments and preferences. For another, its

analyses tend to be confined to the populations of national (and mostly Western) states. (Smith 2009, 74)

Consideration of these factors has not received sufficient attention from researchers of nationalism, partly because of the obvious difficulty and costs of collecting representative responses from different groups, especially in non-Western contexts. While there is a good deal of research on national memorials, both in the landscape and in museums (see for example, Johnson 2003), especially about their emphases and their omissions, few have examined the "iconographies of the mind," the pantheon of achievers that people have placed in their social–psychological profile as an result of the level of education, political socialization, ethnic attachments, community influences, travel ranges, and ideological preferences.

Historical heroes have a strong influence on the formation of local territorial identity, typically persons that took part in its construction. In the context of local identity, we can use the term *genius loci* ("protector of the place"), which, however, can embody a relatively extensive territory memorializing its creator whose biography and whose works have become associated with a particular place (home, homestead, settlement, village, city, the landscape, the terrain). The images of local heroes, unlike other iconic figures, become important for the development of other forms of identity, often deliberately cultivating specific brands. The desire to find a *genius loci* associated with every settlement, every street, and even every building sometimes reaches the point of absurdity. Typically, political, military, cultural, and sports figures (Allen 2013) dominate these street names and of course, they are often changed, sometimes in multiple successive political eras, to reflect the dominant political spirit and national orientation of the day (Light 2004; Young and Light 2001). But often birthplace and icon appear mismatched. The Ukrainian region of Poltava where Nikolai Gogol was born is closely related with his name. In the early 1990s, the "patriotic" zeal gave rise to the intention of some Ukrainian functionaries to close his museums "because he was Russian writer". Former Ukrainian President Kuchma stated that such personalities like Gogol belonged to both Ukraine and Russia, to all humanity (Kuchma 2003).

What is a nation? Though rivers of ink have been expended in the clarification of this concept, claims to nationhood are both historical and ongoing because denigration of the claim and its attendant territorial benefits are central to opposition. Anthony Smith offers a definition that embodies the cultural and psychological values intrinsic in nationhood. For him, the nation is "a named and self-defining human community whose members cultivate shared memories, symbols, myths, traditions and values, inhabit and are attached to historic territories or 'homelands,' create and disseminate a distinctive public culture, and observe shared customs and standardized laws" (Smith 2009, 29). To understand how nations emerge and become self-sustaining, we need to "gage the appeal of different motifs -myths, memories, symbols, values and the like – to various strata of the population, and the reasons for this affinity" (26). This is an approach on which we build in this paper.

The extraordinary circumstances of the implosion of the Soviet Union in the early 1990s gave rise to multiple examples of nationalist mobilizations. Some had been suppressed during Soviet times but remained active (Latvia, Georgia) and were ready to re-blossom after 1991. Others had a brief but ultimately unsuccessful flowering (Chechnya between 1994 and 1999), while others enjoyed a strong local following and foreign patronage resulting in autonomy or independence (including our four study sites). Unlike Latvia, the de facto states could not hark back to an earlier independence that had been squashed during Soviet times, though each had a sense of identity and separation based on ethnic and cultural values and peoples that allowed them to clarify their differences with the parent states from whom they were estranged. Additionally, Abkhazia, South Ossetia, and Nagorno-Karabakh has autonomous status from their parent states, then Soviet republics, till 1991. Given their tenuous status and the absence of recognition of their independence – at least until Russia and a few other states recognized Abkhazia and South Ossetia after the 2008 war with Georgia – the state leaders felt it critical to establish political legitimacy and gain popular support (Bakke et al. 2018). At the same time, in a Janus-like manner characteristic of most nationalisms of "looking back" as well as "looking forward" (Nairn 1975), new countries and new ideologies in an existing polity express their identity in unique ways by commemorating historical and other mythical heroes. Nations are thus "narrated," as Homi Bhabha (1990) underscored, as "a construction of linear narratives of nation, culture, and identity." Thus, people negotiate relationships with a particular nation by constructing narratives that define their boundaries, separating "them" from "others." Billig (1995) drew attention to the unremarked nature of nationalism in its taken-for-granted narratives and symbols and in this sense, our paper is highlighting some of the iconography that underpins the respective storylines. This objective is reached by the state policy of memory adopted officially or de facto in many countries of Central and Eastern Europe. It can be defined as a set of social practices and norms related with the regulation of collective memory. The policy of memory also comprises the attitude to different groups of veterans, access to archives, and funding of certain studies and publications (Miller 2009). As a result, certain events and figures get a strictly defined interpretation, their importance is emphasized and/or exaggerated, while the others are ignored, omitted or became taboo.

Icons in post-Soviet De Facto states

In two open-ended questions, respondents were asked to name up to five prominent political and cultural figures. The text of the questions read (for Nagorno-Karabakh – this republic was substituted in the other three locations). "Please name the five most outstanding political figures in Nagorno-Karabakh or other countries in our time or in the past" and "Please name the five most outstanding cultural persons (writers, composers, artists) in Nagorno-Karabakh, or in other countries in our day or in the past which you most admire." While technically, it was possible that we would be given a list of hundreds of names, in practice, most respondents could only identify two or three names. The average number was 2.7 of the three republics where five names were possible (Nagorno-Karabakh, Abkhazia, and South Ossetia) and less than 2.0 in the location (Transdniestria) where three names could be given to the interviewer. No geographic boundaries were placed on the names offered by respondents. As a result, names from earlier eras such as Tsarist and Soviet times, as well as names from outside the Russian/Soviet world of renowned political leaders (Barack Obama, Winston Churchill, etc.) and famous artists (Picasso, Mozart, etc.) were among the list of personalities from all regions and countries and of all historical periods. The surveys were conducted in 2010 (Abkhazia at 1000 respondents; Transdniestria, 750 respondents; and South Ossetia 500 respondents) and early 2011 (Nagorno-Karabakh, 800 respondents).

Why would de facto states be likely to be more interested in building identities than other states with unquestioned international legitimacy? In a contemporary world where acceptance of a new state on the political map is a very rare event, claims to membership in the international community often resort to the Montevideo Treaty (1933) principles. Its first article reads that "The state as a person of international law should possess the following qualifications: (a) a permanent population; (b) a defined territory; (c) government; and (d) capacity to enter into relations with the other state." Only ratified by 16 countries in the Americas, the treaty does not include a sense of nationhood, but it is certainly claimed by potential members of the international set of states. De facto states are confronted with the widespread support that their parent states (Georgia, Moldova, and Azerbaijan in the case of this paper) have accrued for their position in the international arena. Clem (2014) reviewed these claims in the context of the Ukrainian crisis and the post-Soviet conflicts based on competing entitlements of history and geography. Therefore, claims about legitimate membership in the community of states must be first advocated on the basis of an unquestioned loyalty of the citizens of the putative new state and their identity with it. Small nations will typically underline their long unique histories and residence in a specific territory, and they will emphasize cultural and political distinctiveness where iconic figures feature prominently. If these individuals are locally unique – that is, not shared by other nations – then the identities are further certified. We clarify the provenance of the cultural and political icons

In the presentation of the results of the icons that are named, we consider overall ratios by different eras (Tsarist, Soviet, and post-Soviet), geographic origins (Western and non-Western), and locals. In all cases, it was possible to match the names to well-known individuals. The comparison of these groupings gives significant insights into the nature and complexity of state identity just less than 20 years after the declarations of independence and the cease-fires that allowed them to retain their separatist status. By correlating the groups of names so classified about these lines with socio-demographic characteristics of the respondents, we can gage how national identities are evolving by age group, by ethnic identity,

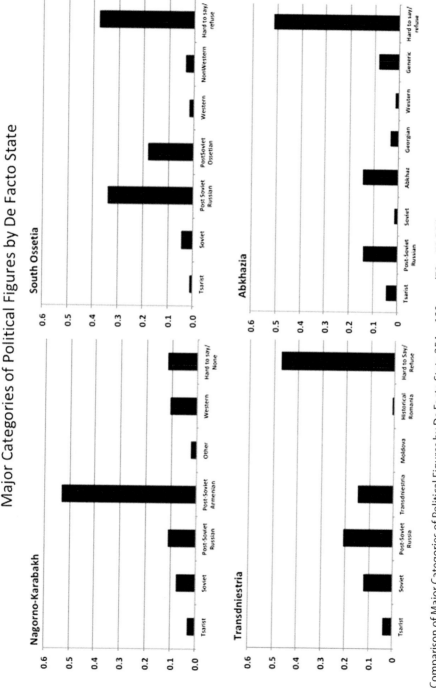

Figure 1. Comparison of Major Categories of Political Figures by De Facto State 254 × 190 mm (72 × 72 DPI).

by educational level, and by views of the state authorities. Because of space limitations, we only show the most important determining factor, which varies by republic, in the graphs and the discussion that follows.

Placing the named individuals into major categories by provenance and era (Tsarist, Soviet, post-Soviet Russian, local (Ossetian, etc.), Western and non-Western) allows us to draw some generalizations for both the political and cultural identifications. For the political categories, a major difference emerged between Nagorno-Karabakh and the other three regions (Figure 1). For all except Karabakh, the ratio of non-responses (don't know or refuse) ranged close to 50% of all possible answers – since five options were possible (three in Transdniestria), this means that the average respondent struggled to give more than a couple of names, though the range was large. For Nagorno-Karabakh, now a homogenous ethnic Armenian region, only 20% of the possible answers were empty.

The comparative graphs in Figure 1 indicate that local political figures, as well as Soviet-era and post-Soviet Russian figures (mostly Russian Prime Minister Vladimir Putin and Dmitri Medvedev, then President) dominated. For Nagorno-Karabakh, the Soviet-era list was dominated by the names of ethnic Armenians who were prominent in the USSR. When these numbers are combined with local political figures, the Karabakh list of icons is the most parochial of the four regions. For the other three regions, locals only constitute about 20%–25% of the names, which reflects both a short history of declared independence from the respective parent state in the early 1990s and the paucity of new national political heroes in the ensuing 20 years, a time of contorted local politics and dominance of the domestic scene by a small handful of individuals. Many of these individuals were involved in the wars of the late Soviet period and remained entrenched in the state apparatus thereafter. Pre-Soviet (from Tsarist times that lasted to 1917), Western and non-Western politicians receive only a handful of nominations, always less than 10% for each of the four republics.

The cultural categories graph (Figure 2) shows that fewer specific names are evident in comparison with the political figures identified; for all republics, a greater representation of earlier (Tsarist era) figures are named. This higher ratio of Tsarist-era persons reflects the classical education of Soviet times with strong representations of persons like writers Alexander Pushkin and Leo Tolstoy visible in all the republics. Once again, Transdniestria has a higher ratio of non-responses compared to the other states, in this case, comprising over 70% of the total. The comparative "don't know" ratios are 35% for both Abkhazia and South Ossetia, and just under 20% for Nagorno-Karabakh. The reason for the dramatic difference between TMR (Transdniestria) and the other republics is suggested by answers to other questions in the questionnaire. The TMR sample was more politically alienated and concerned about economic difficulties, and it may be that such worries reflected a general dissatisfaction with the direction of the state at the time (2010) and negative prospects for the future (O'Loughlin, Toal, and Chamberlain-Creangă

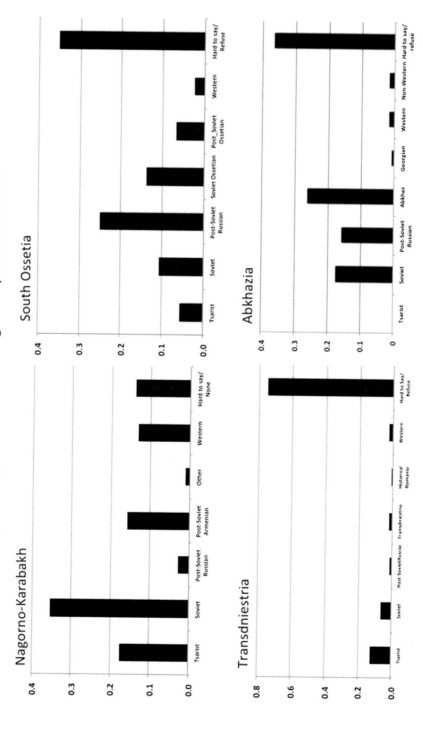

Figure 2. Comparison of Major Categories of Cultural Figures by De Facto State 254 × 190 mm (72 × 72 DPI).

2013). In this social environment, it is not surprising that the people in the sample would not be interested in cultural icons of either the past or the present.

Nagorno-Karabakh and Abkhazia showed a higher ratio of local cultural figures than the other two states, though the names for Abkhazia were significantly different between the four different nationalities. Abkhaz cultural icons were disproportionately named by members of that nationality, while ethnic Georgians gave very few names of any era or provenance. As with the political categories, for the homogenous population of the NKR, Armenians were disproportionately represented in all categories – Tsarist era, Soviet times, post-Soviet years, and even among those identified as Western names. In all de facto states, very few Western or non-Western names were encountered, and the few that were mentioned were typically brought forward by the youngest generation, age 18–30, who are more likely to be exposed to external stimuli via access to mass media and the Internet.

The overwhelming impression of these data on political and cultural figures is that Western and other external influences were unimportant in comparison to the dominance of the Soviet legacy. Second, identities in the new republic that were engaged in both state- and nation-building were strengthened by a high level of identity of national icons in the ethnically homogenous republics of Nagorno-Karabakh and South Ossetia, and in the mixed nationalities of Abkhazia (except for ethnic Georgians) as local persons who have achieved fame in either the cultural or political realms. Transdniestria stands apart in this regard, where iconic names showed strongest attachments to Russia and to that country's political and cultural traditions. In other articles, we have reported the strong level of support for annexation to Russia in Transdniestria (about 50%) and in South Ossetia (about 80%), whereas opinions are quite evenly split on this option versus the independence option in the other two republics (Toal and O'Loughlin 2016). Third, Russian political and cultural figures are prominent in all four republics, and given the economic and security dependence of the three republics on Russia (Karabakh is significantly economically and militarily dependent on Armenia), it is no surprise that the important cultural and political persons from the external patron, Russia, feature so prominently in Transdniestria, South Ossetia, and Abkhazia.

The comparative ratios for the four republics clearly indicates the greater preponderance of local political (that is, Armenian) icons in Nagorno-Karabakh compared to the other locations, while on the cultural side the ratios are lower and quite mixed. In general, respondents could identify more prominent political figures than cultural ones where the "don't know" ratios are over one-third in Abkhazia, South Ossetia, and Transdniestria. The Soviet legacy is strong, though mixed with both national-level (e.g., Josef Stalin and Anna Akhmatova) and local-level luminaries (e.g., Aram Khatchaturyan for Karabakh) evident in the names provided. The prominence of the latter category indicates that the nation-building does not necessarily rely on recent decades to assert the distinctiveness of the nation, and this is especially true when that distinctiveness was repressed before independence. Historical claims to a singular identity prove longevity and consistency.

A comparison of political and cultural icons for individual states reveals both similarity in the presence of Tsarist and Soviet persons in the usual names offered and differences according to the uniqueness of the individual areas. Among the most common names on the cultural side were classic writers like Alexander Pushkin and Leo Tolstoy, and on the political side, Vladimir Lenin and Vladimir Putin. A noteworthy element in all states, though less so in Transdniestria, was the identification of the icon figures who were prominent in the larger Tsarist and Soviet worlds such as General Alexander Suvurov for the TMR, Aram Khatchaturyan for Karabakh, the conductor Valery Gergiev for South Ossetia, and the writer Dmitry Guliya from Abkhazia. A brief tour through the most commonly identified names emphasizes these common and particular themes.

Nagorno-Karabakh

Respondents from Nagorno-Karabakh, as well from other unrecognized republics, recalled many more political figures than cultural ones, particularly those politicians whose names feature in mass media on a daily basis. As the political list shows, Karabakhis were focused primarily on contemporary events in Armenia (to which they are closely tied economically, politically, and socially) and in their republic. At the top of most respondents' lists were then- (2011) and now-NKR President, Bako Sahakyan, as well as two former heads of the republic – Robert Kocharyan and Serzh Sargsyan, both of whom then went on to be elected presidents of Armenia. The first president of Nagorno-Karabakh, Arkady Ghukasyan, was also in the top 10, and the other most prominent politicians were contemporary Armenian leaders, including candidates at recent presidential elections.

The only important non-Armenian/Karabakhi political figures were Vladimir Putin (fourth place) and Josef Stalin (ninth place). Almost all politicians mentioned in Nagorno-Karabakh were living at the time of the survey, except for Stalin. This distribution of political names is evidence of the strongly ethnic character of Armenian and Karabakhi politics, of a deeply traumatized collective consciousness (including the 1915 genocide during the last years of the Ottoman empire and continued hostility from neighboring Turkey and Azerbaijan), and of a highly politicized identity inherited from the USSR that feeds ethnic nationalism. Under the conditions of a besieged fortress, recent fighting in 2016 and a continued risk of a new war with Azerbaijan, contemporary political life is reduced to the struggle for the continued independence of Nagorno-Karabakh and the attainment of political and economic independence of Armenia, still strongly dependent on Russian support. At the same time, paradoxically, the domination of contemporary Armenian or Soviet/Armenian political and cultural figures is evidence of a certain weakness of the NKR's policy of identity building. Because of the ethnic homogeneity after the flight and cleansing of Azerbaijanis in the early 1990s, the

NKR state could not or did not need to use deeper historical strata of collective memory in political mobilization. Post-Soviet Ukraine, a mixed-ethnic state by contrast, paid much attention to compiling a list of politically relevant historical figures, with some of them declared national heroes, others as traitors and Stepan Bandera viewed as a hero and villain by different groups at different times (Liebich and Myshlovska 2014; Wylegała 2017).

A respondent's age (reflecting the period of political socialization) is an important factor determining the views of Karabakhi respondents about political leaders (Figure 3(a)). There is a statistically significant and expected difference in the popularity of Soviet political figures among young, middle-age, and senior respondents. Post-Soviet Armenian political leaders enjoy a high level of knowledge among all age groups, though they are disproportionately mentioned by the youngest respondents (18–35). Like the cultural category (Figure 3b), politicians from the West are also named more often by younger people. Importantly, the frequency

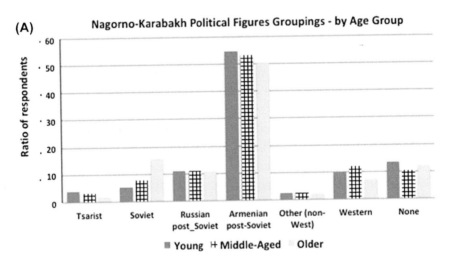

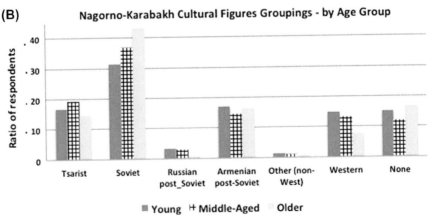

Figure 3. Comparison of Political and Cultural Figures in Nagorno-Karabakh by Age Group 254 × 190 mm (72 × 72 DPI).

of Russian post-Soviet politicians' mentions (mainly of Vladimir Putin) does not vary by age groups.

The top ranks of cultural figures in the representations of the respondents from Nagorno-Karabakh were dominated by ethnic Armenians in all categories. The only non-Armenian among 15 top cultural figures was Alexander Pushkin. Here and elsewhere, the high ranking of this Russian poet (1799–1837), whose poems were studied in every Tsarist/Soviet school, was a remnant of the Soviet past. The short cultural list is evidence of close relationships between Armenia (and Nagorno-Karabakh) and the Armenian diaspora: it included, for instance, Charles Aznavour (born Shahnour Vaghenag Aznavourian), a French singer of Armenian background who was popular in the former Soviet Union and particularly in Armenia because of his strong moral and material support of his historical motherland after the destructive earthquake in December 1988 and the most difficult years of transition 1988–1993. Similarly, the Armenian background was important for Karabakhi respondents in the case of Andre Agassi, the tennis player. Armenia is now the only post-Soviet country with an ethnically homogenous population. Obviously, in this specific ethno-social environment, the Armenian nation is interpreted in ethnic terms, as a closed and exclusive community based on ethnic (that is, biological kinship) and solidarity in the face of external threats (Shnirelman 2003).

The second feature of the Karabakhi world vision and social representations is the opposite side of the same coin. It was based in pride about Armenians who became famous not only in the small country but in the wider Soviet Union as a whole or even worldwide. Prominent was the French singer Aznavour, and this is also the case of the Soviet/Armenian composer, Aram Khachaturyan, who lived and died in Moscow. He was the leading Armenian figure for almost a quarter of the Karabakhi respondents. A third feature of Karabakhi social representations and identity was evidence of a living Soviet legacy and close relations with post-Soviet states, especially Russia. No cultural figure named lived and worked earlier than the nineteenth century, and most of them became famous in the Soviet period. This is noteworthy in the case of Armenians, who strongly proclaim their ancient roots and the Biblical past (converting to Christianity in the early fourth century). The influence of pop culture was unsurprisingly combined with a strong ethnic character. Many respondents included in the list of cultural figures such young pop singers as Razmik Amyan and Arame (born and raised in Russia).

South Ossetia

In South Ossetia, the peaceful Soviet years are remembered with nostalgia, with Soviet authorities now perceived as an efficient counterbalance to Georgian nationalism. Arguments of the Georgian Government and its public position about the negative consequences of Russian imperial policy in the south Caucasus find little appeal among South Ossetians (or indeed Abkhaz).

Accomplishments of native Ossetians at the Soviet Union or Russian scale in cultural, military, political, and other fields has always triggered feelings of national pride. Like Nagorno-Karabakh, the historical depth of prominent figures' list in South Ossetia is low; only the names of Kosta Khetagurov (poet) and of Russian classical writers suggest that the history of Ossetia did not begin only in the twentieth century (Figure 4(a)). This recent dating of figures is particularly striking in light of the typical trend of nationalizing states to revisit national history and to represent the titular ethnic group as old as possible, reinventing or introducing a new pantheon of ancient national heroes into the collective historical memory, especially those who are depicted as fighters for independence and national liberation against an "eternal enemy" (e.g. Turkic peoples, in the case of Armenia and Nagorno-Karabakh).

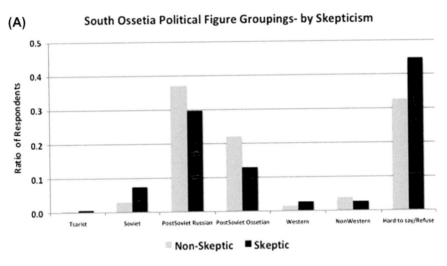

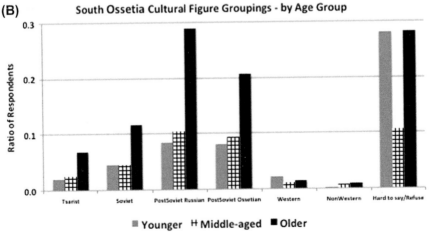

Figure 4. Comparison of Political and Cultural Figures in South Ossetia by Respondent Views of the Authorities 254 × 190 mm (72 × 72 DPI).

For South Ossetia, significant differences are evident between supporters of the Eduard Kokoity government in power at the time of the survey in 2010 and those who were skeptical of it (Figure 4(a)). The difference also correlates to differences between generations, as younger people socialized after the collapse of the Soviet Union can remember many fewer political and cultural figures. This result is certainly a result of the degradation of education during the years of the war and isolation.

A large number of respondents could not or did not want to name prominent cultural figures or named fewer than five of them (Figure 4(b)). The most popular cultural figure was Kosta Khetagurov, an Ossetian poet and activist of the nineteenth-century democratic movement, glorified and transformed into the main national Ossetian hero in the early Soviet period. His poems and biography are studied in all Ossetian schools. Khetagurov was followed by Valery Gergiev, the art director of the prestigious Mariinsky Opera and Ballet Theater in Saint Petersburg. Vasily Abaev, the third most mentioned, was a Soviet/Ossetian linguist and philologist specialized in the studies of the Ossetian language and folklore, the author of the main normative dictionary of the Ossetian. The composer Ilya Gabaraev and writer Nafi Djusoev became known in the Soviet period.

Only two figures of Russian high culture, Alexander Pushkin and Leo Tolstoy, reached the list of the top 15 personalities named by South Ossetian respondents, though listed only by few of them. Some respondents also remembered Russian Soviet and post-Soviet pop singers like Alla Pugacheva. But because of a high variety of mentions, the percentage of post-Soviet Russian figures is high. Geographical proximity and close relations with North Ossetia, a part of the Russian Federation, explain also the much higher share of Russians in the list compared to Nagorno-Karabakh.

The general conclusion is the same as in the case of Nagorno-Karabakh; representations of South Ossetians are highly politicized and ethnicized, a result of the perceived discrimination against the residents of the autonomous oblast in Soviet Georgia, of the Georgian policy of minorities' cultural assimilation, of difficult relations between Ossetians and Georgians in Tsarist times, and of course, of conflict during the period of the post-socialist transition. Social life, including culture, is considered through the prism of the conflict. Not a single Georgian name in South Ossetia (Figure 4(b)), and Azerbaijanian name in the case of Nagorno-Karabakh, appears on the list of prominent cultural figures. People are particularly prone to erect cultural barriers in those countries in which center-periphery and dominance–submission relations are observed.

Transdniestrian Moldovan Republic (TMR)

The hierarchy of the leading politicians in the representations of Transdniestrians was rather simple: the members of the Russian ruling "tandem" (at the time of the survey) Vladimir Putin and Dmitry Medvedev, ranked respectively first and third in the number of mentions, while local leaders – Igor Smirnov (President in 2010) and

Evgeny Schevchuk (his successor) – were the second and fourth (Figure 5(a)). But the distance between Putin and Smirnov, founding father of Transdniestria, was very considerable. Other leaders making a part of top 10 politicians represented exclusively different periods of Soviet/Russian history: Stalin, Lenin, Peter the Great, Leonid Brezhnev, General Suvorov, and Mikhail Gorbachev. While this is obvious evidence of the cultural and political gravitation of Transdniestria to Russia, in our parallel simultaneous survey on the other side of the river, Putin gathered the largest number of mentions also in left-bank Moldova proper (about 30%), which was much more than any Moldovan and Romanian politician.

The Transdniestrian pantheon of cultural heroes was, firstly, a mirror of the cultivated multi-ethnic character of the TMR (Figure 5(b)). In this republic three main ethnic groups – Moldovans, Ukrainians, and Russians – make up almost equal parts of the population (about 30% each). Their languages are official and all of them are languages of instruction in schools, though Russian schools and the Russian

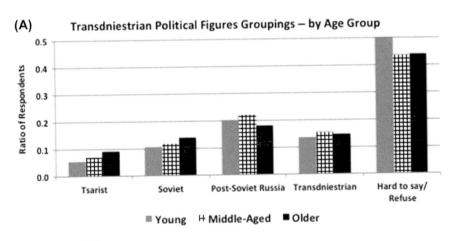

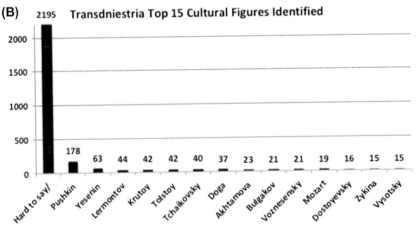

Figure 5. Comparison of Political Figures in Transdniestria by Age Group and Identification of the Top15 Cultural Figures in Transdniestria 254 × 190 mm (72 × 72 DPI).

language dominate. The official rhetoric of the state declares that the national objective is the consolidation of the Transdniestrian political nation whose members share the same values and ideals. Our survey showed that this position is supported with practically no difference between local Moldovans, Ukrainians, and Russians in the set of cultural figures mentioned. As for politicians, this difference was also rather weak: it is only possible to notice a slightly higher "rating" of Transdniestrian politicians among Ukrainians and a bit lower one among Russians. Moldovans were slightly more likely to recall Soviet political leaders while naming contemporary Russian politicians less frequently.

The surveys conducted simultaneously in Transdniestria and in Moldova demonstrated that the Orthodox church and the shared Soviet past are the only common features of their respective societies. They radically differ in the use of languages, historical development, political preferences, and geopolitical orientations. It was clearly seen in the composition of their pantheons of national cultural figures and political heroes, which were very different in our survey. Among cultural icons, common figures were represented by "neutral" Russian and foreign classical writer and composers of the past (Pushkin, Lermontov, Tolstoy, Dostoevsky, Shakespeare, Bach, etc.), as well as a small number of Russian pop singers and composers. In Transdniestria, the share of Russian cultural figures made up 22% of all mentioned names. In Moldova, their ratio was much smaller at 12%, though it is still higher than the share of Romanian cultural icons.

Further statistical analysis of the TMR data on icons revealed that there was no statistically significant difference in the icons identified among three main ethnic communities – by education, by urban or rural residence, by the extent of the personal experiences of the short war between Transdniestria and Moldova in summer 1992, or by the evaluation of the government's political actions. The views of a Transdniestrian Moldovan were not different from a Transdniestrian Russian, but they have very little in common with the opinions of a Moldovan from Moldova across the river Dniester. The TMR population showed a very low level of cultural awareness compared to Nagorno-Karabakh and Abkhazia.

For political figures, the relevant common feature of Trasndniestria and Moldova were mentions of a number of Soviet and Russian political leaders, which did not depend strongly on the ethnic background of respondents. Since 1991, there has been no stable consensus in the Moldovan elite whether their country should be incorporated into Romania or remain independent. This strategic uncertainty is directly reproduced in the contradictory and unstable set of symbolic cultural figures. An obvious material illustration is the Alley of Classics (*Aleea Clasicilor*) in Chișinău, Moldova's capital city, of monuments to prominent persons. Only 6 of the 27 were born within the contemporary boundaries of Moldova; about half are Romanian. Some Moldovan culture icons, like the contemporary writer Grigore Vieru, reject the very existence of the Moldovan language and the Moldovan nation. However, the results of our Moldovan survey confirmed the conclusions made from numerous other national surveys: reunification with Romania has not become a Moldovan national ideal.

A strikingly high number of respondents (47%) could not or refused to name any prominent cultural figure in Transdniestria (Figure 5(b)). The hierarchy of local cultural symbols is not shaped yet – partly because of the ethnic heterogeneity and partly because the republic had been independent for less than 20 years. Except for Evgeny Doga, a Moldovan composer well known in the Soviet Union about 50 years ago for his soundtracks for popular movies, none of the cultural figures mentioned by Transdniestrians was related with this region. Interestingly, no Western figure is on the list

Abkhazia

Like the TMR, Abkhazia is a heterogeneous ethnic republic of Abkhaz, Georgians, Russians, and Armenians. For their political icons, Abkhazian citizens listed mostly contemporary "strong men" in both Abkhazia and Russia. In the top rank (shared with Vladimir Putin) was independence fighter and First President Vladislav Ardzinba, followed by Sergey Bagapsh (president at the time of the 2010 survey). The respondents listed many names from the Soviet past, including Josef Stalin. For political leaders, political solidarity between ethnic Abkhaz, Armenians, and Russians are mirrored in the shared high number of mentions of Abkhazian politicians. In stark contrast, ethnic Georgians did not list such persons (Figure 6(a)). While Soviet political leaders were named more frequently by Russians, contemporary Russian leaders (Putin, Medvedev, Lavrov, etc.) were identified slightly more by Abkhaz than by Russians and Armenians. These high rankings were a clear appreciation by Abkhaz of the support of Russia to Abkhazia during the "5 Days War" in August 2008 with Georgia.

The set of cultural icons admired by Abkhazians reflects the mixed-ethnic structure of the republic and is stratified along ethnic divides (Figure 6(b)). A large proportion of Georgians/Mingrelians did not want to or could not give their opinions. The cultural list bears a strong marker of the Soviet legacy with a high representation of writers who were popular in that era. It is also affected by the proximity to the big patron and neighbor, Russia; both high and mass Russian culture was well represented in the list of names. The list combines ethnic Abkhaz, Soviet/Russian and, to a much lesser extent, Armenian figures with very few Georgian figures on it.

The distribution of cultural icons mentioned by Abkhazian respondents by historical periods and origin shows that respondents referred predominantly either to Abkhaz ethnic figures like the Soviet/Abkazian writer Dmitry Guliya or Soviet singers and composers. Moreover, most Abkhazian figures can be also considered Soviet, such as Guliya who officially was called a classic writer of Abkhazian literature well before the end of the Soviet Union. As in other unrecognized republics, Abkhazians remember the figures from the recent past or from the contemporary period.

Figure 6. Comparison of Political and Cultural Figures in Abkhazia by Nationality 254 × 190 mm (72 × 72 DPI).

Abkhaz were more prone to recall ethnic Abkhazian and Soviet cultural figures, and to a lesser extent, the classic artists from the Tsarist period. Surprisingly, so did also Russians. The views of Armenians and Georgians (the small minority which answered this question) were more balanced. One can notice a higher share of Tsarist and, respectively, Armenian and Georgian figures. Western figures have a similar low rate of recall among all major ethnic groups. Therefore, there certainly exists a certain cultural basis common for all cultural groups related mainly with the Soviet past, but distinct ethnic identities and preferences are also quite visible.

Conclusions

This study was carried out less than two decades after the declaration of independence of the de facto republics, and the mixed domestic and foreign patron (that is, Russian) nature of the icons reflected the development and promotion of key individuals in the infant years of the new states. Over time, as the state authorities crystallize their messages and the state mandates educational texts for all schools, the emphasis can be expected to shift to local traditions and iconic figures. The example of Transdniestria in publishing a national atlas and a memorial book of the 1992 conflict as well as new stamps and banknotes shortly after separation from Moldova was an early foray by that state in state- and nation-building that can be expected to be replicated in the other republics. Geography and history are important subjects that complement literature and the humanities in the making of new nations.

Leading writers on nationalism such as Anthony Smith have long emphasized the mobilizing role of the representations and (re)discoveries of the glorious past and of great national heroes as a part of the symbolic capital of nationalist ideology. Their successors in the study of nationalism analyze its dissemination through mass media and the system of education shaping ethnic and territorial identities, the perception of neighbors and of the world as a whole. The titular ethnic group and its identity are typically epitomized as existing since time immemorial and always based on the same components (language, religion, common territory, and a clear and unbroken ancestral line). The titular group is normally pictured by nationalist intellectuals as integral and consolidated, miraculously surviving despite the countless conspiracies and active efforts of enemies, near and far. A decisive role in the creation, reinforcing, and the diffusion of social representations and imaginations belongs to the state expressing the political interests of those who control it.

Culture and cultural codes are intrinsic components of ethnic/territorial identity. The formation of the unique hierarchical system of deeply venerated national heroes is an organic part of identity building. An ethnic or a political nation needs to identify itself with this pantheon; external observers also identify a nation with its prominent persons. In personifying an "imagined community," to use Anderson's (1983) phrase, they make it an important part of the symbolic capital of a nation. They have an important impact on an individual's vision of the world and serve to justify the rights of an ethnic group to a territory in which it claims exclusive privileges. The rights to a territory were especially important in the Soviet era after the 1920s, with its strictly delimited and hierarchically organized national–territorial autonomies that led to battles of identities and numerous unsettled conflicts after the disintegration of the Soviet Union.

The results of the surveys in the de facto states reveal that these battles had a different nature among political elites and ordinary citizens in the four republics. Irreconcilable discussions about the ancient past are the preserve of intellectuals and politicians. Mass representations on culture and politics are focused only on more or less contemporary symbolic figures representing the ethnic nation, though they still include a few Russian classical writers and composers known in all the post-Soviet spaces as well as a number of recent pop stars. As we have shown in this article, public opinion on the place of key cultural and political figures in the local histories is a good measure of identity and inter-ethnic relations. Some de facto republics (especially Nagorno-Karabakh) are more advanced in such promotions around a clear ethnic definition of the new state that builds on a well-defined set of icons. At the other end of the scale, Transdniestria had the weakest display of knowledge and coherence around a clear set of political and cultural icons, reflecting its "in-between" history and current status as an aspirant Russian republic.

To our knowledge, this study is the first attempt to document the familiarity of residents of new states with iconic national figures that the residents identify. Gottmann's forays into the iconography underpinning identity and territorial claims, central to political geography, were mostly forgotten in the discipline. State decisions on the relative importance and standing of icons are readily seen on banknotes, coins, stamps, street names, statues and memorials, and in school textbooks. Individuals obviously are aware of these public displays, but the extent to which they coalesce in their opinions around certain key individuals is a gage of nation-formation and public education success. The inputs to nation-formation (coins, statues, etc.) are easily identified and explained, but the scale of its reception on the output side, as measured by icon identification, is rarely measured. In this study, we hope to have partly rectified that neglect and in doing so, return attention to Jean Gottmann's enduring legacy in political geography.

Acknowledgments

The surveys were carried out by Levada Center Moscow for Abkhazia, Khasan Dzutsev of Russian Academy of Sciences Vladikavkaz branch for South Ossetia, Gevorg Poghosyan of the National Academy of Sciences (Yerevan, Armenia) for Nagorno-Karabakh and Elena Bobkova of Tiraspol State University for Transdniestria. Gerard Toal (Virginia Tech) was a partner in the survey design and accompanying field work. Valery Dzusati (Arizona State University) helped us to identify Ossetian political and cultural figures. We dedicate this paper to the memory of Alexei Grazhdankin of Levada Center Moscow, whose untimely death in summer 2017 deprived the research community on public opinion on Russia and surrounding regions of a valued colleague, a decent person, and a steadfast worker. He made our research on the de facto states both feasible and trustworthy. Ralph Clem's comments and Nancy Johnston Place's keen editorial eye significantly improved the manuscript.

Disclosure statement

No potential conflict of interest was reported by the authors.

Funding

This research was funded by the National Science Foundation program on Human and Social Dynamics [grant number 0827016].

References

Allen, Dean. 2013. "'National Heroes': Sport and the Creation of Icons." *Sport in History* 33: 584–594.

Anderson, Benedict. 1983. *Imagined Communities: Reflections on the Origin and Spread of Nationalism*. New York: Verso Books.

Bakke, Kristin, Andrew M. Linke, John O'Loughlin, and Gerard Toal. 2018. "Dynamics of State-Building after War: External-internal Relations in Eurasian de facto States." *Political Geography* 63: 159–173.

Barth, Fredrik, ed. (1969) 1998. *Ethnic Groups and Boundaries: The Social Organization of Culture Difference*. Oslo: Universitetsforlaget. Republished and translated ed. Long Grove, IL: Waveland Press.

Bhabha, Homi, ed. 1990. *Nation and Narration*. London: Routledge.

Billig, Michael. 1995. *Banal Nationalism*. Thousand Oaks, CA: Sage.

Bruneau, Michel. 2000. "De l'icône à l'iconographie, du religieux au politique: Réflexions sur l'origine byzantine d'un concept gottmanien." [From Icon to Iconography, from Religious to Political: Reflections on the Byzantine Origin of a Gottmanian Concept.] *Annales de Géographie* 109: 563–579.

Champion, Tony. 1995. "Obituary: Iona Jean Gottmann, 1915-1994." *Transactions of the Institute of British Geographers* 20: 117–121.

Clem, Ralph S. 2014. "Dynamics of the Ukrainian State-Territory Nexus." *Eurasian Geography and Economics* 55: 219–235.

Gottmann, Jean. 1951. "Geography and International Relations." *World Politics* 3: 153–173.

Gottmann, Jean. 1952. *La Politique des Etats et leur Géographie* [The Politics of States and their Geography]. Paris: Armand Colin.

Gottmann, Jean. 1980. "Spatial Partitioning and the Politician's Wisdom." *International Political Science Review /Revue Internationale de science politique* 1: 432–455.

Hobsbawn, Eric, and Terence Ranger, eds. 1983. *The Invention of Tradition*. New York: Cambridge University Press.

Johnson, Nuala C. 2003. *Ireland, the Great War and the Geography of Remembrance*. Cambridge: Cambridge University Press.

Johnston, R. J. 1996. "Jean Gottmann: French Regional and Political Geographer extraordinaire." *Progress in Human Geography* 20: 183–193.

Kolstø, Pål, ed. 2005. *Myths and Boundaries in South-Eastern Europe*. London: Hurst & Company.

Kolstø, Pål. 2006. "The Sustainability and Future of Unrecognized Quasi-States." *Journal of Peace Research* 43: 723–740.

Kuchma, Leonid. 2003. *Ukraina – ne Rossia (Ukraine is not Russia)*. Moscow: Vremia.

Liebich Andre, and Oksana Myshlovska. 2014. Bandera: Memorialization and Commemoration. *Nationalities Papers* 42: 750–770.

Light, Duncan. 2004. "Street Names in Bucharest, 1990–1997: Exploring the Modern Historical Geographies of Post-Socialist Change." *Journal of Historical Geography* 30: 154–172.

Ma, Eric K. W., and Anthony Y. H. Fung. 2007. "Negotiating Local and National Identifications: Hong Kong Identity Surveys 1996-2006." *Asian Journal of Communication* 17: 172–185.

Matjunin, Sergei. 2000. "The New State Flags as the Iconographic Symbols of the post-Soviet Space." *GeoJournal* 52: 311–313.

Miller, Alexey. 2009. "Rossia: Vlast' i Istoria (Russia: Authority and History)." *Pro et Contra* 13 (May-August), 6–23.

Murphy, Alexander B. 2002. "National Claims to Territory in the Modern State System: Geographical Considerations." *Geopolitics* 7: 193–214.

Muscarà, Luca. 1998. "Les Mots Justes de Jean Gottmann. (The Exact Words of Jean Gottmann)." *Cybergeo : European Journal of Geography, Politique, Culture, Représentations*, 54. http://journals.openedition.org/cybergeo/5308

Muscarà, Luca. 2005a. *La Strada di Gottmann: Tra Universalismi della Storia e Particolarismi della Geografia (Gottmann's Road: Between the Universalism of History and the Particularities of Geography)*. Rome: Nexta Books.

Muscarà, Luca. 2005b. "Territory as a Psychosomatic Device: Gottmann's Kinetic Political Geography." *Geopolitics* 10: 26–49.

Nairn, Tom. 1975. "The Modern Janus." *New Left Review* 94: 3–29.

O'Loughlin, John, Gerard Toal, and Rebecca Chamberlain-Creangă. 2013. "Divided space, divided attitudes? Comparing the Republics of Moldova and Pridnestrovie (Transnistria) Using Simultaneous Surveys." *Eurasian Geography and Economics* 54: 227–258.

O'Loughlin, John, Vladimir Kolossov, and Gerard Toal. 2014. "Inside the de facto States: A Comparison of Attitudes in Abkhazia, Nagorny Karabakh, South Ossetia and Transnistria." *Eurasian Geography and Economics* 55: 423–456.

Penrose, Jan, 2011. "Designing the Nation. Banknotes, Banal Nationalism and Alternative Conceptions of the State." *Political Geography* 30: 429–440.

Penrose, Jan, and Craig Cumming. 2011. "Money Talks: Banknote Iconography and Symbolic Constructions of Scotland." *Nations and Nationalism* 17: 821–842.

Pointon, Marcia. 1998. "Money and Nationalism." In *Imagining Nations*, edited by Geoffrey Cubitt, 227–254. Manchester: Manchester University Press.

Raento, Pauliina, Anna Hämäläinen, Hanna Ikonen, and Nella Mikkonen. 2004. "Striking Stories: A Political Geography of Euro Coinage." *Political Geography* 23: 929–956.

Shnirelman, Victor. 2003. *Voina pamyati: Mif'i, identichosti i politika v zakavkaz'e* [Wars of Memory: Myths, Identity and Politics in Transcaucasia]. Moscow: Akademkniga.

Smith, Anthony D. 2008. *The Cultural Foundations of Nations: Hierarchy, Covenant and Republic*. Malden, MA: Blackwell Publishing.

Smith, Anthony D. 2009. *Ethno-Symbolism and Nationalism: A Cultural Approach*. London: Routledge.

Smith, Anthony D. 2010. *Nationalism*. Cambridge: Polity.

Terlouw, Kees. 2014. "Iconic Site Development and Legitimating Policies: The Changing Role of Water in Dutch Identity Discourses." *Geoforum* 57: 30–39.

Toal, Gerard, and John O'Loughlin. 2016. "Frozen Fragments, Simmering Spaces: The post-Soviet de facto States." In *Questioning post-Soviet*, edited by Edward C. Holland and Matthew Derrick, 103–126. Washington, DC: Woodrow Wilson International Center for Scholars.

Verdery, Katherine. 1999. *The Political Lives of Dead Bodies: Reburial and Postsocialist Change*. New York: Columbia University Press.

Wylegała, Anna. 2017. "Managing the Difficult Past: Ukrainian Collective Memory and Public Debates on History." *Nationalities Papers* 45: 780–797.

Young, Craig, and Duncan Light. 2001. "Place, National Identity and Post-socialist Transformations: An Introduction." *Political Geography* 20: 941–955.

The decline and shifting geography of violence in Russia's North Caucasus, 2010-2016

Edward C. Holland ⓘ, Frank D.W. Witmer ⓘ and John O'Loughlin ⓘ

ABSTRACT

A spatial analysis of the geography of insurgency in the North Caucasus of Russia from 1999 through the end of 2016, focused on the period since 2010, corroborates other work on the incidence of violence in the region. A sharp drop in the absolute number of conflict events over the past half-decade occurred as violence diffused from Chechnya in the mid-2000s and is attributable to a range of domestic and international factors. Domestically, the decline is broadly linked to the securitization of the region around the 2014 Winter Olympics in Sochi, the return to the use of the Kremlin power vertical as a system of political management after an interlude focused on economic development as a mitigation strategy, and the wider adoption of harsh management tactics at the regional and republic scales. Internationally, potential insurgents have left Russia to fight in the Middle East and Ukraine. Using a conflict-event data-set ($N = 18,960$) from August 1999 through the end of 2016 and focusing on the period since the creation of the North Caucasus Federal District in January 2010, the paper identifies a set of notable trends within the decline and shift in violence. Key findings include a percentage increase in arrests carried out by Russian security services, a decline in retaliation across conflict actors, and the failure of federal subsidies to contribute to declines in violence in the region. The long-term prospects for continued insurgency in the North Caucasus, specifically in light of the collapse of the Islamic State and Russia's domestic challenges, remain uncertain and should acknowledge the recent decline in violence in the region.

A mountainous region populated by a diversity of peoples mostly practicing Islam rather than Orthodoxy, the North Caucasus has long presented a challenge to the Russian state (Jersild 2002). This challenge has been particularly acute since 1991. The first war in Chechnya – one of "the wars of Soviet succession" that pitted autonomous regions against union republics – was followed by a period of de facto independence for the republic and the renewal of conflict in 1999. This second decade of violence in post-Soviet Russia's North Caucasus was characterized by the rise of an Islamist insurgency, the adoption of extrajudicial tactics by the Russian state, the outsourcing of conflict suppression to the Kremlin's local allies in Chechnya, and the diffusion of violence away from that republic.

Since 2010 political violence in Russia's North Caucasus has both declined and shifted geographically. Previous explanations for this decrease and shift have pointed to a host of factors both international and domestic (Aliyev 2015; Souleimanov and Petrtylova 2015; Souleimanov 2017a). Internationally, the further globalization of Islamist movements through the rise of ISIS has led to the internal fracturing of the North Caucasus Islamist insurgency and provided an alternate venue for these fighters (Ratelle 2016). Domestically, the securitization of the region around the 2014 Winter Olympics in Sochi deflected any discontent associated with ongoing economic stagnation, the continued use of the Kremlin's power vertical as a system of political management, and the broader adoption of harsh tactics of violence suppression and management at the regional and republic scales (Klimenko and Melvin 2016; Youngman 2016). On the surface, it appears that at the end of his third term as Russia's president, Vladimir Putin has finally achieved his "historic mission" of bringing a semblance of peace to the North Caucasus (Taylor 2007).

This paper uses a large-N data-set of conflict events ($N = 18,960$) from August 1999 through the end of 2016 to test explanations for these conflict dynamics. Corroborating other work on the incidence of violence in the region carried out by the *Kavkazskii Uzel* (Caucasian Knot) news service (http://www.kavkaz-uzel.eu/) and the human rights group Memorial (https://www.memo.ru/en-us/), we document a sharp drop in the absolute number of conflict events and a marked shift in their geographical distribution since the start of the decade. The ebb and flow of the conflict – with a high number and concentration of events in Chechnya until the mid-2000s followed by a diffusion of violence first to Ingushetia and then Dagestan and Kabardino-Balkaria – is now characterized geographically by a region-wide decrease in conflict events, extending trends noted in O'Loughlin, Holland, and Witmer (2011). In further exploration of the reasons behind this, we present a series of summary statistics from the data-set, generate surface event maps for the region as a whole, explore tit-for-tat violence between state actors and rebels, and test the effects of shifting monetary subsidy regimes by the federal center to the *rayoni* (the Russian equivalent of counties) and cities of the North Caucasus.

In considering the reasons behind the decline in political violence and the changes in its spatial character in the region, this paper builds on work in geography on the contextual nature of such violence as occurring in space and time. The compilation of large-N data-sets with geolocated conflict events – most notably the Armed Conflict Location Event Data-set (ACLED; Raleigh et al. 2010) – has facilitated the evaluation of insurgent and government action-reaction (O'Loughlin and Witmer 2011, 2012; Linke, Witmer, and O'Loughlin 2012). The disaggregation of event data from the country scale to the local scale also allows for in-depth regional analysis of conflict processes. This approach is particularly informative in the North Caucasus, where the fracturing of the Islamist insurgency has meant that it has been generally difficult to construct a unified narrative of insurgent strengthening or weakening. As violence waned in Chechnya during the middle of the last decade, violence increased first in Ingushetia and then in Dagestan and Kabardino-Balkaria (O'Loughlin, Holland, and Witmer 2011). The aim of the present study is to go beyond absolute event counts at both the regional and republic scales in the North Caucasus to develop a comprehensive picture of why violence declined and shifted, the dynamics of this process (tit-for-tat reactive violence), and the causes of the changes in the violence distributions.

The paper proceeds as follows. We first offer a brief history of Russia's long war, summarizing the evolution of the conflict in the North Caucasus since 1999. Our particular focus in this section is on the period since 2010 – following the creation of the North Caucasus Federal District (NCFD) as a separate region in Russia's territorial organization – and the subsequent decline in violence in the North Caucasus more broadly. We examine the pattern of violence as an action-reaction between rebels and military/police forces before we offer an empirical test of political violence in the North Caucasus, focusing on subsidies as a predictor of such violence. More financial subsidies from Moscow are expected to reduce the incentive to engage in violence. The concluding section offers a reevaluation of Russia's position in the North Caucasus and speculates about the success of the state in conflict management as domestic and international circumstances that condition violence endure.

Russia's long war in the North Caucasus

From its beginnings as a nationalist conflict pitting a Soviet-era third-tier autonomous republic (Chechnya) against a second-order union republic (Russia), the conflict in the North Caucasus has evolved in the past two decades into an Islamist insurgency against the central state and its associated institutions.[1] The first war, from December 1994 until August 1996, exposed the weakness of the Russian state; Lieven (1999, 1) described it as a harbinger of "the end of Russia as a great military and imperial power." In fact, the North Caucasus served as a proving ground for Russian counterinsurgency tactics, the evolution of state policy, and the organization of the power vertical (Kramer 2005; Melvin 2007; Taylor 2007; Lyall 2009; Sakwa 2010; Ware 2011).

Conflict in Chechnya resumed following two key events in August 1999, the incursion by Chechen rebels led by Shamil Basayev and Ibn-al-Khattab into southwestern areas of neighboring Dagestan and the elevation of Vladimir Putin to the position of prime minister of the Russian Federation. "For Putin, the North Caucasus was reflective of the larger problems facing post-Soviet Russia," including the potential for disintegration, the increased influence of foreign actors, and the overall weakness of the state (Dannreuther and March 2008, 99). The September 1999 apartment bombings that occurred in Moscow and the southern cities of Buinaksk (in Dagestan) and Volgodonsk (in Rostov oblast) further galvanized public opinion against the Chechen insurgency and in support of Putin's leadership, though the perpetrators of the bombings have never been positively identified.

The active phase of renewed fighting in Chechnya began in October 1999. In the period from the start of the second war through March 2000, the rebels lost control of the republic's main urban areas including the capital of Grozny (Zürcher 2007; Lyall 2010). The insurgency's tactics shifted in response to a guerrilla campaign targeting Russian military forces. The Islamist faction, led by Basayev, endorsed suicide bombing and carried out hostage takings both in the region and in Moscow. Russian forces used sweep operations – known as *zachistki* – in Chechnya and Ingushetia to detain suspected insurgents, their affiliates, and civilians (Gilligan 2009); the military also targeted the leadership of various factions within the insurgency, killing the president of the separatist Chechen Republic of Ichkeria, Aslan Maskhadov, in 2005 and Basayev the next year.

Violence spread from Chechnya proper to the republics of Dagestan, Ingushetia, and Kabardino-Balkaria beginning in 2007, linked to economic weakness, high unemployment among young men, and the lack of appropriate channels for voicing political concerns (Sagramoso 2007; O'Loughlin, Holland, and Witmer 2011). Ingushetia saw a consistent rise in violence during the tenure of Murat Zyazikov as the republic's president from 2002 to 2008. Violence in Dagestan increased notably beginning in 2009, and the republic has been the site of the highest absolute number of violent events in the North Caucasus since 2010. In Kabardino-Balkaria, militancy is often attributed to the heavy-handed tactics by state forces used against conservative Muslims beginning in 1999; the October 2005 attack on the capital of Nalchik ushered in a more active phase in the republic's insurgency (Souleimanov 2011).

In addition to the diffusion of conflict at the region scale, the awarding of the 2014 Winter Olympics to the resort town of Sochi in the western Caucasus spurred further concern over regional stability. Moscow's securitization of the North Caucasus in the lead-up to the 2014 Winter Olympics in Sochi demonstrated a firm commitment to preventing any sort of terrorist attack against targets associated with the Games. Early planning for the Sochi Olympic Games starting in 2007 endorsed the premise that the region was generally safe and security could be handled by regional agencies until 2012. This approach was revised following the proclamation of the Caucasus Emirate (CE) by Dokku Umarov in October 2007 and terrorist acts in the region and beyond in 2009, 2010, and 2011 (Zhemukhov and Orttung 2014).[2]

In response to the increase in violence in the North Caucasus, Russia's then-president Dmitry Medvedev established the North Caucasus Federal District in January 2010. It was carved out of the existing Southern Federal District and included the national republics of Dagestan, Chechnya, Ingushetia, North Ossetia, Kabardino-Balkaria, and Karachay-Cherkessia, along with Stavropol *krai*. Upon its creation, Alexander Khloponin, formerly the governor of Krasnoyarsk *krai* and Chairman of Norilsk Nickel, was appointed presidential envoy to the district. During his tenure as governor of Krasnoyarsk, Khloponin secured substantial outside investment in the region, thanks to its wealth of natural resources and the presence of Norilsk Nickel (Ware 2011). At the time of his appointment to the envoy post, Medvedev endorsed Khloponin's record in terms of socioeconomic development, which was viewed as "absolutely necessary (*kraine neobkhodimo*) in the North Caucasus" (Medvedev 2010). Appointments made at the republic level also emphasized the economy over security. Arsen Kanokov had been Kabardino-Balkaria's head since 2005, in part on the basis of prior business successes; he led a holding company established in 1991 with interests in banking, investments, and construction. Kanokov was able to continue this success as head of Kabardino-Balkaria; gross domestic product there increased threefold between 2005 and 2012. But his tenure was also marred by infighting among the republic's elites, a series of high-profile corruption cases, and the spike in violence noted previously (RFE/RL 2013).

The decline in violence in the North Caucasus

Russia's strategy in the North Caucasus has further evolved since 2010. The tactic of endorsing economic development has evolved into a carrot-and-stick approach that incorporates revisions to force structures and changes in political leadership in tandem with continued subsidization. Moscow has recognized that force necessarily complements economic growth and support as the key to securing stability in the region; that is, the "carrot" alone does not work. The "hard security" regime that marks this revised approach by the Russian government was enacted prior to and continued after the Olympics. It relies on the close coordination between local and regional authorities and counterinsurgent techniques that target both insurgents and their support networks (Klimenko and Melvin 2016). For example, Moscow responded with decisiveness to a series of attacks in Russia's south (in the city of Volgograd) in the months leading up to the Games, in October and December 2013.

Souleimanov (2016) identifies three strategic implementations by the Russian state with respect to security in the North Caucasus since the start of Putin's third term: the application of tactics – including the targeting of insurgents' families and the destruction of homes – designed to weaken popular support for the insurgency; the increased use of specially trained counterinsurgency forces; and the successful infiltration of insurgent cells by pro-Russian actors. For example, individuals who provide support for the insurgency – such as food and medication – have been pursued and arrested in increasing numbers across the region (Fuller 2015). The upshot, according to Souleimanov (2016), is that these tactics in combination with the exodus of potential fighters to Syria (see discussion below) "have led to a dramatic weakening of the North Caucasus insurgency" (see also Souleimanov 2017a).

Through the use of special forces and the placement of moles in insurgent groups, pro-Russian actors in the North Caucasus have successfully eliminated key figures in the insurgency at both the regional level and in republic-scale *jamaats* (Islamist communities that draw together like-minded believers in opposition to the Russian state; on the Kabardino-Balkaria *jamaat*, see Shterin and Yarlykapov 2011). Russian security services have had success in eliminating key leaders in the Caucasus Emirate; for example, Dokku Umarov, who suggested that insurgents target the Olympics with "maximum force" in a June 2013 video, died in September of that same year from a reported poisoning (Arnold and Foxall 2014, 8). His death was first leaked in January 2014 and subsequently confirmed by the Caucasus Emirate in March 2014 – notably, after the Games had closed. Umarov was replaced by Aliaskhab Kebekov, who himself was killed in an April 2015 operation by Russian security forces near Buinaksk, Dagestan. His replacement, Magomed Suleymanov, was killed in turn in August 2015, further contributing to the leadership vacuum for the Caucasus Emirate that is a sign of the insurgency's weakening (Fuller 2015; Youngman 2016).

This strategy of eliminating the insurgency's leadership has been complemented by the targeting of support networks, a practice long endorsed by Ramzan Kadyrov in Chechnya (Souleimanov and Jasutis 2016).[3] Repressive measures employed by the Russian state in the North Caucasus are increasingly selective in targeting rebels and their support networks (Zhirukhina, forthcoming). A report by Williams and Lokshina (2015) for Human Rights Watch documented the destruction of homes in the village of Gimry, Dagestan during a 2013 counterterrorism operation as evidence of the wider adoption of policies of collective responsibility and collective punishment as endorsed in Chechnya. Ramazan Abdulatipov, Dagestan's head from 2013 until October 2017, ended dialogue with more moderate elements in the Salafist community in the republic, an approach that had been endorsed by his predecessor, Magomedsalam Magomedov. During the second half of 2013, Dagestani authorities stepped up their raids of mosques and cafes where Salafists met; this practice continued during the holy month of Ramadan in 2014 (Williams and Lokshina 2015). Drawing on information from *Kavkazskii Uzel*, Williams and Lokshina's (2015) report also indicates that Salafists have been the target of extra-judicial killings that have been insufficiently investigated by Russian authorities. Islamists in Dagestan have been managed through a so-called "prophylactic list" that includes the names of individuals with suspected ties to Islamist networks – although many are placed on the list without clear justification (Mayetnaya 2017).

This evolution in tactics has been complemented by leadership changes at the regional and republic levels. Sergey Melikov, a general in the Russian military and previously deputy chief of staff in the Russian Interior Ministry, replaced Khloponin as presidential envoy to the NCFD in May 2014. Melikov is from the North Caucasus – he is an ethnic Tabasaran, one of Dagestan's 14 principal ethnic groups – and spent his military career in the region (Vatchagaev 2014). Melikov was subsequently reassigned in July 2016 to the newly created National Guard and replaced as presidential envoy by Oleg Belaventsev, a career naval officer who previously held the plenipotentiary position for the Crimean Federal District (Fuller 2016b). Belaventsev is an outsider to the North Caucasus, with an uneven record in working with local elites in Crimea. The emphasis on security rather than economy is also evident in some republic-level appointments, such as the replacement of Kanakov by Yuri Kokov in Kabardino-Balkaria in December 2013; Kokov has a pedigree from the security services, serving as a Colonel-General in the Interior Ministry.

The renewed emphasis on security experience in the power vertical from the Kremlin underscores the general ineffectiveness of the top-down approach to economic management previously endorsed by the Russian state. The "Strategy for the Socioeconomic Development of the North Caucasus Federal District Until 2025," a development program that aims to improve the region's economy, reduce unemployment, and integrate the region's economy into national and international networks, has thus far been a failure (Holland 2016). Budgetary transfers from Moscow to the North Caucasus republics remain high and have increased over the past half decade (see Figure 7 below); as an example, in 2015 subsidies from the federal center accounted for 81% of Chechnya's budget (Fuller 2016c).

The Russian government has two categories for grants (*dotatsii*) to the federation's regions. The first category includes federal transfers determined at a fixed rate dependent on expenditures and tax revenues. Chechnya has received significant financial support under a second, more nebulous category that is intended to address any shortfalls in regional budgets not sufficiently covered by the first set of transfers. Two other types of federal transfers supplement these grants; subsidies (*subsidii*) and subventions (*subventsii*). *Subsidii* can be understood as matching grants from the federal center to the regions, while *subventsii* are provided to cover expenses at the regional level that are federal responsibility (see Alexeev and Chernyavskiy 2017). The outlay of subsidies under Putin complements political centralization in the Russian state, while transfers to the North Caucasus and Chechnya specifically have allowed for the reconstruction of infrastructure destroyed during the wars and a relative equalization of economic capacity across regions to minimize outmigration. The long-term viability of this support is questionable, as promised outlays to the North Caucasus's development program for 2017 were reduced substantially in 2016 from RUB 30 billion (about USD 500 million) to RUB 12 billion (about USD 200 million) (Dzutsati 2016).

The region's economic problems remain significant, and leaders with business experience have done little to address key structural issues. The republic economies have been unsuccessful at diversifying into potential growth sectors such as agriculture, tourism, and oil and natural gas (Holland 2016). Corruption further contributes to the misuse of funds and the failure of the "Strategy" to gain traction in terms of economic development. The republics' combined debt is approximately

RUB 67 billion (about USD 1.2 billion), while corruption cases rose approximately 12% in 2015 from the previous year (Fuller 2016a). Corruption in the political system has other tangible consequences; it leads, for example, to alternative forms of governance, most prominently the endorsement of Sharia law and the adoption of Islamic education in parts of the North Caucasus (Klimenko and Melvin 2016). More succinctly, Dzutsati (2016) writes that "higher poverty is likely to push some people toward greater radicalism."

International conditions also serve to explain the decline in violence in the North Caucasus over the past half-decade. Souleimanov and Petrtylova (2015) argue that participation by North Caucasians in Syria and Iraq has diverted the focus of potential jihadists away from the region, itself a boon for Russia's anti-Islamist efforts. However, internal divisions among the region's jihadists likely curtailed the outflow of fighters from the North Caucasus to the Middle East beginning in 2014 (Souleimanov 2017a). Russian citizens fighting with ISIS include individuals from Chechnya and Kabardino-Balkaria in addition to Dagestan, although the estimates of the total number of Russian citizens who have traveled to the Middle East vary widely and the precise number is difficult to determine. Recent estimates put this figure at approximately 2,000 (other figures range from a low of 800 to a high of 5,000; see International Crisis Group 2016; Sokolov 2016). In November 2017 remarks to members of the Russian Defense Ministry, Valery Gerasimov, Chief of the General Staff, put the number of Islamic State fighters originating from Russia and killed in the Middle East at 2,800; a further 1,400 from other members of the Commonwealth of Independent States have been killed (Ministry of Defence of the Russian Federation 2017). Combatants from the region can also be found on both sides of the conflict in Ukraine – supporting Russian proxies fighting for the people's republics of Donetsk and Luhansk and on the Kyiv government's side fighting to reintegrate the eastern territories. Walker (2015) quotes a pro-Ukraine Chechen fighter: "That [the war in Syria and Iraq] is not a Chechen war. This, here in Ukraine, is a war for Chechens. If we defeat Russia here, we are closer to freeing our homeland."

Regardless of its precise impact on the number of fighters in the North Caucasus, the rise of the Islamic State resulted in the internal fracturing of the region's Islamist movement at a structural level. Key defections from the core leadership of the CE to the Islamic State include Rustam Asilderov and Aslan Byutukavev, who led the Emirate's Dagestan and Chechnya branches, respectively (Youngman 2016). The Islamic State further weakened the position of the CE in July 2015 when it announced the establishment of an official branch in the North Caucasus. Youngman (2016, 2) summarizes the situation in the region as follows: "by early 2016 it [the Islamic State's Caucasus branch] was clearly the stronger party, with the IK [Imirat Kavkaza, aka CE] leaderless and struggling to survive, at least within the North Caucasus itself."

The endgame in Syria and Iraq matters to the future trajectory of the Russian state, and the collapse of the Islamic State in late 2017 has the potential to destabilize the North Caucasus, thanks to a flood of returning fighters (Ratelle 2016). The effect on domestic conflict of returning fighters is case-specific; many of the foreign fighters who joined the war in Afghanistan during the 1980s stayed on after the conflict ended, in turn limiting the return migration of experienced fighters to their countries of origin. The consequences of involvement by fighters from the North Caucasus in the Syrian civil war remains uncertain; "If allowed to return to their homeland, North Caucasian Jihadists – a committed and experienced force of hundreds of fighters with extensive contacts with jihadists worldwide – may pose an enormous threat to Russia's internal security" (Souleimanov 2014; 154). Ratelle (2016) argues that two waves of fighters from the North Caucasus have joined the war in Syria: a first group who left for Syria from 2011 to 2013 because the opportunities for fighters in the North Caucasus were foreclosed, and a second set who preferred to join the international jihadist networks established in the Middle East. Writing at a time before the fracturing of ISIS, Ratelle suggested that the former group was more likely to return to the North Caucasus. Though precise numbers are difficult to determine, Wright (2017) suggests that about 10% of Russian citizens fighting in the Middle East had returned home by October 2017.

Conflict events in the North Caucasus

A number of organizations track political violence in the North Caucasus. *Kavkazskii Uzel* has compiled statistics on casualties resulting from armed conflict and terrorism-related violence in the North Caucasus Federal District since 2010. According to these data, in 2010 roughly 1,700 people in the North Caucasus were killed or injured. By 2016, the total number of casualties had declined to 287 (*Kavkazskii Uzel* 2016, 2017).[4] Over the past seven years, Dagestan has been the site of the highest number of casualties, although this absolute total is trending downward like the region as a whole, from 659 casualties in 2010 to 204 in 2016.[5] The International Crisis Group (2016, 1) draws on *Kavkazskii Uzel*'s data in their March 2016 report on the region and concludes that the year 2014 "saw a remarkable reduction in violence," a trend that continued in 2015.

Memorial, a leading Russian human rights organization, aggregates news reports to track the number of security personnel killed or injured in combat with insurgents or terrorist attacks occurring in the region (see Toft and Zhukov 2012 for a prior use of these data in conflict analysis). Likewise, this figure has trended down since 2006, although an uptick in Ingushetia in 2008–2009 and Dagestan and Kabardino-Balkaria in 2010–2011 is noted (Memorial 2016). Parallel to the data from *Kavkazskii Uzel*, Memorial's June 2016 report unambiguously states, "beginning in 2012, militant activity has been steadily declining" across the North Caucasus (2016, 6).

The data-set analyzed here is similar in its general aims of documenting violence over time and across the region to those collected by *Kavkazskii Uzel* and Memorial, but ours is more comprehensive in its time frame and geographic detail.

Conflict events are coded in the data-set since the beginning of August 1999, which coincides with the invasion of southwestern Dagestan by forces affiliated with Shamil Basaev and the de facto start of the second Chechen War. There are 18,960 unique events in the data-set, with each entry detailing the date of the event, its best geographic location (the nearest settlement, district, or republic), information about actors, casualties, and a brief textual synopsis. Events were gathered from wire reports and news stories available through Lexis-Nexis's academic search service. The data-set has been used previously for regional analysis (i.e. O'Loughlin and Witmer 2011, 2012; Linke and O'Loughlin 2015) and for cross-national examination of violence in relation to terrain in the wider Caucasus region (Linke et al. 2017). For this article, we updated the data-set to include events from August 2011 through December 2016, inclusive of the seven federal territories of the NCFD. While the data-set corroborates other sources that aggregate violent events in the North Caucasus, the analysis presented here goes beyond absolute event counts to consider how interactions between the state and the rebels have evolved over the past half-decade by focusing on geography, actors and re-actors, and the causal explanation of violence in small geographic areas.

Figure 1 summarizes the absolute event counts by month over the entirety of the data-set, with the creation of the North Caucasus Federal District noted (January 2010). In analyzing the conflict dynamics in the region, prior work using this data-set and other micro-scale data has identified three important trends. First, as discussed above, the diffusion of violence from Chechnya to neighboring republics began in 2007 (O'Loughlin, Holland, and Witmer 2011). Second, violence has been characterized by action-reaction on the part of the Russian military and insurgents (O'Loughlin and Witmer 2012); linking public opinion polling to the incidence of violence, Linke and O'Loughlin (2015, 122) report that violence in the North Caucasus "is associated with lack of trust and a preference for ethno-territorial separation" among survey respondents. While motivations to conflict

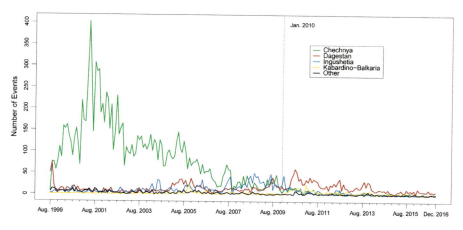

Figure 1. Monthly conflict event totals by republic/*krai*, 1999-2016. January 2010 marks the creation of the North Caucasus Federal District (NCFD).

remain difficult to pin down precisely, the ideas of retribution, the absence of trust associated with corruption, and the continued salience of ethnic identity all contribute as explanatory factors. We formally examine below the pattern of tit-for-tat violence in the region. Third, selective counterinsurgency tactics (e.g. arrests) used by the Russian state have been more successful than indiscriminate tactics (e.g. *zachistki*) in suppressing acts of violence carried out by nationalists affiliated with Aslan Maskhadov and the more secular Chechen Republic of Ichkeria; this distinction does not hold for Islamist violence, however (Toft and Zhukov 2015; see also Lyall 2009).

Shifting aims and narratives: arrests and civilians as targets

Our data-set codes arrests as conflict events when detention is clearly indicated in the news reports. There are 2,613 unique arrests coded in the data-set – 13.8% of the total number of events. This count almost certainly underreports the number of arrests that have been carried out in the North Caucasus region over the past decade and a half, and a pair of caveats should be noted. First, when rebel actors are injured but not killed in engagements with police and military forces, they are detained and subject to criminal prosecution. This outcome is not reported at the time of the event; rather, criminal proceedings occur months or even years after the initial incident and thus were not coded in the data-set. Second, while the data include unique events for *zachistki*, the scale of these operations means that more individuals were detained than a single coded event accurately represents. Despite these caveats, we posit arrests as a proxy for the shifting practices of the Russian state in managing the insurgency (Souleimanov 2017a; Zhirukhina, forthcoming).

As a proportion of total events, arrests have increased markedly over the past seven years (see Figure 2 for arrests as a percentage of total events by key republic for each year, 1999–2016). In Chechnya, for example, more than half of the events recorded in 2015 were arrests (51.8%; 14 out of 27 total events); more generally, Chechnya under Ramzan Kadyrov has had a higher percentage of arrests in

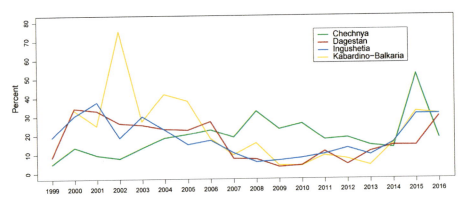

Figure 2. Arrests as percentage of all events by republic by year, 1999–2016.

comparison to the other regions of the North Caucasus. Dagestan, as previously noted, is adopting this securitization model, and 2016 witnessed a sharp increase in the proportion of arrests to the total number of conflict events, although this model is also being applied in Ingushetia and Kabardino-Balkaria – arrests as a percentage of total events top 30% in both of these republics in 2016. As expected, the shift began in 2014 with the securitization of the region at the time of the Sochi Olympics, and these tactics have continued to be applied in the management of the conflict. This active management of potential threats has carried over into day-to-day life in the region. On multiple occasions over the past three years, police detained Muslim worshippers at Salafist mosques in Dagestan's capital, Makhachkala, and other cities; those held were fingerprinted, had their documents photocopied, and were required to complete a questionnaire (see also Mayetnaya 2017).

Both *Kavkazskii Uzel* and Memorial previously indicated that the number of those killed or injured by violence in the region has fallen – either in aggregate or for security personnel as a proxy measure. Figure 3 further explores the decline in violence through the measure of civilian casualties, presenting the aggregate number of civilians killed or injured in comparison to total casualties for the entire region. We note three periods when civilian casualties were notably high: August–December 1999, the early stages of the second Chechen war, when Russian forces entered Chechnya from the north and proceeded toward Grozny and the mountains; 2003–2004, inclusive of the hostage crisis at Beslan, North Ossetia; and 2010–2013, during which period violence was highest in Dagestan and rebels targeted police and military actors in congested urban spaces such as Makhachkala (e.g. see Reuters 2012). During more active phases of fighting (i.e. 2000–2002, when violence peaked in Chechnya), civilian casualties were notably lower, with most fighting occurring between government forces and rebels. Since the recent high point in 2010–2013, the proportion of civilian casualties has declined; inclusive

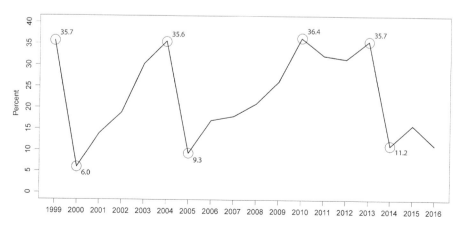

Figure 3. Civilian casualties as percentage of all casualties by year, 1999–2016.

of 2014, the percentage of civilian casualties has been at or below the 15% mark each of the last three years.

In part, the decline in civilian casualties is attributable to the shifting rhetoric of the CE's leadership on the targeting of civilians. Souleimanov (2011, 164–165) distinguishes between the CE's aims within the region and beyond: while attacks within the region "are usually implemented so as to avoid the loss of civilian lives" (though collateral casualties do occur), in Moscow and elsewhere indiscriminate violence is carried out with "the goal of as many civilian deaths as possible." However, since the January 2011 Domodedovo airport attack, such violence has been rare and includes the series of bombings in Volgograd in late 2013 and the April 2017 bombing of the St. Petersburg metro, which was carried out by a Russian citizen of Uzbek descent born in Kyrgyzstan. Russian authorities have also been successful at killing rebels or individuals identified as such; according to our data, nearly 60% of the reported casualties in the North Caucasus in 2016 were comprised of rebels who were killed in fighting with state forces (141 out of 240 total casualties).

The geography of the decline in violence

Following the decline in the absolute number of conflict events at both the regional and republic scales, actions carried out by rebels and military actors in the region have also experienced a corresponding drop in the past half-decade. Presenting the data for selected years, Figure 4(a) and (b) display potential surface maps for conflict events carried out by rebel and military/police actors, respectively, across the region. Potential surface maps show spatial distributions by considering both the distance between points and the magnitude of the observation (Warntz 1964; Frolov 1977). For violence, these maps show the influence of conflict across space as a function of distance between locations. The aim of these maps is to show how violence in one location influences violence in a neighboring location, as well as to show overall distributions (see O'Loughlin, Holland, and Witmer 2011 for a prior application of potential surface maps). The left column shows the annual potential surface for rebel events and the right column shows military/police events at five-year intervals.[6] Military/police violence was initially more intense and widespread than rebel violence. The diffusion in violence from Chechnya and Grozny to neighboring republics is especially clear when comparing the 2000 maps to 2010.

The decline in violence is further displayed by mapping the absolute change in event counts per 1000 people at the *rayon* level across the two selected periods, August 1999–December 2009 and January 2010 to December 2016 (Figure 5). This map clearly shows steep declines in violence for Chechnya and Ingushetia, small changes for Stavropol' *krai*, and notable increases in violence for Kabardino-Balkaria and Dagestan. We examine potential drivers of these changes in our *rayon*-scale statistical models below.

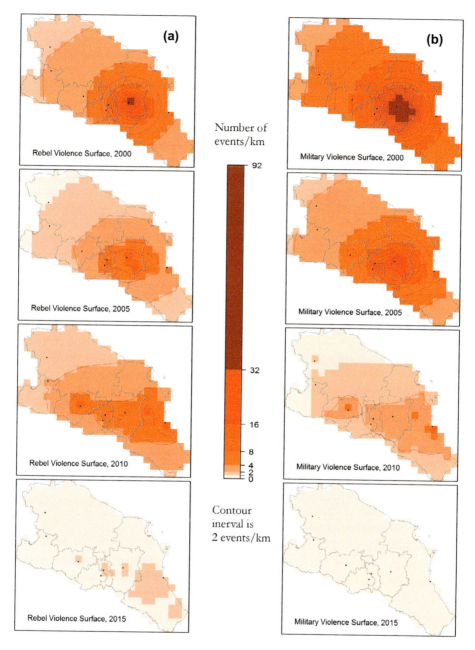

Figure 4. Violence potential surfaces for select years; (a) rebel events on the left, (b) military/police events on the right.

Modeling violence in the region: action-reaction and the role of subsidies

In this section, we present results from two sets of statistical models. The first tests action-reaction behavior between rebels and government forces to see if the tit-for-tat violence intensity has declined after the creation of the NCFD. Our second set of models examines the role of subsidies in predicting the decline in violence

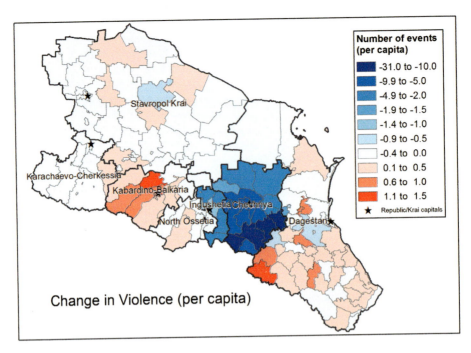

Figure 5. Change in the number of violent events per 1000 people from the first period, August 1999–December 2009, to the second period, January 2010–December 2016. Source: Authors.

while controlling for spatial autocorrelation and other known factors connected to the geographic distribution of violence.

Action–reaction models

For the action–reaction models, we create a spatio-temporal data-set of violence by overlaying a 10 × 10 km set of grid cells (1,787 in total) on our violent events data-set and aggregating violence by event type (rebel, military/police, arrest) to each grid-week. To test the tit-for-tat relationship over multiple spatial and temporal periods, we calculate the first- and second-order spatial lags (queen contiguity) for each grid cell and then temporally lag them for one through five time periods. This allows us to test the possible effects of violence that occurred in the prior month in surrounding territory. This type of spatial Granger analysis has been used to study reciprocity in Iraq (Linke, Witmer, and O'Loughlin 2012) and builds upon the simpler temporal Granger causality analysis (Granger 1969; Freeman 1983).

The simplest form of the model estimates the direction and strength of the relationship between action and subsequent reaction as follows:

$$Y_t = \beta_1 Y_{t-1} + \beta_2 X_{t-1} + \epsilon$$

where the influence of actor X at the prior time period, $t-1$, against actor Y at time t is determined by the coefficient β_2 after controlling for prior actions by Y at $t-1$ and allowing for an error term of unspecified influences, ϵ. Our spatial version of this model extends the temporal lags from one to five and also includes first- and second-order neighboring violence. We estimate violent event counts using a negative binomial generalized linear model available from the R "MASS" package.

The plots of the β coefficients for just the reciprocity terms in our regression models, with control terms omitted for clarity, are shown in Figure 6. The darker circles (filled or open) in each plot are for models where rebel violence is the dependent variable, Y. Lighter circles in Figures 6(a) and 6(c) show results from models where military/police violence is being predicted, while lighter circles in Figures 6(b) and 6(c) are for arrest models. We estimate and present results for all republics/*krai*, for Chechnya, and for Dagestan for both the early period, August 1999–December 2009 (Figures 6(a) and 6(b)), and later period, January 2010–December 2016 (Figures 6(c) and 6(d)).

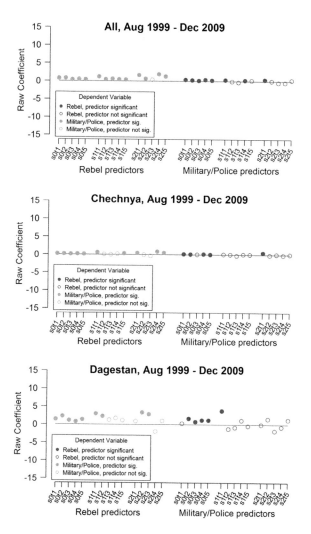

Figure 6a. Action–reaction models (controls omitted) for military/police violence during the first period, August 1999–December 2009 for all grid cells, and Chechnya and Dagestan subsets.

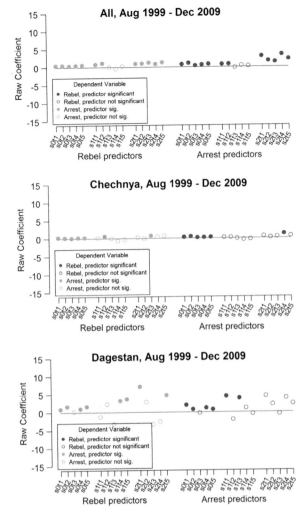

Figure 6b. Action–reaction models (controls omitted) for arrests during the first period, August 1999–December 2009 for all grid cells, and Chechnya and Dagestan subsets.

Overall, the reciprocity relationships are stronger for the early period and weaken for the second period with fewer coefficients significant. This suggests that the retaliatory nature of the violence has declined along with the overall intensity of the violence. Presumably, this shift corresponds to the change in tactics pursued by the Russian state and the fracturing of the insurgency noted previously. Comparing rebel vis-à-vis arrest violence to rebel vis-à-vis military/police violence indicates a somewhat stronger action-reaction relationship for the military/police events than the arrest events. These differences are greater for the second time period.

The tit-for-tat pattern of violence found in the North Caucasus is generally weaker than that shown for violence in Iraq (Linke, Witmer, and O'Loughlin 2012). This is likely due in part to differences in the intensity and scale of the violence. In

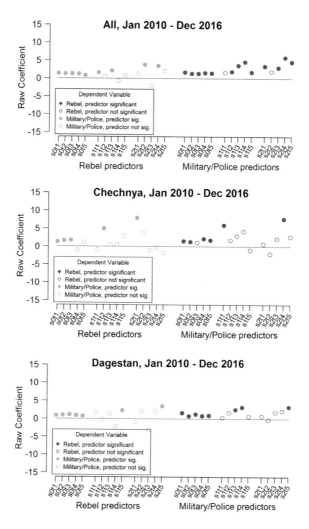

Figure 6c. Action-reaction models (controls omitted) for military/police violence during the second period, January 2010–December 2016 for all grid cells, and Chechnya and Dagestan subsets.

Iraq, the data-set consisted of nearly 400,000 events precisely geocoded by the U.S. military, whereas our North Caucasus data-set has 16,928 events with sufficient locational precision to be used in this kind of action-reaction analysis. For instance, the Iraq data-set is geocoded to street-level precision, enabling city-scale reciprocity relationships to emerge. The North Caucasus data-set is geocoded at the village level, which may explain why the second-order spatial lag reaction relationship is sometimes stronger than for nearby violence in some of the all-republic models (Figures 6b–6d). This distance of about 25 km may also reflect a natural spacing of villages along connecting roadways.

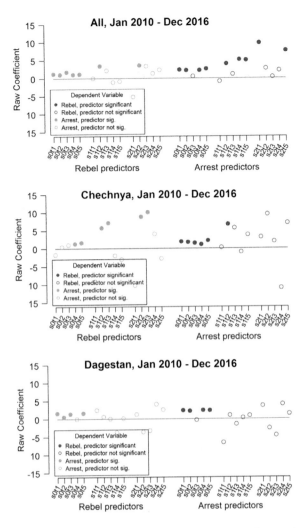

Figure 6d. Action-reaction models (controls omitted) for arrests during the second period, January 2010–December 2016 for all grid cells, and Chechnya and Dagestan subsets.

Change in violence models

In our predictive model of the change in violence across the North Caucasus, we focus on the role of federal subsidies. In particular, we test the hypothesis that federal subsidies reduce the level of violence as more government assistance may raise the opportunity costs of engaging in violence. This expectation draws on Fearon's (2008) argument about the economic inducements for rebels, while Child (2017) extends the argument to consideration of licit (often government-sponsored) and illicit (shadow economy and militant activities to raise money) undertakings. If those who would potentially engage in violence against the state and its forces in the region believe that state services are improving due to more subsidies from Moscow, their motivations for attacking the state would be lowered. In the wider realm of COIN (counterinsurgency) studies, mostly about U.S. military operations in Iraq and Afghanistan, the evidence for this argument is mixed. A review of the literature by Chou (2012) indicates that governmental (including U.S. aid) spending in Afghanistan does not statistically reduce the level of violence. Small targeted projects, such as village-level infrastructural projects, however, seem to be more effective in persuading potential fighters to avoid taking up arms (Berman, Shapiro, and Felter 2011). More general funds on a larger scale do not seem to have any impact (Berman et al. 2011).

We formally test the effect of subsidies on violence by estimating a set of statistical regression models. The outcome variable is change in the number of violent events per 1000 people (Figure 5) as measured at the *rayon* administrative unit between the first period, August 1999–December 2009, and second period, January 2010–December 2016. Our key predictor variable is change in subsidies per capita from 2010 to 2016 (Figure 7) as reported at the *rayon* and city scale in the North Caucasus region by the Russian Federal State Statistics Service, Goskomstat (http://www.gks.ru).[7]

We also include several control variables for factors that have been shown to influence violent conflict distributions. Metrics such as titular percentage (proportion of the nominal ethnic group), employed percentage, and urban percentage are included from census data published by Goskomstat. Additionally, we include the mean forest cover percentage calculated for each *rayon* for the year 2000. Forest cover is sometimes associated with violent conflict, typically under the

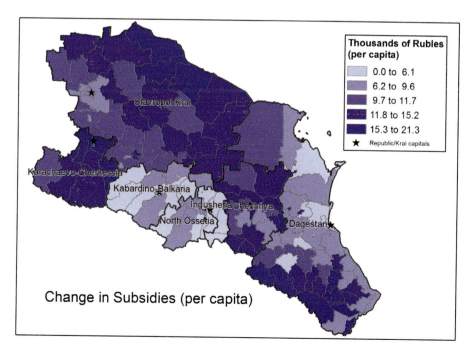

Figure 7. Change in federal subsidies (measured in thousands of Rubles per capita) from 2010 to 2016.[8] Source: Goskomstat, http://www.gks.ru/

theory that rebels use forests as hideouts from which to strike (Linke et al. 2017). These data are derived from Landsat imagery where forest is defined as closed canopy vegetation of at least 5 m in height (Hansen et al. 2013). Lastly, a distance to road measure is included based on the Global Roads Open Access Data Set (CIESIN 2013). To calculate the mean distance to road for each *rayon*, a 1 × 1 km raster layer was created where each pixel represents the distance to the nearest road. From this grid, the mean distance to road was generated for each administrative unit.

The first model we estimate is the simple linear model (Table 1). In this model, only our control variable, forest cover, is statistically significant and its negative sign indicates that areas with higher percentages of forest cover saw greater declines in violence; for instance, many of the districts in Chechnya are heavily forested and experience large declines in violence. Our variable of interest, change in subsidies, is not statistically significant, though it does have the correct sign, with increases in subsidies associated with decreases in violence (for similarly ambiguous results on the efficacy of subsidies in the region, see Alexseev 2013). This model should not be given too much weight, however, since it violates one of the assumptions of linear

Table 1. Models for change in violence per 1000 people.

	Linear model			Spatial lag model			Multilevel model		
	Estimate	SE	*p*-val.	Estimate	SE	*p*-val.	Estimate	SE	*p*-val.
Fixed part									
Constant	0.103	1.483	0.94	1.328	1.117	0.24	1.638	1.557	0.29
Change in subsidies (per cap)	−0.105	0.084	0.21	−0.079	0.063	0.22	0.002	0.073	0.98
Titular percentage	1.136	1.198	0.35	−0.082	0.906	0.93	−2.644	1.715	0.13
Employed percentage	2.640	6.132	0.67	1.227	4.617	0.79	−0.479	4.290	0.91
Urban percentage	0.444	1.332	0.74	−0.121	1.003	0.90	−0.486	0.923	0.60
Forest cover percentage	−0.111	0.027	0.00	−0.056	0.021	0.01	−0.035	0.021	0.09
Distance to road (km)	0.167	0.380	0.66	−0.081	0.286	0.78	−0.411	0.268	0.13
Spatial lag of violence				0.694	0.068	0.00	0.502	0.134	0.00
Random part									
Chechnya							−5.456	0.697	0.00
Dagestan							0.512	0.420	0.22
Ingushetia							−0.201	0.980	0.84
Kabardino-Balkaria							1.125	0.851	0.19
Karachaevo-Cherkessia							1.123	0.888	0.21
North Ossetia							0.935	0.931	0.32
Stavropol *krai*							1.961	0.501	0.00
Model diagnostics									
Log-likelihood	−395.69			−368.69			−352.48		
AIC	807.39			755.39			724.95		
Residuals (s.d.)	4.27			3.30			2.84		

Notes: Number of districts = 138. SE = Standard error.

regression in that the errors are not independent (Anselin 1988). In particular, they exhibit spatial dependence with a statistically significant Moran's I for the regression residuals, meaning we must reject the null hypothesis of uncorrelated error terms. In evaluating a spatial error versus spatial lag model to address the problem, we use the robust Lagrange multiplier diagnostics for spatial dependence (Anselin et al. 1996) that indicate a preference for the spatial lag model (p-value = 0.00014) over the spatial error model (p-value = 0.037).

The second model in Table 1 shows the results for this properly specified spatial lag model (using R package "spdep"). Note that the signs for several of the non-significant variables change when this spatial lag term is introduced, although the subsidies metric remains stable. For this model, the Lagrange multiplier test for residual autocorrelation is not statistically significant, indicating the spatial dependence in the model has been sufficiently addressed.

Lastly, we estimate a multilevel model (R package "lme4") using republic/*krai* level random effects (Table 1). This enables us to explicitly model the nested structure of the administrative units and capture any unexplained factors within each republic/*krai*, such as changes in leadership and republic-level policy toward combatants and their affiliates. Often, these random effects are not reported and viewed simply as controls in the model (Witmer et al. 2017), but here, we are interested in evaluating the effects substantively since our other variables generally perform poorly. For instance, the fixed part of the model did not capture the steep decline of violence in Chechnya clearly visible in Figure 5, but this effect is captured by the large and statistically significant random effect coefficient for Chechnya. Each subsequent model reduces the magnitude and significance of the forest cover factor, indicating it is not as influential as the initial linear model suggested. The model diagnostics, log-likelihood, AIC, and residuals standard deviation all indicate the multilevel spatial model performs best in modeling the change in violence. In the final model, the change in subsidies remains non-significant and confirms the relationship that has been seen in other studies for Afghanistan and the North Caucasus. As the "carrot" of economic development and "winning hearts and minds" programs failed to have the desired effects of violence reduction, by the time of the Sochi Olympics, the Russian government was committed to the "stick" approach of police and military operations.

Why would the results of research on the relationship between subsidies and violence be inconsistent? One possibility is that rebels are motivated by ideological motivations rather than economic ones. Nationalist and religious movements would appear to be more likely to reject the possibility of the loyalty of their members being bought by the state. A second possibility is the ineffectiveness of the largess from the government due to its capture by local political leaders who implement the distribution of funds. Corruption is an endemic problem in many conflict zones and the North Caucasus is no exception. Foxall (2014, 49) notes that

> While the Russian government has, since 2000, allocated significant federal subsidies to the [North Caucasian] republics, these have had no tangible impact on the economic

situation in the region, where economic dislocation is widespread. Rather, they appear to have created a culture of dependency among regional leaders.

Republic presidents can distribute the Moscow subsidies according to their personal and local preferences so that the expected positive effect on the level of violence is weakened.

Conclusion

The key finding in this paper is that violence in the North Caucasus has declined substantially since the beginning of the decade in 2010. But violence still occurs. August and September of 2016 saw the initiation of four counter-terrorist operations by the Russian government, three in Dagestan and one in Kabardino-Balkaria. In the last of these operations, six locally based Islamists were killed in a firefight with security forces in Makhachkala (Aliyev 2016). Dzutsati (2016) also identifies a spike in violence in the region in the fall of 2016, writing that, "recent trends in the North Caucasus indicate that low-level insurgency-related violence is likely to continue to plague the region despite regular triumphant statements of Russian officials." Souleimanov (2017b) reaches a similar conclusion for Chechnya, as "sporadic attacks against the Kadyrov regime, and local law enforcement associated with it, will likely recur in the years to come."

Has Russia at last established a strategy for the North Caucasus that works to reduce violence in the region? The proportional rise in arrests reported in our analysis, complemented by the "hard security" tactics that targeted networks of support and placed military personnel in leadership positions at the republic and regional scales have corresponded to a shift in violence away from Chechnya and an overall decline. Also influencing the decline in violence in the region is the rise of the Islamic State and the simmering conflict in eastern Ukraine. Both conflicts have served as alternate venues for fighters from the region, although their precise impact is difficult to evaluate using our data-set. Questions remain about the effectiveness of Russian policy in the North Caucasus if Islamists with fighting experience return to the North Caucasus from abroad or the center's capacity for subsidizing the region remains curtailed (Souleimanov 2014; Dzutsati 2016; Ratelle 2016). For now, the alliance of the Kremlin and the Kadyrov regime in Chechnya is tamping down militant attacks from the heartland of Caucasian rebellions. It is unclear how long this arrangement will remain *in situ*.

In turn, whether the decrease in political violence seen over the past seven years becomes the new normal is contingent on dynamics both internal and external – the very same dynamics that led to the decrease in the first place: the continued success of the power vertical; securitization; economic policy as a counter to corruption and slow growth; and the outcome of events in the Middle East and, to a lesser extent, Ukraine. Although subsidy regimes are not a significant predictor in explaining the decline in violence across the two time periods considered, they remain important to regional economies and have increased across the region during the period under consideration (from 2010–2016; see Figure 7 above). With Russia's continued dependence on oil and natural gas for its federal budget and potential uncertainty in these markets, it is possible that the ability of Moscow to support the North Caucasus will be drawn into question in the near term (though popular discontent with this policy has been more muted recently). And the alternative strategies for economic development in the region – including tourism and agriculture – have not led to real improvement in the region's economic fortunes (Holland 2016). Russia's protracted war in the North Caucasus will continue so long as it is unable to resolve these outstanding issues of management, economic development, and regional integration.

Notes

1. In the nested hierarchy of the Soviet Union's political geography – often likened to a set of *matrioshka* dolls – the Union of Soviet Socialist Republics is the first tier, the 15 constitutive Soviet Socialist Republics the second tier (the countries that gained independence with the Union's collapse in 1991), and the variously named autonomous regions (including the Chechen-Ingush Autonomous Soviet Socialist Republic) the third tier (Bremmer and Taras 1997).
2. The Caucasus Emirate is the self-declared Islamic theocracy that replaced the more nationalist oriented Chechen Republic of Ichkeria following the death of Maskhadov with the aim of consolidating Islamist groups across the North Caucasus into a single organization.
3. Kadyrov has been de facto head of the republic since the assassination of his father, Akhmad, in May 2004 and remains as the key actor in Putin's Chechenization policy (Russell 2008).
4. This number was up slightly from 2015, when *Kavkazskii Uzel* recorded 258 casualties across the district. For academic work similar in aim, see Zhirukhina (forthcoming); the data-set reported in this article extends from 2007 to 2014 and collects information on types of state repression – special operations, arrests, and seizures, among others – from media and official sources.
5. In terms of casualties per capita, Dagestan's casualty rate was 52.6 per million residents in 2015, a figure slightly below that of Kabardino-Balkaria (55.8 per million residents). Both rates are calculated using total population figures from the 2010 Russian census (Dagestan's 2010 population was 2.91 million; Kabardino-Balkaria's was 860,000).
6. Animated gif files showing potential surfaces for every year are available at https://www.colorado.edu/ibs/waroutcomes/caucasus/.
7. See discussion above on the variety of forms, including grants (*dotatsii*), subsidies, and subventions, which transfers to the region take. The data presented here are calculated by summing these three categories and other interbudgetary transfers for the *rayoni* and cities of the federal subunits of the NCFD.
8. Data from 2010 are used for all republics with the exception of Chechnya (the earliest available data for that republic is 2013); 2016 data are used for all republics except Kabardino-Balkaria (2012) and Ingushetia (2014), as 2016 or more recent data are not reported at the subregional scale for either republic.

Acknowledgments

Many undergraduates at the University of Colorado Boulder and Jordan Estes at the Kennan Institute coded and georeferenced the events data. The statements and views expressed herein are those of the authors and are not necessarily those of the Kennan Institute or the Woodrow Wilson Center. Ralph Clem provided insightful feedback that strengthened the final version of the paper. All errors remain ours.

Disclosure statement

No potential conflict of interest was reported by the authors.

Funding

This work was supported by the National Science Foundation Human and Social Dynamics Program [grant number 0433927], [grant number 0827016]; Kennan Institute for Advanced Russian Studies [Title VIII Research Scholar]; University of Colorado Boulder.

ORCID

Edward C. Holland http://orcid.org/0000-0003-3914-0505
Frank D.W. Witmer http://orcid.org/0000-0002-2646-0705
John O'Loughlin http://orcid.org/0000-0002-9053-7885

References

Alexeev, Michael, and Andrey Chernyavskiy. 2017. *A Tale of Two Crises: Federal Transfers and Regional Economies in Russia in 2009 and 2014–2015*. Working paper. Accessed January 17, 2018. https://pdfs.semanticscholar.org/6870/0c66e8ada28f76be0fcf024ff24d50d66dcf.pdf

Alexseev, Mikhail. 2013. "Local versus Transcendent Insurgencies: Why Economic Aid Helps Lower Violence in Dagestan, but not in Kabardino-Balkaria." In *Development Strategies, Identities, and Conflict in Asia*, edited by W. Ascher and N. Mirovitskaya, 277–314. New York, NY: Palgrave Macmillan.

Aliyev, Huseyn. 2015. "Conflict-related Violence Decreases in the North Caucasus as Fighters go to Syria." *Central Asia-Caucasus Analyst*. Accessed January 10, 2018. https://www.cacianalyst.org/publications/analytical-articles/item/13171-conflict-related-violence-decreases-in-the-north-caucasus-as-fighters-go-to-syria.html

Aliyev, Huseyn. 2016. "Revival of Islamist Insurgency in the North Caucasus?" *Central Asia-Caucasus Analyst*. Accessed March 24, 2017. http://cacianalyst.org/publications/analytical-articles/item/13403-revival-of-islamist-insurgency-in-the-north-caucasus?.html

Anselin, Luc. 1988. *Spatial Econometrics: Methods and Models*. Dordrecht, NL: Kluwer Academic.

Anselin, Luc, Anil K. Bera, Raymond Florax, and Mann J. Yoon. 1996. "Simple Diagnostic Tests for Spatial Dependence." *Regional Science and Urban Economics* 26: 77–104.

Arnold, Richard, and Andrew Foxall. 2014. "Lord of the (Five) Rings: Issues at the 2014 Sochi Winter Olympic Games: Guest Editors' Introduction." *Problems of Post-Communism* 61: 3–12.

Berman, Eli, Michael Callen, Joseph H. Felter, and Jacob N. Shapiro. 2011. "Do Working Men Rebel? Insurgency and Unemployment in Afghanistan, Iraq, and the Philippines." *Journal of Conflict Resolution* 55: 496–528.

Berman, Eli, Jacob N. Shapiro, and Joseph H. Felter. 2011. "Can Hearts and Minds be Bought? The Economics of Counterinsurgency in Iraq." *Journal of Political Economy* 119: 766–819.

Bremmer, Ian, and Ray Taras, eds. 1997. *New States, New Politics: Building the Post-Soviet Nations*. New York, NY: Cambridge University Press.

CIESIN (Center for International Earth Science Information Network). 2013. *Global Roads Open Access data set. Version 1 (gROADSv1)*. Palisades, NY: NASA Socioeconomic Data and Applications Center.

Child, Travers. 2017. "Reconstruction and Conflict: Losing Hearts and Minds." *Center for Economic Policy Research*. Accessed January 10, 2018. http://voxeu.org/article/reconstruction-and-conflict-losing-hearts-and-minds.

Chou, Tiffany. 2012. "Does Development Assistance Reduce Violence? Evidence from Afghanistan." *Economics of Peace and Security Journal* 7: 5–13.

Dannreuther, Roland, and Luke March. 2008. "Chechnya: Has Moscow Won?" *Survival* 50: 97–112.

Dzutsati, Valery. 2016. "Russian Policy in the North Caucasus Remains in Flux." *Eurasia Daily Monitor*. Accessed March 24, 2017. https://jamestown.org/program/russian-policy-north-caucasus-remains-flux/

Fearon, James. 2008. "Economic Development, Insurgency, and Civil War." In *Institutions and Economic Performance*, edited by E. Helpman, 292–328. Cambridge, MA: Harvard University Press.

Foxall, Andrew. 2014. *Ethnic Relations in Post-Soviet Russia: Russians and Non-Russians in the North Caucasus*. London: Routledge.

Freeman, John R. 1983. "Granger Causality and the Time Series Analysis of Political Relationships." *American Journal of Political Science* 27: 327–358.

Frolov, Yuri S. 1977. "Mapping of the Population Potential." *Soviet Geography: Review and Translation* 18: 110–123.

Fuller, Liz. 2015. "Caucasus Emirate Weakened by Death of New Leader, but not Defunct." *RFE/RL Caucasus Report*. Accessed March 26, 2017. http://www.rferl.org/a/caucasus-emirate-new-leader-death/27193195.html

Fuller, Liz. 2016a. "Why is the Death Toll Tumbling in The North Caucasus?" *RFE/RL Caucasus Report*. Accessed March 26, 2017. http://www.rferl.org/a/insurgency-north-caucasus-terrorism-isis/26840778.html

Fuller, Liz. 2016b. "Putin Names New Envoy for North Caucasus." *RFE/RL Caucasus Report*. Accessed March 26, 2017. http://www.rferl.org/content/caucasus-putin-names-new-envoy-/27888978.html

Fuller, Liz. 2016c. "Chechen Leader Kadyrov Challenges Moscow Over Budget Subsidies." *RFE/RL Caucasus Report*. Accessed January 8, 2018. https://www.rferl.org/a/chechnya-kadyrov-challenges-moscow-budget-subsidies/28123822.html

Gilligan, Emma. 2009. *Terror in Chechnya: Russia and the Tragedy of Civilians in War*. Princeton, NJ: Princeton University Press.

Granger, Clive W. 1969. "Investigating Causal Relations by Econometric Models and Cross-Spectral Methods." *Econometrica* 37: 424–438.

Hansen, Matthew C., Peter V. Potapov, Rebecca Moore, Matt Hancher, Svetlana Turubanova, Alexandra Tyukavina, David Thau, et al. 2013. "High-Resolution Global Maps of 21st-Century Forest Cover Change." *Science* 342: 850–853.

Holland, Edward C. 2016. "Economic Development and Subsidies in the North Caucasus." *Problems of Post-Communism* 63: 50–61.

International Crisis Group. 2016. *The North Caucasus Insurgency and Syria: An Exported Jihad?* Brussels: International Crisis Group.

Jersild, Austin. 2002. *Orientalism and Empire: North Caucasus Mountain Peoples and the Georgian Frontier, 1845–1917*. Montreal: McGill-Queen's University Press.

Kavkazskii Uzel. 2016. "Infographics. The Statistics of the Number of Victims in the North Caucasian Federal District Regions for a Period of Six Years." *Kavkazskii Uzel* [Caucasian Knot]. Accessed March 26, 2017. http://www.eng.kavkaz-uzel.eu/articles/34546/

Kavkazskii Uzel. 2017. "Infographics. Statistics of Victims in Northern Caucasus for 2016 under the data of the Caucasian Knot." *Kavkazskii Uzel* [Caucasian Knot]. Accessed January 10, 2018. http://www.eng.kavkaz-uzel.eu/articles/38325/

Klimenko, Ekaterina, and Neil John Melvin. 2016. "Decreasing Violence in the North Caucasus: Is an End to the Regional Conflict in Sight?" *Stockholm International Peace Research Institute.* Accessed March 24, 2017. https://www.sipri.org/commentary/blog/2016/decreasing-violence-north-caucasus-end-regional-conflict-sight

Kramer, Mark. 2005. "The Perils of Counterinsurgency: Russia's War in Chechnya." *International Security* 29: 5–63.

Lieven, Anatol. 1999. *Chechnya: Tombstone of Russian Power*. New Haven, CT: Yale University Press.

Linke, Andrew M., and John O'Loughlin. 2015. "Reconceptualizing, Measuring, and Evaluating Distance and Context in the Study of Conflicts: Using Survey Data from the North Caucasus of Russia." *International Studies Review* 17: 107–125.

Linke, Andrew M., Frank D. W. Witmer, and John O'Loughlin. 2012. "Space-Time Granger Analysis of the War in Iraq: A Study of Coalition and Insurgent Action-Reaction." *International Interactions* 38: 402–425.

Linke, Andrew M., Frank D. W. Witmer, Edward C. Holland, and John O'Loughlin. 2017. "Mountainous Terrain and Civil Wars: Geospatial Analysis of Conflict Dynamics in the Post-Soviet Caucasus." *Annals of the American Association of Geographers* 107: 520–535.

Lyall, Jason. 2009. "Does Indiscriminate Violence Incite Insurgent Attacks? Evidence from Chechnya." *Journal of Conflict Resolution* 53: 331–362.

Lyall, Jason. 2010. "Are Coethnics More Effective Counterinsurgents? Evidence from the Second Chechen War." *American Political Science Review* 104: 1–20.

Mayetnaya, Yelizaveta. 2017. "On Russia's Extremism Watch List: 'Is It Just Because We Wear the Hijab?'" *RFE/RL*. Accessed June 29, 2017. https://www.rferl.org/a/daghestan-extremism-watch-list-hijab/28529530.html?ltflags=mailer

Medvedev, Dmitry. 2010. "V Rossii obrazovan novyi federal'nyi okrug – Severo-Kavkazskii [A New Federal District is Formed in Russia – The North Caucasusian]." *Kremlin.ru*. Accessed March 26, 2017. http://www.kremlin.ru/events/president/news/6664

Melvin, Neil J. 2007. "Building Stability in the North Caucasus: Ways Forward for Russia and the European Union." *Stockholm International Peace Research Institute*. Accessed January 6, 2018. https://www.sipri.org/sites/default/files/files/PP/SIPRIPP16.pdf

Memorial. 2016. *Counter-Terrorism in the North Caucasus: A Human Rights Perspective. 2014 – First Half of 2016*. Moscow: Memorial Human Rights Centre.

Ministry of Defence of the Russian Federation. 2017. "Remarks by Chief of General Staff of the Russian Federation General of the Army Valery Gerasimov at the Russian Defence Ministry's board session (November 7, 2017)." Accessed January 29, 2018. http://eng.mil.ru/en/news_page/country/more.htm?id=12149743@egNews

O'Loughlin, John, and Frank D. W. Witmer. 2011. "The Localized Geographies of Violence in Russia's North Caucasus, 1999–2007." *Annals of the Association of American Geographers* 101: 178–201.

O'Loughlin, John, and Frank D. W. Witmer. 2012. "The Diffusion of Violence in the North Caucasus of Russia, 1999–2010." *Environment and Planning A* 44: 2379–2396.

O'Loughlin, John, Edward C. Holland, and Frank D. W. Witmer. 2011. "The Changing Geography of Violence in Russia's North Caucasus: Regional Trends and Local Dynamics in Dagestan, Ingushetia, and Kabardino-Balkaria." *Eurasian Geography and Economics* 52: 596–630.

Raleigh, Clionadh, Andrew Linke, Håvard Hegre, and Joakim Karlsen. 2010. "Introducing ACLED: An Armed Conflict Location and Event Dataset." *Journal of Peace Research* 47: 651–660.

Ratelle, Jean-François. 2016. "North Caucasian Foreign Fighters in Syria and Iraq: Assessing the Threat of Returnees to the Russian Federation." *Caucasus Survey* 4: 218–238.

Reuters. 2012. "Twin Bomb Attacks Kill 12 in Russia's Dagestan." *Reuters*. Accessed January 10, 2018. https://www.reuters.com/article/us-russia-dagestan-blast/twin-bomb-attacks-kill-12-in-russias-dagestan-idUSBRE8430E620120504

RFE/RL. 2013. "Kabardino-Balkaria President Leaves Office Halfway Through Second Term." *RFE/RL Caucasus Report*. Accessed March 26, 2017. http://www.rferl.org/content/caucasus-report-kabardino-balkaria-leader-leaves-office/25193318.html

Russell, John. 2008. "Ramzan Kadyrov: The Indigenous Key to Success in Putin's Chechenization Strategy?" *Nationalities Papers* 36: 659–687.

Sagramoso, Domitilla. 2007. "Violence and Conflict in the Russian North Caucasus." *International Affairs* 83: 681–705.

Sakwa, Richard. 2010. "The Revenge of the Caucasus: Chechenization and the Dual State in Russia." *Nationalities Papers* 38: 601–622.

Shterin, Marat, and Akhmet Yarlykapov. 2011. "Reconsidering Radicalisation and Terrorism: The New Muslims Movement in Kabardino-Balkaria and its Path to Violence." *Religion, State & Society* 39: 303–325.

Sokolov, Denis. 2016. "Can the North Caucasus Adapt to Political Change?" *Open Democracy*. Accessed March 24, 2017. https://www.opendemocracy.net/od-russia/denis-sokolov/can-north-caucasus-adapt-to-political-change

Souleimanov, Emil. 2011. "The Caucasus Emirate: Genealogy of an Islamist Insurgency." *Middle East Policy* 18: 155–168.

Souleimanov, Emil A. 2014. "Globalizing Jihad? North Caucasians in the Syrian Civil War." *Middle East Policy* 21: 154–162.

Souleimanov, Emil. 2016. "The North Caucasus Insurgency: Weakened but not Eradicated." *Central Asia-Caucasus Analyst*. Accessed March 24, 2017. http://cacianalyst.org/publications/analytical-articles/item/13400-the-north-caucasus-insurgency-weakened-but-not-eradicated.html

Souleimanov, Emil Aslan. 2017a. "A Failed Revolt? Assessing the Viability of the North Caucasus Insurgency." *The Journal of Slavic Military Studies* 30: 210–231.

Souleimanov, Emil. 2017b. "Attacks in Chechnya Suggest Opposition to Kadyrov is Far from Eradicated." *Central Asia-Caucasus Analyst*. Accessed March 24, 2017. http://cacianalyst.org/publications/analytical-articles/item/13436-attacks-in-chechnya-suggest-opposition-to-kadyrov-is-far-from-eradicated.html

Souleimanov, Emil Aslan, and Grazvydas Jasutis. 2016. "The Dynamics of Kadyrov's Regime: Between Autonomy and Independence." *Caucasus Survey* 4: 115–128.

Souleimanov, Emil Aslan, and Katarina Petrtylova. 2015. "Russia's Policy toward the Islamic State." *Middle East Policy* 22: 66–78.

Taylor, Brian D. 2007. "Putin's 'Historic Mission': State-Building and the Power Ministries in the North Caucasus." *Problems of Post-Communism* 54: 3–16.

Toft, Monica Duffy, and Yuri M. Zhukov. 2012. "Denial and Punishment in the North Caucasus: Evaluating the Effectiveness of Coercive Counter-Insurgency." *Journal of Peace Research* 49: 785–800.

Toft, Monica Duffy, and Yuri M. Zhukov. 2015. "Islamists and Nationalists: Rebel Motivation and Counterinsurgency in Russia's North Caucasus." *American Political Science Review* 109: 222–238.

Vatchagaev, Mairbek. 2014. "Appointment of General Melikov to Replace Khloponin Points to Kremlin Bid to Subdue Dagestani Insurgency." *Eurasia Daily Monitor*. Accessed March 24, 2017. http://www.jamestown.org/single/?tx_ttnews%5Btt_news%5D=42375&no_cache=1#.V59M0aJGjis

Walker, Shaun. 2015. "'We Like Partisan Warfare.' Chechens Fighting in Ukraine – On Both Sides." *The Guardian*. Accessed January 10, 2018. www.theguardian.com/world/2015/jul/24/chechens-fighting-in-ukraine-on-both-sides

Ware, Robert. 2011. "Has the Russian Federation been Chechenised?" *Europe-Asia Studies* 63: 493–508.

Warntz, William. 1964. "A New Map of the Surface of Population Potentials for the United States, 1960." *Geographical Review* 54: 170–184.

Williams, Daniel, and Tanya Lokshina. 2015. "Invisible War: Russia's Abusive Response to the Dagestan Insurgency." *Human Rights Watch*. Accessed March 26, 2017. https://www.hrw.org/report/2015/06/18/invisible-war/russias-abusive-response-dagestan-insurgency

Witmer, Frank D.W., Andrew M. Linke, John O'Loughlin, Andrew Gettelman, and Arlene Laing. 2017. "Sub-National Violent Conflict Forecasts for Sub-Saharan Africa, 2015–2065, Using Climate-Sensitive Models." *Journal of Peace Research* 54: 175–192.

Wright, Robin. 2017. "ISIS Jihadis Have Returned Home by the Thousands." *The New Yorker*. Accessed January 9, 2018 https://www.newyorker.com/news/news-desk/isis-jihadis-have-returned-home-by-the-thousands

Youngman, Mark. 2016. "Between Caucasus and Caliphate: The Splintering of the North Caucasus Insurgency." *Caucasus Survey* 4: 194–217.

Zhemukhov, Sufian, and Robert W. Orttung. 2014. "Munich Syndrome: Russian Security in the 2014 Sochi Olympics." *Problems of Post-Communism* 61: 13–29.

Zhirukhina, Elena. Forthcoming. "Protecting the State: Russian Repressive Tactics in the North Caucasus." *Nationalities Papers*. doi: https://doi.org/10.1080/00905992.2017.1375905

Zürcher, Christoph. 2007. *The Post-Soviet Wars: Rebellion, Ethnic Conflict, and Nationhood in the Caucasus*. New York, NY: NYU Press.

Clearing the Fog of War: public versus official sources and geopolitical storylines in the Russia-Ukraine conflict

Ralph S. Clem

ABSTRACT

Military action undertaken by the Russian Federation against Ukraine in 2014 has had enormous geopolitical ramifications. This resulted in what is almost certainly a permanent change in sovereign territory, with the former gaining and the latter losing the strategic Crimean peninsula. But Russia's moves also set in motion a violent conflict in the Donbas region of eastern Ukraine. Although the United States and the NATO alliance have advocated a geopolitical storyline that attributes blame for this to Russia, close scrutiny of the evidence they have adduced in this regard fails to establish this culpability conclusively. However, by utilizing data collected and analyzed in the public realm, it is possible to determine with more certainty that, in certain places and at given times, Russia was indeed the aggressor. The rapidly increasing amount of public-sourced information globally and the growing sophistication of analytical methods by non-governmental groups presages more complete understanding of such conflicts without reliance on official information.

Introduction

Russia's seizure and subsequent annexation of Ukraine's Crimea region in early 2014 ushered in heightened tensions between Moscow on the one hand and the European Union (EU) and the United States on the other. But it is Russia's alleged complicity in the subsequent spread and intensification of fighting to the Donbas region of eastern Ukraine[1] in the summer of 2014 – the subject of this paper – that poses an even greater risk to peace and security in Europe, inasmuch as that crisis has already resulted in thousands of casualties, well over 1.5 million internally displaced persons, and widespread destruction.

Most analyses and commentaries regarding the current Russo-Ukrainian conflict in the Donbas center around the question of Russia's motives in undertaking these moves and/or the appropriate responses to be taken in return by the United States, the EU, individual European states, and the North Atlantic Treaty Organization

(NATO) (Götz 2015; Laruelle 2016; Marten 2015; Tsygankov 2015). As such, it is usually taken as a point of departure that Russia has indeed, by whatever means, violated Ukraine's territorial sovereignty, most obviously by its capture of Crimea. As will be detailed below, official and diplomatic statements by the EU, NATO, and the United States Government (USG) assert that Russia has in fact attacked eastern Ukraine as well in the months following the annexation of Crimea. Indeed, the USG and NATO have adduced evidence purporting to prove Russian aggression in the Donbas specifically, evidence that has been widely accepted uncritically by established journalists and by most foreign affairs commentators.

Russia, as is well known, has vigorously pushed back against these allegations of wrongdoing in eastern Ukraine and has consistently denied official involvement in the ongoing hostilities there; indeed, as early as April 2014 Russia's ambassador to the United Nations, Vitaly Churkin, accused the Ukrainian government of attacking civilians in the Donbas and portrayed Russia's actions as purely humanitarian in nature ("Kiev must stop war" 2014). Others have suggested that Russia is engaged in what some have called "information warfare" to both disguise and enhance its operations on the ground in Ukraine (Pomerantsev 2014; Snegovaya 2015).

Thus, it behooves us, in light of the seriousness of the situation, to delve more deeply into what exactly is the evidence brought forward officially that might lend credence to claims of Russian malfeasance. Further, we should ascertain if unofficial (or public-sourced) information is available that might shed additional light on what is a prime example of Clausewitz's maxim that "war is the realm of uncertainty," and what the implications of those data sources might be for geopolitical analysis writ large.

Adducing facts and attributing hostile acts

Governments present information to shape discourse consistent with the state's goals in the conduct of its international relations. As Gearóid Ó Tuathail (Gerard Toal) put it, in the furtherance of their geopolitical agendas states construct "storylines" concerning events: "Storylines are sense-making organizational devices tying the different elements of a policy challenge together into a reasonably coherent and convincing narrative" (2002, 617). Governments, as well as international and non-governmental organizations, work furiously to establish and propagate their storylines by providing information through official channels and via print and digital media. But when it comes specifically to attributing hostile acts to other states as part of a geopolitical storyline, official sources almost always stop short of providing everything they know out of concern for the sanctity of their sources and the methods by which they analyze and report data (Clem 2017a). The need also to achieve some degree of agreement among different national intelligence or security agencies can create a bureaucratic imperative to "sanitize" or politicize information to the point that it becomes relatively uninformative or even misleading. In the context of US intelligence, the failure to disrupt the 9/11 attacks and the mistaken conviction about purported Iraqi weapons of mass destruction are extreme, but instructive cases (Betts 2007).

Absent the kind of unambiguous, convincing proof – a "gold standard," as it were – that official sources could provide regarding, say, the origin of kinetic attacks, state perpetrators are allowed a degree of deniability that shields them from reprisals or sanctions and seriously weakens the credibility of deterrence (Eichensehr 2017; Lindsay 2015). Further, this anonymity allows an aggressor state to present a more compelling denial of complicity in hostile acts through its own storylines.

What might that dispositive "gold standard" look like in cases where evidence of armed conflict is adduced? That a reported event shown in satellite or ground level imagery occurred at the time and place stated would be a good starting point. That a vehicle, aircraft, or missile is in fact of the type or model indicated, and that imagery thereof is sufficiently detailed to make that determination and place it at a particular location is another. Or that holes in the ground that are purported to be artillery or rocket impact craters caused by firing across a border are in fact of the size and shape known to be caused by such impacts, and that the axis of attack can be determined by forensic analysis, likewise might be relevant.

Acquiring and analyzing data from, say, satellite imagery that would meet such a standard has long been almost entirely within the control of governments and, again, governments have been loath to release such information to the public. But the advent over the last two decades of increasingly more pervasive, persistent, and effective public sources of information now allows for a better understanding of geopolitical events using data gathered solely from these unofficial sources, what Corey Hinderstein (2014) calls "societal verification."[2] As Sean Larkin (2016) noted, "…the market-driven explosion of surveillance sensors and data analytics will bring an unprecedented level of transparency to global affairs." Indeed, to an appreciable degree, that is already the case. When combined with rapidly expanding networks of telecommunications users, the public collection, analysis, and dissemination of types of data hitherto usually closely held by governments is rapidly transforming and enhancing the public's ability to not only observe, but to confirm, what states have, or have not, done on the ground.[3]

Briefly, there are four major technological changes that, taken together, have enabled the very rapid expansion of public-sourced data analysis; the first three involve technologies that originated in the military or government sector and migrated after a time into the civilian world. First, the Global Positioning System (GPS) allows for precise, almost instantaneous geospatial location anywhere on Earth (Milner 2016). Second, satellite imagery is now widely available for purchase in the commercial sector with much finer resolution (i.e. shows more detail on the ground) and more frequent imaging of places (Florini and Dehqanzada 1999). Third, the Internet ties together over half the world's population at increasingly higher transmission speeds, enabling the very rapid transmission of information right around the planet (Westcott 2008). The fourth, the availability of privately generated ground-level digital imagery, including video, owes to the proliferation of wireless telephony, smart phones, and social media platforms. Other innovations, such as exponential increases in data storage capacity, more capable software, and powerful algorithmic search engines further enhance the capability to process and analyze information gathered from private sector surveillance systems – such as commercial satellite imagery – or by individuals or groups outside government.

Given that there is now available a constant and massive flow of information through unofficial channels, what ought to guide the analysis and dissemination of those data by individuals, non-state groups, or non-governmental organizations in the furtherance of transparency? A central tenet of official intelligence analysis is that the efficacy of a product is enhanced by the incorporation of data from multiple sources; that is, the interpretation of satellite imagery might be more relevant if ground-level photography or video of an event were available as a complement. Often referred to as "all-source analysis" in official intelligence parlance (Fingar 2012), this practice should also apply to research on national security issues or events – geopolitics – conducted in the public realm. Further, the use of one type of source might also cue the collection of data by other means (Caddell 2017). Finally, the fusion of information from different sensors, together with date, time, and location data frequently embedded in imagery ("geo-tagged"), permits not only a more complete analysis, but also the validation of the provenance and accuracy of raw data; regrettably, the manipulation of imagery of various kinds is rife, and even experienced journalists or analysts have been misled by state and non-state actors (Chivers 2013; Powers and O'Loughlin 2015; Sienkiewicz 2014).

Russia and Ukraine: the descent into war

The case in point here involves purported military action taken by one state, Russia, against another sovereign state, Ukraine. Although the armed conflict phase of the Russia-Ukraine crisis begins with Russia's invasion of Crimea in February 2014, the focus of this paper will be on the much narrower time frame of June-August 2014, by which time the fighting had shifted to eastern Ukraine and during which the most intense combat of the war took place. It is not my intent here to take up the related and important question of whether these actions constitute "acts of war" in terms of international law, as those are essentially political judgments made by the interested parties. Nor do I wish to engage on the perceived "special" nature of this conflict, often described as "hybrid war" or "stealth war" (on this subject, see Kofman and Rojansky 2015; Rácz 2015, Ch. 2). Certainly the scale of fighting in the period June-August 2014 in eastern Ukraine would appear to any reasonable person to be warfare; that is, it involved large numbers of troops in typical military formations, accompanied by armor, artillery, and air strikes.

Secondly, the focus here will be on whether the Russian state itself committed hostile military acts in eastern Ukraine, specifically in Donetsk and Luhansk oblasts where the fighting has taken place. This analysis only indirectly addresses the issue of non-state actors in the Russo-Ukraine war; that is, the separatist groups operating in eastern Ukraine under the rubrics of the Donetsk and Luhansk People's Republics (the DNR and LNR). Whereas the separatists are undoubtedly important actors in the conflict, and their existence allows the Russian state to obscure its role in the military dimension of the crisis, the focus here is on the *direct* action of states (i.e. the Russian Federation) in armed conflict rather than involvement through proxies. As noted above, because of the time frame focus here, we leave aside the earlier Russian invasion and subsequent annexation of Crimea; important as it is, the subject has been extensively analyzed elsewhere (see Mankoff 2014; Menon and Rumer 2015, Ch. 2; Sakwa 2015, Ch. 5; Wilson 2014, Ch. 6). Further, I do not take up here the most noteworthy event that occurred in the conflict in eastern Ukraine: the downing of Malaysia Airlines Flight 17 (MH17). Although this horrific tragedy has competing narratives and very different interpretations of the forensic evidence involved, it has become a *cause célèbre* purely in its own right and has already been the subject of extensive coverage and both official and public analyses (Clem 2017b; Toal and O'Loughlin 2017).

The issue at hand is how the storyline portraying Russian actions in June-August 2014 in eastern Ukraine has been shaped by two of the most important external actors in this conflict, the USG and NATO, via the release to the media of selected intelligence products.[4] That official evidence will be compared to what has become available from the innovative use of the trove of publicly derived data from social media and other non-governmental sources, principally satellite imagery, and ground level video and imagery (including geo-tagged "selfies"). Thus, the paper focuses first on evidence released by the USG and NATO purportedly verifying actual Russian attacks against Ukraine and other military actions on sovereign Ukrainian territory (such as the presence of Russian troops and equipment in Ukraine). Not included here are any of the observations on the security status in Ukraine filed by the Special Monitoring Mission (SMM) of the Organization for Security and Co-operation in Europe (OSCE); a review of the daily reports from SMM monitors in the field, while generally descriptive of the situation on the ground, revealed no data relevant to the question of actions of the Russian state itself in this conflict during the period covered in this paper (OSCE 2014). Official and public-sourced evidence relating to three separate aspects of Russian military intervention in Ukraine within the time frame June-August 2014 will be analyzed here: (1) the cross-border transfer of tanks and other military equipment directly from Russia to the pro-Russian separatists in Ukraine; (2) rocket and artillery attacks from inside Russia into Ukraine; and (3) the cross-border introduction of Russian army units into Ukraine.

Background: Russia and the battles for the Donbas

If the Russian conquest of Crimea was accomplished by stealthy means and involved little, if any, loss of life, the situation in the Donbas very quickly morphed from civil unrest to actual war with large numbers of military and civilian casualties and widespread destruction of towns and cities and economic infrastructure. Once separatist fighters occupied territory and prevented government forces from restoring Kyiv's writ–arguably marked by the rebel seizure of Slavyansk in April 2014–what had been a fairly chaotic sequence of hit-and-run skirmishes and guerilla-style actions escalated into pitched battles involving armored vehicles and artillery ("War in the Donbass" 2015; Wilson 2014, 136–139). By June 2014 it had become clear that the Ukrainian government faced a tenacious and increasingly stronger and better-equipped enemy; any notion that it might be otherwise was dispelled by the intense battle near Krasniy Liman in Donetsk Oblast engaging thousands of troops on both sides, backed by armor and artillery (Reuters 2014). On July 1, following a brief ceasefire the Ukrainian government, employing both regular army formations and fighters from the "volunteer battalions" or private militias, launched a major offensive against separatist forces and gained significant territory in both Donetsk and Luhansk oblasts, to the extent that the pro-Russian rebels were driven into pockets around the cities of Donetsk and Luhansk proper. However, sections of the border with Russia remained under separatist control, allowing Russia to resupply or equip rebel fighters or even infiltrate Russian troops (either individually as volunteers or as regular army units) into Ukraine. Moscow, perhaps anticipating the overthrow of the pro-Russian government of Viktor Yanukovych, had carefully laid the groundwork for both supporting the DNR and LNR and/or preparing for direct action against Ukraine if necessary by initiating a series of large-scale military exercises as early as February 2014. These exercises involved tens of thousands of troops, many of which were deployed along regions bordering Ukraine. From that time, ever-larger concentrations of Russian troops and huge numbers of tanks, armored personnel carriers, and various types of artillery (including multiple rocket launchers) were massed inside Russian territory just across the Ukrainian frontier.

With the survival of the DNR and LNR hanging in the balance as the Ukrainian government offensive progressed in July 2014, there is broad consensus among academics and members of the foreign affairs commentariat that Russia was simply not prepared to allow the Ukrainian government to crush the separatist insurrection in the Donbas, and accordingly, took direct military action itself to insure the survival of the DNR and LNR regardless of the consequences (Menon and Rumer 2015, Ch. 3; Wilson 2014, 142–143; Rácz 2015, 75). Again, how those actions –transferring armor and other heavy weaponry across the Russo-Ukrainian border; conducting cross-border rocket and artillery attacks against Ukrainian units operating inside Ukraine; and, sending units of the Russian army into Ukraine – might be verified by official and public-sourced data is the purpose here. What is clear regardless, is that in August 2014, massive counterattacks were launched against the Ukrainian government's offensive and in short order devastated the Ukrainian army and the volunteer battalions. Within a matter of weeks the territory previously gained by government forces was lost and, importantly, as were other areas to the south within Donetsk Oblast as far as the Sea of Azov right up to the outskirts of the vital port and industrial center of Mariupol. The absolute nadir for the Ukrainian government cause was the virtual annihilation of a large concentration of their troops and equipment near the town of Ilovaisk (see Judah 2014), after which debacle Kyiv agreed to an internationally brokered ceasefire ("Minsk I") effective September 5, 2014, bringing to a tenuous end this phase of the war.

Official and public-sourced storyline evidence

From the time when it became obvious that Russia was intent on invading, occupying, and annexing Crimea, the USG and the NATO alliance (not to mention European national governments and the EU) have made literally dozens of statements, press briefings, information releases, and diplomatic demarches condemning Moscow's actions against Ukraine and demanding that Ukraine's territorial sovereignty and integrity be respected. When that respect was not forthcoming, as is well known, a series of economic and other sanctions were imposed to alter Russia's behavior, sanctions that have had little, if any effect on turning Russian public opinion against the Putin government (Frye 2017) and have not, as of this writing, brought an end to the fighting in the Donbas. As events unfolded through the summer and into the fall of 2014, the official storyline emanating from Washington and Brussels continued to place the blame for the violence in eastern Ukraine on Russian meddling and later on Russian aggression. But Moscow simultaneously engaged in an extremely effective narrative of its own[5] via a public relations campaign to disavow these alleged actions against Ukraine under various guises, including the characterization of Russian Federation military personnel fighting inside Ukraine as "volunteers." More commonly, Moscow simply denied any contention that Russian troops or equipment were directly involved at all; in early 2015 Russian Foreign Minister Sergei Lavrov stated: "I say every time: if you allege this so confidently, present the facts. So before demanding from us that we stop doing something, please present proof that we have done it" (Baczynska 2015). To provide evidence backing up their storyline, it would thus obviously fall to the USG and NATO to provide sufficient official information relating to Russia's alleged actions.

Tanks across the border in June 2014

From the outset of the heavy fighting in the Donbas, there was intense focus on the question of how the separatist forces acquired their growing and increasingly more lethal arsenal (Ferguson and Jenzen-Jones 2014). In most cases, at least early on, the answer was most likely from equipment and munitions taken from fleeing or surrendering Ukrainian troops or looted from Ukrainian government arms depots. Later, however, it appeared that more sophisticated weapons and ammunition of all types were being supplied by Russia. The first serious instance wherein Russia's complicity in this conflict might be definitively established via the release of official information was the alleged movement of three tanks and multiple launch rocket systems (MLRS) across the border from Russia into Ukraine in June 2014. Although seemingly insignificant when compared to the scale of fighting later in the conflict, this event, which was publicly announced and denounced by the United States government on June 12, with a more detailed release two days later by NATO, received extensive media coverage, more or less uncritically, by the Western press (e.g. Kramer and Gordon 2014a; Marcus 2014). The acquisition of heavy armor by the separatists, even in such small numbers, was a very serious matter; the aphorism "If you want to make an impression, send a tank" is apt. To back up their storyline, NATO provided both annotated commercial satellite imagery and ground-level photography to, as they put it, "…inform the debate regarding recent events in the border region of Russia and Ukraine" (Miller 2014). The imagery suggests that the tanks moved from a staging area inside Russian territory into Ukraine, the implications of which would, "If these latest reports are confirmed mark a grave escalation of the crisis in eastern Ukraine in violation of Russia's Geneva commitments."[6] As one might imagine, these reports were quickly and strongly denied by Russia.

As dramatic as the imagery released by NATO might seem at first glance, it does not conclusively demonstrate that the tanks observed on the Russian side of the border in the satellite imagery are the ones shown in videos from inside Ukrainian territory. Specifically, images from the commercial satellite provider Digital Globe are presented with annotations indicating that ten main battle tanks (MBTs) of a type not identified are observed in a military staging area in the Rostov Oblast of the Russian Federation (which adjoins eastern Ukraine) on June 11, with three of the ten loaded on tank transporters (used to move tanks over longer distances by road). The presumption is that the three MBTs on the transporters subsequently moved toward the Ukrainian border and eventually were off-loaded and are later found moving through cities in eastern Ukraine. In the public-sourced images from inside Ukraine included in this official release, the tanks are identifiable as T-64 MBTs, but bear no national markings. Although the State Department also made reference to "Internet video" showing the MLRSs moving through the separatist-controlled city of Luhansk, no video footage or stills of those were officially adduced (Kramer and Gordon 2014a).

We are thus left with a fairly substantial but clearly circumstantial case wherein the official evidence presented fails to validate the assertion put forth by the USG and NATO concerning Russian complicity in this alleged escalation of the fighting in eastern Ukraine that the introduction of armor and other heavy weaponry across the Russo-Ukrainian border would connote. Two questions emerge from this inconclusive presentation. First, why did the USG and/or NATO not provide stronger evidence of the provenance of the MBTs that would more clearly establish the "chain of custody" from one side of the border to the other? Such evidence would do much to substantiate their claim. Possible answers are (a) that USG/NATO intelligence did not have higher resolution imagery or other types of intelligence that would allow for the identification of the suspect T-64 MBTs on the Russian side of the border; (b) that they had that information but were precluded from releasing it out of concern that sources or methods relating to its acquisition and analysis would be compromised; or (c) that they had that information but preferred not to release it for other reasons.

The other question that arises from the USG/NATO briefing is the unsophisticated use of the public-sourced imagery showing the T-64s inside Ukraine. Specifically, the imagery from a public-source video showing T-64s ostensibly in the Ukrainian city of Makiivka does not precisely correlate the location with overhead imagery of the purported location (one is left to attempt to reconcile the two images, which is difficult owing to the misalignment of the photos and the absence of directional cues). The second public-source image of a T-64 in Snizhne, Ukraine, in the NATO information release has no locational corroboration whatsoever. Further, there are other photos of that same tank available on the Internet that would have allowed an analyst to identify that specific location by reference to landmarks identifiable with street level or overhead imagery even absent any embedded geo-tagged metadata, and quite possibly to ascertain the time of day depicted in the photo. Given these shortcomings in the official release, it is no surprise that within 48 hours Russian Internet bloggers attacked the evidence as superficial.

In summary, information about the June 2014 Russian tank incursion released to the public by official sources does not show conclusively that the MBTs imaged inside Russia were transferred to the pro-Russian separatists in eastern Ukraine. That is to say, it does not meet the "gold standard." Claiming that it does on the basis of the evidence presented, as is the case in the transcript below of the US Department of State's Press Briefing for June 13, 2014, is disingenuous:

> Question [Matthew Lee, Associated Press]: Can we go back to Ukraine? There were reports yesterday on these three tanks that you were talking about, which you said was an escalation. There were reports that these in fact were Ukrainian tanks that had been kind of ripped off by the separatists.

> Ms. Harf [Marie Harf, Deputy Spokesperson, Department of State]: Nope.

> Question: You have convincing evidence –

> Ms. Harf: Yes.

> Question: – or whatever, that they were –

> Ms. Harf: They have acquired heavy weapons and military equipment from Russia, including Russian tanks. Yes.

Question:	Well – but these three tanks were driven across the border –
Ms. Harf:	Yes, yes.
Question:	– from – okay.
Ms. Harf:	They were somehow pulled out of the Russian warehouses, someone taught them how to use them, and they were sent from Russia to Ukraine.
Question:	So you don't buy the stolen from the Ukrainians?
Ms. Harf:	Nope, not at all.
Question:	Definitely Russian tanks? Okay.
Ms. Harf:	Yeah. (US Department of State 2014a)

In this case, however, other public-sourced information, specifically the videos and photos of the T-64 tanks driving through what are supposed to be Ukrainian cities, is helpful in understanding what actually happened in this case, but is also not dispositive. Mark Galeotti (2014) provided a timely and appropriately cautionary initial analysis on June 14 pointing to the fact that NATO did not identify the MBTs on the Russian side of the border as T-64s, and that there was at least the possibility that the T-64s seen in the public-sourced video inside Ukraine could have been captured by the separatists from the Ukrainian army or from depot storage in Ukraine. Ukrainian bloggers, who no doubt have their own agendas, posted videos of the T-64s moving through two towns and correlated the video with Google maps (something NATO did not do, or did not release to the public), but they misidentified the tanks. Later, however, Adam Čech and Jakub Janda (2015) conducted a thorough search of public-sourced video to ascertain that the Donbas separatists did not possess tanks prior to the June 12 incident, lending credence to the transference from Russia hypothesis.

However, this particular public-sourced information, specifically the videos and photos of the T-64 tanks driving through what are supposed to be Ukrainian cities, is not helpful in determining what actually happened in this case absent a more detailed independent analysis to establish the venues from which they were taken or to provide any clues as to the provenance of the vehicles. Nor was the inclusion of these public-sourced data in the official NATO press release on the June 2014 tank incident rendered in sufficient detail to establish that the tanks in question were in fact inside Ukraine. This underscores the relative paucity of public-sourced data available in the Russia-Ukraine context at that time (June 2014). As will be seen, this will change dramatically beginning with the events of July-August 2014.

Cross border rocket attacks in July and August 2014

As noted above, the July 1 Ukrainian offensive against separatist forces in eastern Ukraine began well. Key rebel strongholds such as Slavyansk were captured and government troops advanced steadily with the goal of sealing the border

POLITICAL GEOGRAPHIES OF THE POST-SOVIET UNION

with Russia to the east and southeast of Luhansk, which would have seriously hampered Russian efforts to resupply the separatists. One of the first indications of how difficult this would be came on July 11, 2014 when a motorized rifle unit of the Ukrainian army was severely mauled by a barrage of artillery rockets near the village of Zelenopillya in Luhansk Oblast (close to the border with Russia), incurring heavy casualties and the loss of many of its vehicles. Most reports at the time suggested that pro-Russian separatists carried out the attacks; indeed, the Ukrainian side claimed to have destroyed the rocket launchers with air strikes, which would certainly indicate that the launchers were inside Ukraine (BBC 2014). However, preliminary public analysis of this event and subsequent rocket attacks in southeastern Ukraine began with the release of reporting by Michael Weiss and James Miller (2014) to the effect that some of these attacks – indeed, perhaps the July 11 incident – were actually launched from inside Russian territory in the direction of Ukraine. Weiss and Miller, both of whom reported on the Syrian civil war, used videos taken of rocket launches; by geo-locating the camera positions with reference to ground-level imagery, they were able to establish that rockets were fired from a point in the vicinity of a town in Russia near the Ukrainian border toward Ukrainian territory. What this analysis does not show, however, are the vectors between these rocket launches and specific impact areas inside Ukraine, which renders them inconclusive as to Russian culpability. The United States and NATO were quick to point to these attacks and to denounce them, building them into their evolving storyline in which Russia was supposed to be taking a more direct role in the Donbas fighting. To emphasize their concern that these attacks added a new and more serious dimension to the conflict in eastern Ukraine, a series of public presentations and statements by top USG leadership were made to publicize the Russian actions. This exchange with White House Press Secretary Josh Earnest on July 25, 2014 makes the point:

> Question [Bill Plante, CBS News]:   Is there firing across the [Russia-Ukraine] border in either direction?

> Mr. Earnest: We have seen in the last couple of days, according to some social media reports but also to some intelligence assessments that have been released by the intelligence community, reports that there has been firing of Russian heavy weapons from the Russian side of the border at Ukrainian military personnel. We have detected that firing, and that does represent an escalation in this conflict. I know that the Pentagon and the State Department both talked about this a little bit yesterday, but it only underscores the concerns that the United States and the international community has about Russian behavior and the need for the Putin regime to change their strategy (White House Press Briefing 2014).

It is instructive that the referenced State Department press briefing the day prior featured a heated exchange between the Department's spokesperson (Marie Harf) and a reporter concerning the fact that the Department alleged "…that Russia is firing artillery from within Russia to attack Ukrainian military positions," but that no tangible evidence to support that claim would be forthcoming owing to the need to protect sources (US Department of State 2014b). The Supreme Allied Commander Europe (SACEUR), US Air Force General Phillip Breedlove, posted on Twitter that "I am deeply concerned by this latest video that appears to show Russia engaging in military action against Ukraine" and attached one of the YouTube videos showing rocket launches from Russian territory; SACEUR did not offer any official evidence to substantiate his concerns (Breedlove 2014).

On July 27, 2014, however, the US Department of State did release a set of four satellite images that claim to show the cross-border rocket attacks from Russia targeted against Ukrainian army units inside Ukraine (Frizell 2014). Two of the four graphics, annotated on images from Digital Globe, juxtapose what appear to be firing sites for multiple rocket launchers and artillery on the Russian side of the border and impact crater fields on the Ukrainian side. These images, while certainly compelling, are not irrefutable. The Russo-Ukrainian border is shown, which provides a visual cue to back up the contention that the attacks were cross-border. However, the specific location of the images is not provided; only a small reference map is included which makes it difficult to verify. Second, and more importantly, no assessment is provided of the alignment of impact craters inside Ukraine that would provide the azimuth of attack back to the firing site inside Russia. The other two images are of what are alleged to be rocket and artillery attacks within Ukraine; presumably, the point is that the artillery units are from the Russian army operating within Ukrainian territory, but this is not stated.

After several months had passed, public-sourced analysis on the Russian cross-border rocket artillery attacks against the Ukrainian army in July-August 2014 was forthcoming, and it would provide a far more exhaustive study of the subject than the official sources cited above (Case 2015). The first such report, originated in December 2014 (see the 2015 revised version) from the blogger Sean Case, and – although termed "initial" – it certainly was more convincing than the NATO release (Frizell 2014). Subsequently, an even more-detailed report was assembled by a team of analysts (including Case) headed by Eliot Higgins (one of the pioneers of public investigative journalism from the Syrian civil war) (Bellingcat 2015). Higgins and his associates painstakingly compiled satellite imagery of both rocket launch sites on the Russian side of the border and impact crater fields on the Ukrainian side, and a further analysis of videos taken of rocket launches from Russian territory. In addition to a much more detailed analysis of the suspected launch sites than that released by official sources, the Bellingcat report scrutinized over 1,300 impact craters within Ukraine to derive trajectories and then trace those back to firing locations within Russia. From these data they ascertained to a high degree of certainty – much more of a "gold standard" than the official storyline version – that in several specific incidents over the period July 14 to August 8, 2014 the Russian army attacked Ukrainian government forces maneuvering inside the sovereign territory of Ukraine.

Russian army units enter Ukraine

Arguably the most flagrant Russian violation of Ukraine's territorial sovereignty would be the crossing of the former's military units into the latter's territory. Actual Russian army units–not individuals or small groups of special operation troops–operating inside Ukraine to engage Ukrainian forces might account for the dramatic reversal of Ukrainian fortunes on the battlefield in August 2014, as it is difficult to imagine that the separatists themselves, even with Russian supplies and advisors, would have been able to achieve such a feat of arms. In late August NATO accused Russian army artillery units of providing fire support to separatist forces both across the border and from *within* Ukraine, which would, if confirmed, represent a significant escalation of Russian involvement in the fighting (Gordon 2014). But it was a dramatic new thrust in southeastern Donetsk Oblast from the Russian border along the littoral of the Sea of Azov, capturing the Ukrainian town of Novoazovsk and advancing to the outskirts of the key city of Mariupol, that heightened suspicions that the Russian army had itself now entered the fight (Kramer and Gordon 2014b).In response to this newest dimension to the conflict, during State Department press briefings the USG made clear that it accused Russia of introducing units of its armed forces into the war. This exchange from August 28, 2014:

Question [Bradley Klapper, Associated Press]: Yesterday, you decried what you said were incursions and for all intents and purposes an apparent invasion. That looks all too real today, and the action has gotten much more serious, with direct contact between Russian and Ukrainian forces. What are you going to do about it? Because you didn't really outline anything specific you were going to do yesterday.

Ms. Psaki [Jen Psaki, Spokesperson, Department of State]: Well, let me first say that what we're seeing, not just over the last couple of days but certainly weeks and even months, is a pattern of escalating aggression in Ukraine from the Russians and Russian-backed separatists. And it's clear that Russia has not only stepped up its presence in eastern Ukraine and intervened directly with combat forces – armored vehicles, artillery, and surface-to-air systems – and is actively fighting

Ukrainian forces as well as playing a direct supporting role to the separatist proxies and mercenaries. (US Department of State 2014c)

Pursuant to this assertion that Russia was now "intervening directly with combat forces" in Ukraine, on August 28, 2014 NATO released a set of five annotated images from Digital Globe claiming to "…provide additional evidence that Russian combat soldiers, equipped with sophisticated heavy weaponry, are operating *inside Ukraine's* sovereign territory" (NATO 2014, emphasis added). As was the case with prior NATO imagery releases, these images to some degree fall short of the "gold standard." First, three of the images purported to show Russian military equipment *inside Russia*, albeit near the Ukrainian border; clearly, from these three images it is not possible to deduce that Russian units are operating inside Ukraine. A fourth image (Figure 1) claims to show Russian self-propelled artillery (with support vehicles) moving east to west inside Ukraine, with the equipment alleged to be Russian because the vehicles are in an area where the Ukrainian army had yet to penetrate.

However, as before with NATO-released imagery, precise locations are not given, so that contention becomes more difficult to check. Likewise, because the specific types of vehicles were not identified, one cannot rule out that they could belong to Ukrainian army units or was captured equipment being operated by the separatists. The fifth image, which according to NATO shows Russian self-propelled

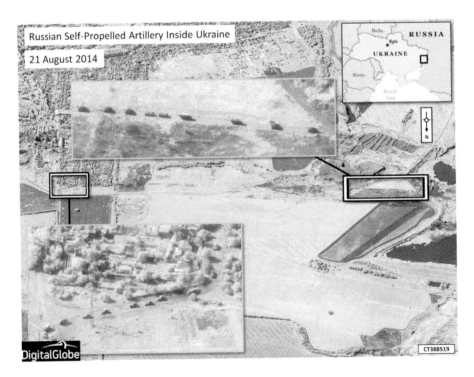

Figure 1. NATO imagery. Source: NATO 2014.

artillery inside Ukraine, again bears no specific location information, so it is not possible to conclude that this equipment is actually inside Ukraine.

In contrast to the sterile and unconvincing official data, dramatic evidence of large-scale movements of Russian military equipment into and inside Ukraine during this period began appearing on social media as the Russian operation in Ukraine progressed (for example, see, Ukraine@war 2014). Convoys of up to 100 Russian vehicles, including tanks, armored personnel carriers, self-propelled and towed artillery, surface-to-air missile systems, and troop transports were video recorded moving through locations geo-located within Ukraine (Kolona voisk Rossii 2014). Although the process is labor intensive, public analysts have acquired the skill sets necessary to identify individual vehicles and use ground-level views for specific geo-location (Higgins 2015). Linking those photos to satellite imagery in the unofficial analyses provides a more comprehensive and compelling portrayal of Russian army activity in the eastern Ukraine theater of war than anything officially released.

Russia attacks Ukraine: storylines and evidence

According to the official US/NATO storyline, Russia escalated its military involvement in eastern Ukraine in a sequence from a huge buildup of forces along the Russo-Ukrainian frontier; the cross-border transfer of armor and heavy weaponry; to cross-border artillery and rocket attacks; and lastly, to the insertion of units of the regular Russian army into Ukraine. Based on a careful analysis of officially released evidence, is it possible to build a fairly persuasive circumstantial case that over the period June through August 2014 the Russian government steadily ramped up its engagement in the fighting in the Donbas? In none of these cases did information officially released from either the United States Government or NATO *definitively* confirm any of these parts of the official storyline. One must assume that more definitive official analyses were not forthcoming owing to concerns about making public the full extent of intelligence gathering and analysis capabilities of the United States and/or the NATO alliance; because the United States and NATO lack the ability to conduct such analyses; because it was assumed that the material officially released would be convincing and stand up to scrutiny; or, finally, that the United States and its allies did not desire that the full extent of Russian involvement in the fighting in eastern Ukraine be definitively known as it might sway public opinion toward providing more advanced weapons to the Ukrainian government. As regards the first point specifically, as was stated previously, the overwhelming tendency of the epistemic sovereign state is to withhold information from the public or to characterize it in ways sufficiently vague as to cloak the full extent of the state's knowledge about a given situation. In some cases this is a reasonable approach; in other cases, especially when public sources are used in making the state's case, the necessity for secrecy is less obvious.

Both journalists and scholars attempting to verify elements of the official storyline constantly confront the security classification wall behind which policy-makers operate, resulting in less-than-satisfactory explanations of whatever actions are taken, or not taken, by governments. As David M. Herszenhorn and Peter Baker of the *New York Times* note with reference to reporting on the conflict in eastern Ukraine, "… the United States has been hesitant to make its intelligence public…" which means that in many cases evidence is unattributed because "… officials with access to classified intelligence assessments, [must speak] on the condition of anonymity" (2014). With reference to the June 2014 tank issue, Jonathan Marcus of the BBC, after describing the NATO release on that incident, stated that: "The evidence shown comes from Nato military sources and is *not necessarily conclusive*. But, despite Russian denials, it is *strongly suggestive* of the narrative that Nato is setting out" (2014, emphasis added). Interestingly, Marcus (2014) also writes that "A senior Nato officer told me, however, that military satellite material covering the same locations gives added evidence of Russian involvement." Just what that adds to an assessment of the evidence actually presented, allowing one to move from "strongly suggestive" to "conclusive," is not made clear. Despite the occasionally expressed frustration at not having more compelling evidence from official sources, the tendency among the professional media is to defer to military and governmental authority.

But that cloak of secrecy, related in general terms earlier in this paper, is now being lifted in the specific case of the conflict in eastern Ukraine through the proliferation of public-sourced data and more sophisticated non-official analysis. A main catalyst in this regard was the intensive work and significant investment in "digital labor" involved in the public investigation of the MH17 incident, said cause drawing in the experienced Syrian civil war analysts and new talent (Sienkiewicz 2015). The public-sourced reporting on Russian activities in eastern Ukraine now manifests previously unseen scope and depth of analysis, including the tracking of particular types of Russian weapons and military vehicles (see Ferguson and Jenzen-Jones 2014), front-line reporting by new media organizations (in particular VICE News), the geo-location of "selfies" taken by Russian soldiers, geo-located texts posted to social media sites, information concerning the deaths of Russian soldiers killed in action in Ukraine, and even more detailed dissection of videos and satellite imagery. The totality of the public-sourced effort dwarfs any data released to date by the USG and/or NATO. Further, some of this non-official work product, in addition to being increasingly cited by journalists, has been curated via gatekeepers and also synthesized into report format by foreign affairs institutes. These overview analytical pieces are very comprehensive and, for the most part, technically sophisticated beyond anything coming from official sources (see Czuperski et al. 2015; Miller et al. 2015). It can also be the case, however, that such reporting aims to influence government policy and often advocates for an agenda tied to the priorities of funding sources.

Conclusions

Despite their strongly worded assertions that Russia committed acts of aggression against Ukraine, the actual evidence released by the United States and NATO, although certainly timely, does not conclusively verify that these acts took place. Furthermore, these official releases are hamstrung by the perceived need to cloak certain details in secrecy despite the fact that imagery and other data are readily available from commercial sources. On the other hand, public source investigations, although longer in gestation, provide much more persuasive evidence that Russian aggression did indeed occur. This incongruity raises obvious questions about the depth of commitment of the external governmental/inter-governmental parties to pursue vigorously the case against Russia, perhaps (understandably) to avoid direct military conflict with that country; the culture of protecting sources and methods involved in gathering and analyzing official intelligence even when open sources are being used; or to tamp down calls to provide more substantial military assistance to Ukraine. This last, which raises the possibility of yet more Western involvement in a conflict that might escalate to a more serious level, is a possibility that most individual Western governments have been reluctant to entertain.

Actions such as those undertaken by Russia against Ukraine go to the heart of the viability of the extant spatial structuring of polities worldwide, engaging fundamental geopolitical concepts such as sovereignty and legitimacy (Clem 2014). As noted above, the rapid expansion of publicly generated data such as those adduced in the Ukraine crisis and elsewhere, allows the critiquing of official storylines and attributing blame in a geopolitical context beyond that previously possible. These official storylines, hitherto difficult to parse, are more likely now to be challenged without having to resort to breaches of official security. As a result, analysis conducted wholly in the public realm will almost certainly enhance the study of geopolitics and at the same time allow for greater transparency into the actions of states, especially those involving territorial conflict.

Notes

1. The Donbas region of eastern Ukraine (sometimes Donbass) is defined here as Donetsk and Luhansk oblasts.
2. She refers here to advances made in public-sourced reporting on nuclear non-proliferation, but the same rationale applies to other national security issues as well.
3. The distinction is made here between so-called Open Source Intelligence (OSINT) that refers to the analysis of public information within intelligence agencies (which is often not available to the public), and public-sourced information, which is gathered, analyzed, and disseminated by non-governmental or commercial entities, including private citizens.
4. As Alasdair Roberts (2006, 130–131) makes clear, NATO functions much as a state would in maintaining control over classified information shared among its members.

5. For a much more extensive discussion of the geopolitical narratives involved, see Toal 2017.
6. "Geneva commitments" references an agreement reached in that city in April 2014 between US Secretary of State John Kerry and Russian Foreign Minister Sergei Lavrov "…to de-escalate tensions and restore security…" in Ukraine (Gordon 2014). Obviously this agreement was overtaken by events.

Acknowledgements

An earlier version of this paper was presented at the 11th Annual Danyliw Research Seminar on Contemporary Ukraine at the Chair of Ukrainian Studies, University of Ottawa, Canada, October 22–24, 2015. The author wishes to thank participants at that conference and Gerard Toal for comments on previous versions of this paper. The author remains fully responsible, of course, for any shortcomings.

Disclosure statement

No potential conflict of interest was reported by the author.

References

Baczynska, Gabriela. 2015. "Russia Says No Proof It Sent Troops, Arms to East Ukraine." Reuters, January 21. Accessed December 14, 2017. https://www.reuters.com/article/us-ukraine-crisis-lavrov/russia-says-no-proof-it-sent-troops-arms-to-east-ukraine-idUSKBN0KU12Y20150121

BBC. 2014. "Ukraine Conflict: Many Soldiers Dead in Rocket Strike." July 11. Accessed September 10, 2017. http://www.bbc.com/news/world-europe-28261737

Bellingcat. 2015. "Origin of Artillery Attacks on Ukrainian Military Positions in Eastern Ukraine between 14 July 2014 and 8 August 2014." February 17. Accessed November 17, 2017. https://www.bellingcat.com/news/uk-and-europe/2015/02/17/origin-of-artillery-attacks/

Betts, Richard K. 2007. "Two Faces of Intelligence Failure: September 11 and Iraq's Missing WMD." *Political Science Quarterly* 122: 585–606.

Breedlove, General Philip M. 2014. July 14. Accessed November 17, 2017. https://twitter.com/PMBreedlove/status/489762254670544896

Caddell, Joseph. 2017. "Discovering Soviet Missiles in Cuba: How Intelligence Collection Relates to Analysis and Policy." *War on the Rocks*, October 19. Accessed October 19, 2017. https://warontherocks.com/2017/10/discovering-soviet-missiles-in-cuba-intelligence-collection-and-its-relationship-with-analysis-and-policy/

Case, Sean. 2015. "Smoking GRADs: Evidence of 90 Cross-border Artillery Strikes from Russia to Ukraine in Summer 2014." July 16. http://mapinvestigation.blogspot.com/2015/07/smoking-grads-evidence-of-90-cross.html

Čech, Adam, and Jakub Janda. 2015. *Caught in the Act: Proof of Russian Military Intervention in Ukraine*. Brussels: Wilfried Martens Centre for European Studies. Accessed November 15, 2017. https://www.martenscentre.eu/publications/caught-act-proof-russian-military-intervention-ukrain

Chivers, C. J. 2013. "New Study Refines View of Sarin Attack in Syria." *New York Times*, December 28. Accessed November 7, 2017. http://www.nytimes.com /2013/12/29/world/middleeast/new-study-refines-view-of-sarin-attack-in-syria.html

Clem, Ralph S. 2014. "Dynamics of the Ukrainian state-territory nexus." *Eurasian Geography and Economics* 55: 219–235.

Clem, Ralph S. 2017a. "Intel and Attributing Bad Acts: The United States needs to do Better." *War on the Rocks*. January 24. Accessed November 17, 2017. https://warontherocks.com/2017/02/intel-and-attributing-bad-acts-the-united-states-needs-to-do-better/

Clem, Ralph S. 2017b. "MH17 Three Years Later: What Have We Learned?" *War on the Rocks,* July 18. Accessed November 17, 2017. https://warontherocks.com/2017/07/mh17-three-years-later-what-have-we-learned/

Czuperski, Maksymilian, John Herbst, Eliot Higgins, Alina Polyakova, and Damon Wilson. 2015. "Hiding in Plain Sight: Putin's War in Ukraine." Washington, DC, Atlantic Council. Accessed November 17, 2017. http://www.atlanticcouncil.org/publications/reports/hiding-in-plain-sight-putin-s-war-in-ukraine-and-boris-nemtsov-s-putin-war

Eichensehr, Kristen. 2017. "Trump's Dangerous Attribution Message on Russian Hacking – and How to counter it." *Just Security*, January 10. Accessed November 18, 2017. https://www.justsecurity.org/36161/trumps-dangerous-attribution-message-and-counter/

Ferguson, Jonathan, and N. F. Jenzen-Jones. 2014. "Raising Red Flags: An Examination of Arms & Munitions in the Ongoing Conflict in Ukraine." Armament Research Services, November 18. Accessed November 18, 2017. http://armamentresearch.com/publications/

Fingar, Thomas. 2012. "A Guide to All-Source Analysis". Journal of U.S." *Intelligence Studies* 19: 63–66.

Florini, Ann, and Yahya Dehqanzada. 1999. "Commercial Satellite Imagery Comes of Age." *Issues in Science & Technology* 16: 46–52.

Frizell, Sam. 2014. "U.S.: Satellite Imagery Shows Russians Shelling Eastern Ukraine." *Time,* July 27. Accessed November 20, 2017. http://time.com/3042640/satellite-russian-ukraine-shelling/

Frye, Timothy. 2017. "Do Economic Sanctions Cause a Rally around the Flag?" Columbia/SIPA Center on Global Energy Policy. Accessed November 18, 2017. http://energypolicy.columbia.edu/sites/default/files/SanctionsandRallyAroundtheFlagTimFrye0817.pdf

Galeotti, Mark. 2014. "Those Mysterious Tanks in Ukraine." *In Moscow's Shadows,* June 14. Accessed November 18, 2017. https://inmoscowsshadows.wordpress.com/2014/06/14/those-mysterious-tanks-in-ukraine/

Gordon, Michael R. 2014. "Russia Moves Artillery Units into Ukraine, NATO Says." *New York Times*, August 22. Accessed September 4, 2017. https://www.nytimes.com/2014/08/23/world/europe/russia-moves-artillery-units-into-ukraine-nato-says.html

Götz, Elias. 2015. "It's Geopolitics, Stupid: Explaining Russia's Ukraine Policy." *Global Affairs* 1: 3–10.

Herszenhorn, David M., and Peter Baker. 2014. "Russia Steps up Help for Rebels in Ukraine War." *New York Times,* July 25. Accessed November 20, 2017. https://www.nytimes.com/2014/07/26/world/europe/russian-artillery-fires-into-ukraine-kiev-says.html

Higgins, Eliot. 2015. "Confirming the Location of the Same Msta-S in Russia and Ukraine." *Bellingcat*, May 29. Accessed November 18, 2017. https://www.bellingcat.com/?s=msta

Hinderstein, Corey. 2014. "Redefining Societal Verification." In *Innovating Verification: New Tools and New Actors to Reduce Nuclear Risks*. Washington, DC: Nuclear Threat Initiative. Accessed November 18, 2017. http://www.nti.org/analysis/reports/

Judah, Tim. 2014. "Ukraine: A Catastrophic Defeat." *New York Review of Books*, September 5. http://www.nybooks.com/daily/2014/09/05/ukraine-catastrophic-defeat/.

"Kiev must Stop War on Ukrainians." 2014. Rt.com, April 13. Accessed December 13, 2017. https://www.rt.com/news/security-council-ukraine-violence-312/

Kofman, Michael, and Matthew Rojansky. 2015. *A Closer look at Russia's 'Hybrid War'*. Washington, DC: Woodrow Wilson Center, Kennan Cable, No. 7.

"Kolona voisk Rossii dvizhetsya po g. Roven'ki Luganskaya oblast' (Column of Russian forces moving through the city of Roven'ki in Lugansk Oblast)." 2014. Posted July 27. Accessed November 20, 2017. https://www.youtube.com/watch?v=VeWjViTVsTo

Kramer, Andrew E., and Michael R. Gordon. 2014a. "Russia Sent Tanks to Separatists in Ukraine, U.S. Says." *New York Times*, June 13. Accessed November 20, 2017. https://www.nytimes.com/2014/06/14/world/europe/ukraine-claims-full-control-of-port-city-of-mariupol.html

Kramer, Andrew E., and Michael R. Gordon. 2014b "Ukraine Reports Russian Invasion on a New Front." *New York Times*, August 27. Accessed November 20, 2017. https://www.nytimes.com/2014/08/28/world/europe/ukraine-russia-novoazovsk-crimea.html

Larkin, Sean P. 2016. "The Age of Transparency: International Relations without Secrets." *Foreign Affairs* 95: 136–146.

Laruelle, Marlene. 2016. "The Three Colors of Novorossiya, or the Russian Nationalist Mythmaking of the Ukrainian Crisis." *Post-Soviet Affairs* 32: 55–74.

Lindsay, Jon R. 2015. "Tipping the Scales: The Attribution Problem and the Feasibility of Deterrence against Cyber Attacks." *Journal of Cybersecurity* 1: 53–67.

Mankoff, Jeffrey. 2014. "Russia's Latest Land Grab: How Putin Won Crimea and Lost Ukraine." *Foreign Affairs May-June* 60–68.

Marcus, Jonathan. 2014. "Russia and Ukraine's Mystery Tanks." BBC News, June 14. Accessed November 20, 2017. http://www.bbc.com/news/world-europe-27849437

Marten, Kimberly. 2015. "Putin's Choices: Explaining Russian Foreign Policy and Intervention in Ukraine." *The Washington Quarterly* 38: 189–204.

Menon, Rajan, and Eugene Rumer. 2015. *Conflict in Ukraine: The Unwinding of the Post-Cold War Order*. Cambridge, MA: MIT Press.

Miller, Christopher. 2014. "NATO Says Satellite Photos Support Claim that Russian Tanks Entered Ukraine." Mashable.com, June 15. Accessed November 20, 2017. https://www.nato.int/cps/ic/natohq/news_112193.htm

Miller, James, Pierre Vaux, Catherine A. Fitzpatrick, and Michael Weiss. 2015. "An Invasion by Any Other Name: The Kremlin's Dirty War in Ukraine." Washington, DC: Institute of Modern Russia. Accessed November 17, 2017. http://imrussia.org/en/research

Milner, Greg. 2016. *Pinpoint: How GPS is changing Technology, Culture, and our Minds*. New York: W.W. Norton.

NATO (North Atlantic Treaty Organization). 2014. Allied Command Operations. "NATO Releases Satellite Imagery Showing Russian Combat Troops inside Ukraine." August 28. Accessed November 20, 2017. https://www.nato.int/cps/ic/natohq/news_112193.htm

Ó Tuathail, Gearóid (Gerard Toal). 2002. "Theorizing Practical Geopolitical Reasoning: The Case of the United States' Response to the War in Bosnia." *Political Geography* 21: 601–628.

OSCE (Organization for Security and Cooperation in Europe). 2014. "2014 Daily Updates from the Special Monitoring Mission to Ukraine." Accessed November 18, 2017. http://www.osce.org/ukraine-smm/daily-updates

Pomerantsev, Peter. 2014. "Russia and the Menace of Unreality." *The Atlantic*, September 9. Accessed November 15, 2017. https://www.theatlantic.com/international/archive/2014/09/russia-putin-revolutionizing-information-warfare/379880/

Powers, Shawn, and Ben O'Loughlin. 2015. "The Syrian Data Glut: Rethinking the Role of Information in Conflict." *Media, War & Conflict* 8: 172–180.

Rácz, András. 2015. Russia's Hybrid War in Ukraine: Breaking the Enemy's Ability to Resist. Helsinki: Finnish Institute of International Affairs.

Reuters. 2014. "Ukraine Fights Fierce Battle with Rebels." Accessed November 20, 2017. http://ewn.co.za/2014/06/19/Ukraine-forces-fight-fierce-battle-with-rebels

Roberts, Alasdair. 2006. *Blacked Out: Government Secrecy in the Information Age*. New York: Cambridge University Press.

Sakwa, Richard. 2015. *Frontline Ukraine: Crisis in the Borderlands*. London: I.B. Tauris.

Sienkiewicz, Matt. 2014. "Start Making Sense: A Three-tier Approach to Citizen Journalism." *Media, Culture & Society* 36: 691–701.

Sienkiewicz, Matt. 2015. "Open BUK: Digital Labor, Media Investigation and the Ukrainian Civil War". *Critical Studies in Media Communication*. Published online July 3, 2015.

Snegovaya, Maria. 2015. *Putin's Information Warfare in Ukraine: Soviet Origins of Russia's Hybrid Warfare*. Washington, DC: Institute for the Study of War.

Toal, Gerard. 2017. *Near Abroad: Putin, the West, and the Contest over Ukraine and the Caucasus*. New York: Oxford University Press.

Toal, Gerard, and John O'Loughlin. 2017. "'Why Did MH17 Crash?': Blame Attribution, Television News and Public Opinion in Southeastern Ukraine, Crimea and the *De Facto* States of Abkhazia, South Ossetia and Transnistria." *Geopolitics*. doi:10.1080/14650045.2017.1364238.

Tsygankov, Andrei. 2015. "Vladimir Putin's Last stand: The Sources of Russia's Ukraine Policy." *Post-Soviet Affairs* 31: 279–303.

Ukraine@war (blog, now Putin@war). 2014. "Russia redeploys HUGE convoy of armor in Ukraine." September 2. http://ukraineatwar.blogspot.ru/2014/09/russian-moves-huge-convoy-of-armor-into.html

US Department of State. 2014a. Daily Press Briefing, June 13. Washington, DC. Accessed November 20, 2017. https://2009-2017.state.gov/r/pa/prs/dpb/2014/06/227573

US Department of State. 2014b. *Daily Press Briefing, July 24*. Washington DC. Accessed November 20, 2017. https://2009-2017.state.gov/r/pa/prs/dpb/2014/07/229752.htm

US Department of State. 2014c. Daily Press Briefing, August 28. Washington, DC. Accessed November 20, 2017. https://2009-2017.state.gov/r/pa/prs/dpb/2014/08/230798.htm

"War in the Donbass." 2015. Wikipedia. Accessed November 20, 2017.https://en.wikipedia.org/wiki/War_in_Donbass

Weiss, Michael, and James Miller. 2014. "Russia is Firing Missiles at Ukraine." *Foreign Policy*, July 17. Accessed November 20, 2017. http://foreignpolicy.com/2014/07/17/russia-is-firing-missiles-at-ukraine/

Westcott, Nicholas. 2008. *Digital Diplomacy: The Impact of the Internet on International Relations*. Research Report no. 16. Oxford Internet Institute. https://www.oii.ox.ac.uk.

White House Press Briefing. 2014. July 25. Accessed November 20, 2017.http://www.presidency.ucsb.edu/ws/index.php?pid=105478

Wilson, Andrew. 2014. *Ukraine Crisis: What it means for the West*. New Haven, CT: Yale University Press.

Cleavages, electoral geography, and the territorialization of political parties in the Republic of Georgia

David Sichinava ⓘ

ABSTRACT
This article examines the territorialization of party support in the Republic of Georgia as political parties in Georgia try to territorialize by aligning themselves to existing societal cleavages. The article specifically focuses on the case of the United National Movement (UNM), which from its inception in 2001 was led by Georgia's former president, Mikheil Saakashvili, and was the country's governing party from 2004 to 2012. While in power, the UNM enjoyed nationwide support. After being unseated, instead of nationalizing countrywide, the UNM has based its support in national elections on specific areas populated by ethno-linguistic and religious minorities. By analyzing the results of the most recent five national elections and the 2014 national census, the article shows that continuing support for the UNM and the subsequent territorialization of the party is dictated by these existing societal cleavages.

Introduction

In this paper, I explore the territorialization of political parties in the Republic of Georgia. Territorialization of party systems refers to the strategy of a political party to concentrate its support in a particular geographic area. To the contrary, the *nationalization* of political parties describes a situation when voters of a particular political party are spread evenly across a country (Agnew 1987). As Daniele Caramani argued (2004), the gradual disappearance of territorial patterns of party support in Western European societies is associated with the proliferation of the left–right divide in politics. Still, omnipresent regional disparities can be attributed to existing contextual factors, such as the presence of territorially concentrated linguistic, religious, and ethnic minorities (also see Bochsler 2006).

Apart from ideological constraints, different factors push political parties to nationalize. There have been several analyses of this issue, and party nationalization has been linked to the electoral regime (Morgenstern, Swindle, and Castagnola 2009; Golosov 2016b) and economic and social factors (de Miguel 2011), as well as the peculiarities of territorial organization of the polity (Chhibber and Kollman 2004). Additionally, as one stream of research argues, chances of party nationalization are higher if there are limited incentives for participation in regional politics; that is, regional autonomy is absent or malfunctioning (Simón 2013). Party nationalization strategies could be context-dependent, as shown by Agnew (1997).

The intrinsic nature of interrelations between electoral cleavages and party territorialization has been recently explored by de Miguel (2016), who links existing sociocultural cleavages, regional diversity, and the territorialization of political parties. Indeed, as the formation of these cleavages is context- and space-bound (Johnston 2009), voting decision-making is linked to the particular place and its contextual peculiarities (Cox 1969; Agnew 1987, 1996). Though often overlooked (such as by Golosov 2016b), the political geography perspective on voting has been recently invigorated with the highly polarizing results of US presidential elections (Johnston, Jones, and Manley 2016; Johnston, Manley, and Jones 2016; Johnston et al. 2017; Scala and Johnson 2017). However, the study of the spatial patterns of voting beyond the Anglo-American realm has been scarce (Leib, Quinton, and Warf 2011; Shin 2015), especially those investigating the patterns of voting in the post-Communist polities and so-called new democracies (e.g. O'Loughlin, Shin, and Talbot 1996; O'Loughlin, Kolossov, and Vendina 1997; Perepechko, Kolossov, and ZumBrunnen 2007; Maškarinec 2017).

When it comes to Georgia, the relative omission of the country from both specific and comparative studies on Eastern European politics and geography could be attributed to its relatively short history of free and fair elections. However, elections themselves in the country have systematically been labeled as "free and fair" by international elections observer organizations at least since 2008 (e.g. OSCE ODIHR 2008a, 2008b, 2012, 2013). The country is positioned moderately high in global democracy indices and is considered to be an "electoral democracy" by Freedom House (2016). As Tavits (2005) argues, in post-Communist polities, a short exposure to democracy would definitely prevent political parties operating there to align themselves along cleavages and policy-based politics. Therefore, the emergence of spatial cleavages could be attributed to the socio-demographic and cultural composition of the subnational territorial units (Bochsler 2006; Clem and Craumer 2008; O'Loughlin 2001). In the same vein, the most pronounced electoral cleavages in Georgia mirror the territorial cleavages between ethnic Georgians and other ethnicities in the country, as well as politically and economically engaged urban areas versus rural settlements (Sichinava 2015).

In this paper, I investigate a specific case of party support in Georgia by examining the temporal and spatial shifts in support of the United National Movement (UNM), a major political party that governed the country from 2004 until losing legislative elections in 2012. Despite subsequent losses in popularity and the emergence of splinter parties, the UNM still maintains its relatively stable party support.

In the analysis, I proceed as follows. The subsequent section explores the literature on party nationalization/territorialization and the electoral geography research relevant to this study of Georgia. In the data analysis section, I present global and local measures of spatial autocorrelation in order to describe the geography of party support for the UNM. I also present a fixed-effects regression model that predicts the district-level vote share for the UNM through proxy measures of various societal cleavages. I hypothesize that the spatial concentration of party support of the UNM has been aligned to existing societal cleavages. As the pattern has been maintained after the party lost power, it could indicate a territorialization strategy of the UNM instead of adopting a nationalizing strategy.

Literature review

Although it has long been argued that the emergence of nationalized political parties indicates the maturity of political systems (Caramani 2004), as Golosov (2016a, 2016b) attests, this fact is no longer relevant for many contemporary polities. The question of how political parties attempt to stabilize their support nationwide has been a key question for several significant contributions to the political science literature (e.g. by Caramani 2004; Chhibber and Kollman 2004; Jones and Mainwaring 2003; Golosov 2016a).

Students of the nationalization of political systems suggest different explanations for the territorialization and nationalization of voting and party support. The first strand of academic thought has been focused on the effects of institutional design on territorial support for political parties. For example, the devolution of power to regional entities through federal government and/or fiscal decentralization (Chhibber and Kollman 2004; Harbers 2010) has a reverse effect on party nationalization; however, as it has been recently put forward by Simón (2013), the influence of decentralization on the territorialization of political parties largely depends on the extent of incentives granted to the local political actors. A detailed analysis of the influence of electoral systems on party nationalization by Golosov (2016b) shows that systems that allow the "de-personalization" of party votes (e.g. party list proportional systems) strongly contribute to the nationalization of politics, whereas any other system involving single-member constituencies contributes to the territorialization of political parties, although the evidence for this is mixed (e.g. in Moser 1999).[1]

A second broad school of thought focuses on the peculiarities of various non-structural factors. Agents such as ethnic and religious diversity (Ordeshook and Shvetsova 1994; Golosov 2016b), regional economic inequalities (de Miguel 2011), and societal cleavages (Caramani 2004; Bochsler 2006; de Miguel 2016) contribute to the nationalization process, although these explanations could be proven to be complex and context-dependent. For example, in the case of African elections, Wahman (2017) showed that in spite of ethnic diversity that would technically incentivize territorialized voting, incumbent political parties in Africa manage to nationalize effectively. Additionally, Tiemann (2012) convincingly links the nationalization of party systems in Eastern Europe with the characteristics of transformation, specifically the peculiarities of electoral institutional arrangements and local historical cleavages. The role of societal cleavages in the territorialization of politics has been recently advanced by (de Miguel 2016). In her comparative analysis of 382 elections across 60 polities, de Miguel argues that the spatial concentration of ethnic and religious diversity is associated with more territorialized party systems, as is the existence of societal cleavages. This argument sounds especially compelling for the study of electoral geography, where societal cleavages used to be one of the key ontologies explaining spatially distinct patterns of voting behavior (Shin 2015).

The societal cleavage model of Lipset and Rokkan (1967) has been a useful tool for understanding the mechanisms of party formation and voter alignment. Lipset and Rokkan, based on Parsons' model of social systems (Parsons, Bales, and Shils 1953), argued that in Western European societies, party formation has been associated with four key societal cleavages, namely: center–periphery, church–state, employer–employee, and urban–rural dichotomies. Each of these pairs has contributed to the formation of political parties with corresponding ideological strains that have played a major role in stabilizing party systems.

The societal cleavages model has been established based on the experiences of Western European societies and generally neglects the peculiarities of voting beyond Western Europe (Deegan-Krause 2007). On the other hand, the role and the nature of the very cleavages ascribed by Lipset and Rokkan have been criticized, as the ideological dichotomies in the West also change and are more aligned to the post-materialist values (Deegan-Krause 2007). Although as Lipset (2001) himself later recounted, the model describes the attachment to the particular side and its institutionalization into political parties, ensuring the adaptability of the model to new contexts and cleavages. Indeed, the flexibility of the cleavage model yielded contributions from differing contexts such as Tunisia (Van Hamme, Gana, and Rebbah 2014), Turkey (West 2005), and – importantly for the Georgia case – post-Communist polities of Eastern Europe (Kitschelt 1995; Evans and Whitefield 1998; Whitefield 2002; Evans 2006). Overall, voter alignments in post-Communist societies have been formed along "Leninist" (Kitschelt 1995) and pre-Communist sociocultural legacies, including the peculiarities of Communist rule and the transition to democracy (Whitefield and Evans 1999).

Political geographers have been successful in adapting the Lipset–Rokkan cleavage model to the explanation of territorial patterns of voting and party formation (Taylor and Johnston 1979). They also produced one of the pioneering works investigating the peculiarities of voting and territorial organization of party systems in the New Europe (O'Loughlin, Shin, and Talbot 1996; O'Loughlin, Kolossov, and Vendina 1997; O'Loughlin 2001). The political geography approach underlines the role of a particular local context in which voters socialize, become politically engaged, and align themselves to political parties (Cox 1969; Agnew 1987, 1996).

Similarly, local analysis of election data, putting aside the problem of ecological fallacy (Freedman 1999; Pearce 2000; Seligson 2002), allows the detection of local trends in voting data, which is sometimes impossible to do with global statistical analysis (O'Loughlin 2003).

Setting the scene: a political history of the UNM

Contemporary Georgia's party system has been formed as the result of the compounding factors of personality and elite politics (Nodia and Scholtbach 2006) and the semi-authoritarian (Bader 2008), or competitive authoritarian (Wheatley and Zürcher 2008; George 2014), political environment where political parties have been operating since the inception of the independent post-soviet Georgian state. The initial years of the country's independence were dominated by nationalist political groups that emerged as champions in the country's independence movement (Jones 2013). However, political instability in the initial years of independence after 1991 brought civil war, violent ethnic conflicts, and economic downturns that resulted in the malfunctioning political system. For almost a decade, the former Communist leader Eduard Shevardnadze, who certainly contributed to the stabilization and institutional build-up, including the consolidation of the country's fractured party system, ruled the country. However, endemic corruption and overwhelming economic hardships (Jones 2000) soon triggered a peaceful "Rose Revolution" in 2003 that led to Shevardnadze's political demise.

The UNM and its leader, Mikheil Saakashvili, were key actors in bringing Shevardnadze's rule to an end and managed to radically transform the country and its political system. First created as a splinter group of Shevardnadze's "big tent" Citizens' Union of Georgia in 2001 (Chiaberashvili and Tevzadze 2005), the UNM managed to lead this peaceful revolution against Shevardnadze and, after consolidating power in its hands, swept the 2004 parliamentary and presidential elections as well as the 2006 municipal elections. Electoral success energized the UNM to push its reform agenda further. However, these efforts were mostly dedicated to the improvement of administrative institutions (Aprasidze and Siroky 2010) and less focused on the transparency of decision-making (Gallina 2010), which led to the relative marginalization of other political groups (Siroky and Aprasidze 2011). The UNM would rather focus on the improvement of the state's enforcement capacity (Rekhviashvili and Polese 2017) and international ratings, such as the World Bank's Ease of Doing Business Index, in order to ensure quick growth (Schueth 2011). However, the party spectacularly failed to ensure inclusive economic institutions and subsequently excluded large swaths of the country's population from post-Rose revolutionary economic growth (De Waal 2011; Baumann 2012; Gugushvili 2016). Fast modernization and Westernization became a trending mantra for Georgia's ruling political class. The UNM managed to pose as the driving force behind the country's transformation from post-Soviet failed state to the "beacon of democracy," and the alternative to the Russian model of development on the Eastern fringes of Europe (Kupatadze 2012). On top of that, Georgia was deemed as a role model for the democracy promotion project (Jawad 2005), which in turn helped the UNM secure large sums of international aid needed to stabilize the regime (Mitchell 2006).

The UNM's post-revolutionary institutional design hindered wider inclusion of other political forces to formal political institutions (Wertsch 2006; Miriam Lanskoy and Giorgi Areshidze 2008), who resorted to radicalization by staging street protests and refusing to enter the 2008 convocation of Georgian parliament. Political protests staged in the capital city of Tbilisi by opposition groups then led to the violent dispersion of a protest rally on 7 November 2007 and subsequent early presidential and parliamentary elections in the first half of 2008. In the new electoral cycle, the UNM managed to retain power by narrowly avoiding a runoff in early presidential elections, while easily carrying the subsequent parliamentary elections. Despite concerns about the fairness of the electoral campaign as well as the transparency of voting procedures during the election itself (OSCE ODIHR 2008a), both elections were deemed to be democratic (Nilsson and Cornell 2008). The devastating war with Russia in 2008 further contributed to the radicalization of politics (Cornell and Nilsson 2009), which climaxed in continued protest rallies and the occupation of the Tbilisi city center by opposition political groups in early 2009. However, contrary to the demands of the protesters who required an immediate resignation of Mikheil Saakashvili (Harding 2009), the UNM avoided early elections and maintained its power until 2012 parliamentary elections.

The eventual decline of the UNM from power may be explained on the one hand as the result of the emergence of a powerful political figure in the person of billionaire Bidzina Ivanishvili, and on the other hand, to the structural problems in the party's top-down governing style. Ivanishvili, a wealthy tycoon who earned his fortune in Russia in the early 1990s, championed the consolidation of fragmented opposition forces throughout the 2012 parliamentary election campaign. He also managed to compete successfully with the state apparatus in terms of financial and human resources, and when necessary, with compromising misinformation, or *kompromat* (Fairbanks and Gugushvili 2013; Roudakova 2017). The UNM itself failed to maintain a balance of stabilizing forces, such as resource superiority over other political groups and the legitimation of its own regime (Shubladze and Khundadze 2017). The post-2012 development of the UNM reflects the struggle of the party to maintain its support base and regional structure. Mikheil Saakashvili, a flamboyant charismatic founder and the leader of the party has been in exile since leaving his presidential office and of late has dedicated most of his time and effort to Ukrainian politics. Several former leaders such as Vano Merabishvili and Bacho Akhalaia have been jailed for misconduct during their tenure as government officials. Internally, although the party managed to maintain unity in the 2013 presidential and 2016 parliamentary elections, right after the last legislative elections "Movement for Liberty – European Georgia," a faction that splintered off of the UNM, managed to take over the majority of the UNM's seats in the parliament.

Despite the fragmentation and the internal and the external constraints, the party managed to maintain the core of its support base; in the elections, the party's poll numbers have been steady. In the 2013 presidential elections, about 354,000 voters, or about 22% of the total, endorsed the UNM candidate, Davit Bakradze. In the 2016 parliamentary elections, the party garnered 478 thousand votes, or 27% of all votes.

Data and methods

Methods of studying party nationalization generally utilize regression-based approaches, measurements of inequality (e.g. Gini-based indices), and even simple descriptive statistics. (For a comprehensive survey see Bochsler [2010].) Political scientists rarely resort to spatial methods; however, the rationale behind the usage of spatial measures for this purpose of identifying geographic concentrations lies in the very definition of the territorialization/ nationalization thesis. Based on this logic, Tapiador and Mezo (2009) utilize global spatial autocorrelation measures, namely Moran's I, to account for the geographic (de)concentration of party votes in Spain. More recent contributions to the party territorialization/nationalization literature (e.g. Harbers 2016; Ozen and Kalkan 2016) effectively employ spatial and spatio-temporal methods to explore regional patterns of party territorialization in Mexico and Turkey, respectively.

I use both spatial and non-spatial methods in order to explore the peculiarities of electoral support for the UNM.[2] As my variable of interest is the overall performance of the party, I analyze the vote share for the UNM and its presidential candidates. I look at the vote share of the UNM in 2008, 2012, and 2016 parliamentary elections and the vote share of Mikheil Saakashvili and Davit Bakradze, respectively, in the 2008 and 2013 presidential elections.

First, I present exploratory spatial data analysis of precinct-level election results; namely, I calculate local and global spatial autocorrelation.[3] Moran's global spatial autocorrelation (Moran 1950) assesses the overall degree of spatial concentration of a phenomenon and has been already utilized as a measure of the geographic concentration of votes (e.g. in O'Loughlin, Shin, and Talbot 1996). The measure fluctuates between −1 and 1, where positive numbers are associated with geographic concentration of a phenomenon while negative numbers refer to dispersed geographic patterns. As the result, I present the correlogram of spatial lag and Moran's I values which describe the association among spatial concentration and the distance between spatial units. In order to identify statistically significant areas of concentration of high and low votes share for the UNM, I refer to the local indicators of spatial autocorrelation, namely Anselin's local Moran's I (Anselin 1995). As the result, I present a set of maps for each of the five national elections and identify geographic areas of high and low support for the UNM.

In addition to the exploratory spatial data analysis, I also model district-level vote share for the UNM and its presidential candidates in the last five national elections.[4] In order to control for the district and election-specific effects, I utilize a fixed effects model for both electoral districts and the separate elections. Conceptually, the independent variables control for existing societal cleavages (urban/rural, center–periphery), voter mobilization, district magnitude, and a variety of geographic characteristics. Descriptive statistics of these variables are summarized in Table 1.

Measures of nationalization of party systems are often based on the effective number of parties nationally or locally in the electoral unit (e.g. in Jones and Mainwaring 2003; Bochsler 2010; de Miguel 2016; Wahman 2017; Ozen and Kalkan 2016). However, in this analysis, I focus on the strategies of a particular political party; following Harbers (2016), I utilize district-level election outcomes to assess the territorialization process. For societal cleavages, I control for the urban–rural and center–periphery divide in the Georgian electorate, which has significant impact in almost all post-Communist polities (Evans 2006). Lipset and Rokkan (1967) define the center–periphery cleavage as the tension between the nationalizing culture of the political center with ethnically or religiously different peripheries, while the emergence of the urban–rural divide is associated with rising alienation between the city and the countryside in terms of development and the quality of life. From the theoretical perspective on Georgia as a post-Communist polity, these two cleavages are likely to be most relevant, as they often mirror more recent developments, such as the type of the Communist regime and the peculiarities of transition to the democracy and consequently, represent a stable basis for party formation and voter affiliation (Whitefield 2002; Evans 2006).

I operationalize the center–periphery divide using the proportion of Orthodox population in the district and the proportion of native Georgian speakers. The rationale behind introducing the religious dimension of electoral cleavages lies in the peculiarities of Georgian nation building. As Pelkmans (2002, 2005) has argued, the construction of the national identity during the last years of the Soviet Union was largely associated with Orthodox Christianity. Therefore, Muslim populations, especially in their communities in Adjara and Guria regions, have been under constant pressure for religious conversion or are treated as lower class citizens. More broadly for ethnic minorities, their representation in Georgian politics is still limited (Zollinger and Bochsler 2012). However, in the aftermath of the Rose Revolution, the Saakashvili government pushed for better integration of Muslims and more access to resources (George 2008), and in turn, this has been associated with higher support of ethnic minorities in the UNM (George 2014).

The urban–rural cleavage is operationalized with the proportion of urban population in the electoral district, conventionally known as the level of urbanization, and the proportion of those with higher education. A stark urban–rural divide in development has been an endemic characteristic for Georgia. Georgian rural areas were especially affected with the transition to the market economy and since then have been systematically overlooked politically (Jones 2013). Moreover, the neoliberal characteristics of Georgia's economic development led to the increase

Table 1. Descriptive statistics of the variables.

Variable		Mean	Median	Standard deviation	Source
Dependent variable:					
	Vote share for the United National Movement or its candidate	0.41	0.38	0.20	Central Elections Commission of Georgia (CEC)
Predictors					
	Turnout	0.57	0.57	0.10	Central Elections Commission of Georgia (CEC)
	Proportion of population with higher education	0.21	0.17	0.12	2014 National Census of Georgia
	Proportion of white collar workers in the working-age population	0.20	0.14	0.15	2014 National Census of Georgia
	Proportion of orthodox population	0.82	0.94	0.27	2014 National Census of Georgia
	Proportion of native Georgian speakers	0.88	0.98	0.22	2014 National Census of Georgia
	Proportion of urban population	0.39	0.26	0.31	2014 National Census of Georgia
Covariates	Inverse of the district size	0.00004	0.00003	0.00003	Central Elections Commission of Georgia (CEC)
	Distance from international roads (m)	11,510	3312	16,373.04	Department of Road Transportation of Georgia
	Distance from national roads (m)	2024	501.4	4405.639	Department of Road Transportation of Georgia
	Median altitude of the district from the sea level (m)	612	488.2	498.9	Extracted from ASTER digital elevation model using ArcGIS software
	District size and median	48,212	36,103		
	Number of observations	365			

of rural poverty (Gugushvili 2016), and that indeed contributed to the alienation of the rural population toward the UNM. Apart from cleavages, I also control for several other measures, such as voter turnout, which could serve as a proxy for voter mobilization and electoral fraud (as high turnout in similar contexts is sometimes associated with fraud, e.g. in Enikolopov et al. [2013]) and the inverse number of the electoral district size. The latter measure allows control for the large district magnitude inherent to Georgia's electoral system.

Data analysis

Spatio-temporal patterns of UNM vote territorialization

The first part of my analysis refers to the global and local patterns of spatial concentration of the UNM votes. I summarize in Table 2 the values of Moran's global spatial autocorrelation of the precinct-level election outcomes for the last five national elections. In the case of the strongly contested 2008 early presidential elections, the UNM votes can be characterized with territorial concentration, while it is less pronounced in the 2008 parliamentary elections. The 2012 parliamentary election was also highly contested, resulting in the UNM's loss of power. In the 2012 dataset, the value of the UNM votes' spatial autocorrelation again indicates the highest degree of the concentration. For the subsequent elections, the measures of UNM spatial autocorrelation stabilized and stayed at the level of 2008 parliamentary elections – in the 2013 presidential elections, the Moran's coefficient decreased to 0.421, and it reached the value of 0.450 in the case of 2016 parliamentary elections (both significant at $p < 0.001$).

Figure 1 illustrates the dependence of Moran's autocorrelation measures on the spatial lag values (spatial lag is the weighted average of values in the neighboring spatial units). As expected, the dependence of the vote concentration on distance bands hints at the distance–decay nature of the concentration of the precincts with high values. Similar to the single coefficients of spatial autocorrelation described above, results of 2012 parliamentary elections have the most pronounced dependence of spatial autocorrelation values on the distance lags, while the 2008 and 2016 parliamentary elections are less prone to express spatial dependence.

The empirical results show a moderate to high level of spatial autocorrelation of the UNM vote share in the analyzed elections. The values are especially high when it comes to the two most contested and polarized elections – the 2008

Table 2. Moran's global autocorrelation coefficient values for 2008–2016 national elections.

Elections	Moran's I	Monte Carlo simulated p-values	Number of precincts	Average precinct size, voters
Parliamentary 2016	0.450	0.001	3624	966
Presidential 2013	0.421	0.001	3622	967
Parliamentary 2012	0.587	0.001	3602	988
Parliamentary 2008	0.433	0.001	3524	976
Presidential 2008	0.522	0.001	3435	1032

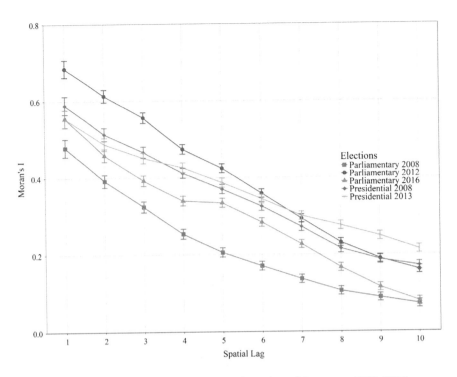

Figure 1. Spatial correlogram of UNM and its presidential candidate votes, 2008–2016.

presidential and the 2012 parliamentary elections. In the first case, the UNM's presidential candidate, Mikheil Saakashvili, narrowly avoided a runoff with Levan Gachechiladze, a candidate endorsed by the opposition. In the 2012 parliamentary elections, albeit the most contested in Georgia's recent political history, overall political polarization was also reflected in the territorial patterns of voting. The high territorial concentration of UNM votes indeed mirrors the party's strategy for territorializing its support in specific geographic areas. Finally, the continuing trend of the concentration of vote share for the UNM also hints to the continuation of the territorialization strategy (Figure 2).

Cleavage dimension of UNM vote territorialization

The second set of analyses explores the specific geographic areas of concentration of the UNM's votes. Anselin's local Moran measurement of spatial autocorrelation allows us to detect the areas of high and low concentration of a particular value as well as spatial outliers; for example, high values surrounded by spatial units holding a low value (Anselin 1995). In the 2008 presidential elections, the main areas of concentration of high votes for Mikheil Saakashvili were the westernmost Samegrelo province, the mountainous southwestern areas of Adjara, territories bordering South Ossetia (for the specific locations of these regions refer to Figure 2), and a wide belt of ethnically and religiously diverse southern Georgia (top leftmost

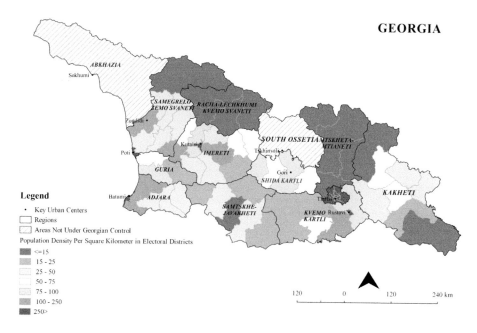

Figure 2. Population density and the indicator map of geographic locations mentioned in this paper. Source: FOI request from the National Statistics Office of Georgia; own work.

panel of Figure 3). The electoral map of UNM support in 2008 parliamentary elections displays fewer areas of concentration (top right panel of Figure 3); however, the party performed especially well in the territory bordering South Ossetia and in the southern fringes of the country with a predominantly Armenian or ethnically mixed population.

The situation was altered in and after the 2012 parliamentary elections. The UNM lost support in the territories adjoining South Ossetia that suffered the most from the 2008 armed conflict between Russia and Georgia. However, the key areas of the "Southern Belt," densely populated northwestern areas of Samegrelo, and parts of mountainous Adjara remained as the cornerstones of support for the UNM. To the contrary, the blue areas of western Georgia denote the concentration of extremely low values for the UNM in the municipality of Sachkhere, birthplace of then-opposition leader, Bidzina Ivanishvili.

Since its fall from power, the UNM has concentrated its support to several key geographic areas. The party still maintains its power base in Samegrelo; areas of Samtskhe in Southern Georgia with mixed Georgian Orthodox, Armenian Apostolic, Catholic, and Muslim populations; and the Kvemo Kartli area in southeastern Georgia, where the dominant ethnic group is Azerbaijani. The concentration of low values of UNM's support are clustered in the Javakheti region populated by ethnic Armenians as well as in selected mountainous areas in western and eastern Georgia.

Overall, the UNM's electoral maps show consistent patterns of party support concentration. The party's key vote base has been located in the densely populated

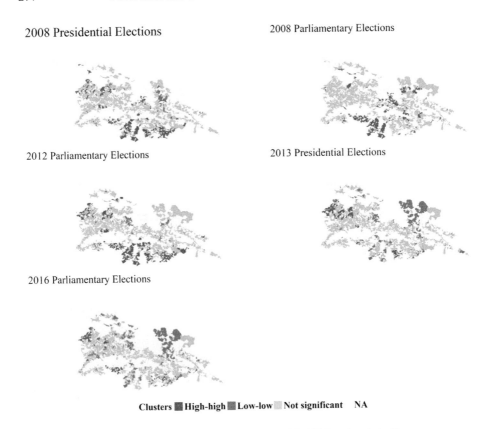

Figure 3. Spatial patterns of UNM vote distribution in 2008–2016 national elections.

regions of Samegrelo and Kvemo Kartli, as well as in Samtskhe, while party support has been recurring in Javakheti and Upper Adjara areas. The ethnic Georgian population of Samegrelo is a linguistic minority speaking a distinct language of the Kartvelian language family. Although it has always been fully integrated into mainstream Georgian society and even considered a vanguard force of Georgian nation building, certain prejudices still exist against Mingrelian-speakers (as attested in an experimental study of Dragojevic, Berglund, and Blauvelt [2018]). Apart from that, in the early 1990s, the Samegrelo region was the epicenter of the civil war between the loyalists of the first president of independent Georgia, Zviad Gamsakhurdia (who was Mingrelian), and the supporters of the military council who ousted him and invited Shevardnadze as the leader. The violence perpetuated by the military council armed groups against the local population has often been framed in terms of a distinct regional identity (Kolsto 1996; Broers 2001).

Two distinct patterns emerge with regard to the areas populated with ethnic Armenian and Azerbaijani minorities. First, in the 2012 watershed parliamentary elections, the areas predominantly populated with ethnic minorities supported the UNM. As George (2014) has attested, this support was consistent with the earlier patterns of voting for the incumbent, even controlling for possible electoral fraud. Indeed, apart from the usual circumstances, higher support of minorities was articulated by the UNM's campaign and actual policies; for example, introducing special educational programs to encourage minority students to apply to Georgian universities (Office of the State Minister 2017). However, since 2013, the pattern has been altered – although it lost power the UNM managed to retain its support among Azerbaijani-populated constituencies, while the visible decline of its vote in Armenian-populated communities could be attributed to the distinct and long-attested patterns of cooptation of local power elites to the ruling political groups in the central government (Gotua 2011).

Global statistical model of UNM vote territorialization

The outcomes of the fixed effects model predicting the district-level vote share for the UNM and its presidential candidates in national elections between 2008 and 2016 are shown in Table 3. By employing the fixed effects model, I control for the district and election specificities that may influence the outcomes beyond the predicted effects that are indicated in Table 3. The results indicate a strong association of the district-level vote share for the UNM and the societal cleavages identified above. When it comes to the center–periphery dichotomy, the ethno-linguistic

Table 3. Results of regression analysis.

	Dependent variable					
	Vote share for the United National Movement or its candidate					
	(1)	(2)	(3)	(4)	(5)	(6)
Total turnout	0.339***	0.318***	0.310***	0.383***	0.310***	0.295***
	(0.067)	(0.067)	(0.063)	(0.067)	(0.067)	(0.074)
Proportion of orthodox population	−0.023	−0.028	−0.026	−0.072**	−0.180***	
	(0.03)	(0.03)	(0.029)	(0.031)	(0.021)	
Proportion of native Georgian speakers	−0.163***	−0.166***	−0.168***	−0.179***		
	(0.036)	(0.036)	(0.035)	(0.038)		
Proportion of population with higher education	−0.375***	−0.395***	−0.372***			
	(0.081)	(0.08)	(0.048)			
Proportion of urban population	−0.008	0.011				
	(0.032)	(0.031)				
Inverse of the district size	−344.758**					
	(170.219)					
Observations	365	365	365	365	365	365
R^2	0.365	0.358	0.357	0.247	0.201	0.043
Adjusted R^2	0.347	0.341	0.343	0.232	0.187	0.03
F statistic	33.904***	39.519***	49.487***	38.997***	44.907***	16.087***
	(df = 6; 354)	(df = 5; 355)	(df = 4; 356)	(df = 3; 357)	(df = 2; 358)	(df = 1; 359)

Note: Standard errors reported in parenthesis.
*$p < 0.1$; **$p < 0.05$; ***$p < 0.01$.

dimension of this electoral cleavage is statistically significant to the $p < 0.01$ level in the initial model (model 1 in Table 3) after control for all covariates. Overall, the concentration of ethnic and religious minorities inside voting districts is associated with a positive change in the vote share of the UNM and its candidates. The proportion of Georgian-speakers in the electoral district is associated with a large negative change in the dependent variable. A percentage point increase in the proportion of Georgian speakers is associated with about a 16 percentage point decrease in the mean proportion of the UNM vote share. The pattern is persistent when we gradually remove covariates from the main model (columns 2–5 in Table 3). When it comes to another measure of the center–periphery cleavage, religious composition of the district becomes significant when the urban–rural divide is removed from the regression model (columns 4–5 in Table 3). Specifically, one percentage point change in the proportion of the Georgian Orthodox population in the electoral district is associated with a −7% point change in the mean proportion of UNM's district-level election outcomes, when not controlling for the urban–rural cleavages. The fact that these two variables are associated with the same phenomenon is illustrated in the change of coefficient of the Georgian Orthodox population when the covariate measuring the proportion of Georgian speakers is dropped (columns 4 and 5).

With regard to the urban–rural cleavage, in the initial model only the characteristics of education are associated with the statistically significant change in the dependent variable. The proportion of higher education holders in the electoral district is associated with a large decrease in the proportion of UNM votes in the electoral districts. Namely, 1% point change in this covariate is associated with almost 38% decline in the mean value of UNM's vote share. Interestingly, in the initial model the effect of urbanization does not exhibit a statistically significant association with the change in the dependent variable.

Turnout values have an overall positive effect in the vote share of the UNM. In the full model (column 1), 1% point change in the turnout is associated with about a 35% increase in the mean district-level vote share for the UNM, *ceteris paribus*. However, removal of other covariates from the regression model leads to a dramatic decrease in R-squared and adjusted R-squared values of the model, indicating that electoral turnout is not a substantial explanation of the variation in the dependent variable (vote of UNM).

Discussion and conclusion

What are the pillars to which political parties in the post-Communist polities align themselves in order to recruit supporters? Although my analysis refers to a specific case of Georgia, the fate of the UNM shows that to a significant extent, cleavage politics in the post-Communist polities is still alive and well. Moreover, these cleavages also contribute to the territorialization of party systems. Overall, the presented analysis showed that the UNM's support in the last national elections has been aligning with urban–rural and center–periphery dichotomies in a consistent manner. The empirical materials also show that high territorial concentration of the UNM's support especially in the highly contested 2012 elections contributed to the demise of the party from power. Despite that fact, the party's support further territorialized along identified cleavages from earlier elections.

The empirical results presented in this paper resonate with findings from an exhaustive study by de Miguel (2016). Spatially concentrated diversity, in the Georgian case, means that ethnic, religious, and linguistic minorities form a necessary context for the nationally oriented political party to territorialize its support. As is known from the political geography literature, ethnic, religious, and linguistic minorities almost always have distinct voting behaviors, especially in the post-Soviet realm (O'Loughlin 2001). This statement holds for the Georgian case as well. The center–periphery cleavage in Georgia is seemingly a later development and is linked to the peculiarities of transition to democracy, as is attested in Whitefield and Evans (1999). Ethnic conflicts of the beginning of the 1990s as well as the characteristics of Georgia's nation building process and the policies toward minorities administered by the central government could be in play. Although the compounding effect of local elites, who successfully navigate the moribund waters of political transition and pledge allegiance to particular political groups, cannot be denied (Gotua 2011; George 2014).

To conclude, the consistency of global and local spatial autocorrelation measures indicate that the patterns are stable. The ethnically, linguistically, and religiously distinct areas of Georgia exhibit a clear pattern of voting behavior which goes beyond the relative simplistic explanation of elite, institutional, or personal politics. As John Agnew wrote in 1996, electoral geography perspectives on voting demonstrate the role of spatial complexity in defining voting patterns; therefore, reducing the analysis to a mere set of simple models would definitely "miss the multiscalar quality of social causation" (144). The Georgian example of the party territorialization process well illustrates how geographic analysis of voting contributes to the understanding of seemingly "global" strategies of the national political class and helps identify the contextual factors influencing voting decision-making process.

Notes

1. The author would like to thank Ralph Clem for pointing out this issue.
2. The data that support the findings of this study are available in Open Science Framework at http://doi.org/10.17605/OSF.IO/CD475.
3. The boundary data were compiled for the National Democratic Institute, an international democracy watchdog, by the Caucasus Research Resource Centers, and Tbilisi State University and was provided to the author free of charge. Precinct-level election results were obtained through the freedom of information request from Georgia's Central Elections Commission. Census materials were provided by the National Statistics Office of Georgia through the freedom of information request.

4. Areas that are under Ossetian and Abkhazian separatist control after the 2008 Russo-Georgian War (Akhalgori, Liakhvi, Kodori electoral districts) are excluded from the analysis to maintain comparison across the elections.

Acknowledgments

The initial drafts of this work were prepared when the author was a Fulbright Visiting Scholar at the Institute of Behavioral Sciences University of Colorado Boulder. The author would like to thank Dr John O'Loughlin for hosting and mentoring at the Institute of Behavioral Sciences, Dr Barbara Buttenfield for the advice on spatial analysis, and Andrew Eaman for the help on processing the geographical data. The author acknowledges insightful comments from Dr John O'Loughlin and Dr. Ralph Clem on this paper. All errors in this manuscript are my own.

Disclosure statement

No potential conflict of interest was reported by the author.

ORCID

David Sichinava 🆔 http://orcid.org/0000-0003-4660-4363

References

Agnew, John. 1987. *Place and Politics: The Geographical Mediation of State and Society*. Boston, MA: Allen & Unwin.

Agnew, John. 1996. "Mapping Politics: How Context Counts in Electoral Geography." *Political Geography* 15: 129–146.

Agnew, John. 1997. "The Dramaturgy of Horizons: Geographical Scale in the 'Reconstruction of Italy' by the New Italian Political Parties, 1992–1995." *Political Geography* 16: 99–121.

Anselin, Luc. 1995. "Local Indicators of Spatial Association – LISA." *Geographical Analysis* 27: 93–115.

Aprasidze, David, and David Siroky. 2010. "Frozen Transitions and Unfrozen Conflicts, or What Went Wrong in Georgia." *Yale Journal of International Affairs* 5: 121–136.

Bader, Max. 2008. "Fluid Party Politics and the Challenge for Democracy Assistance in Georgia." *Caucasian Review of International Affairs* 2: 1–10.

Baumann, Eveline. 2012. "Post-Soviet Georgia: It's a Long, Long Way to 'Modern' Social Protection." *Economies et Sociétés* 46: 259–285.

Bochsler, Daniel. 2006. "The Nationalization of Political Parties. A Triangle Model, Applied on the Central and Eastern European Countries." *CEU Political Science Journal* 1: 6–37.

Bochsler, Daniel. 2010. "Measuring Party Nationalisation: A New Gini-based Indicator That Corrects for the Number of Units." *Electoral Studies* 29: 155–168.

Broers, Laurence. 2001. "Who Are the Mingrelians? Language, Identity and Politics in Western Georgia." *Sixth Annual Convention of the Association for the Study of Nationalities*, New York, April 5–7.

Caramani, Daniele. 2004. *The Nationalization of Politics: The Formation of National Electorates and Party Systems in Western Europe*. Cambridge: Cambridge University Press.

Chhibber, Pradeep, and Ken Kollman. 2004. *The Formation of National Party Systems: Federalism and Party Competition in Britain, Canada, India and the US*. Princeton, NJ: Princeton University Press.

Chiaberashvili, Zurab, and Gigi Tevzadze. 2005. "Power Elites in Georgia: Old and New." In *From Revolution to Reform: Georgia's Struggle with Democratic Institution Building and Security Sector Reform*, edited by P.H. Fluri and E. Cole, 187–207. Vienna: Landesverteidigungsakademie (Austrian National Defence Academy).

Clem, Ralph S., and Peter R. Craumer. 2008. "Orange, Blue and White, and Blonde: The Electoral Geography of Ukraine's 2006 and 2007 Rada Elections." *Eurasian Geography and Economics* 49: 127–151.

Cornell, Svante E., and Niklas Nilsson. 2009. "Georgian Politics since the August 2008 War." *Demokratizatsiya* 17: 251–268.

Cox, Kevin R. 1969. "The Voting Decision in a Spatial Context." *Progress in Geography* 1: 81–117.

De Waal, Thomas. 2011. *Georgia's Choices: Charting a Future in Uncertain times*. Washington, DC: Carnegie Endowment for International Peace.

Deegan-Krause, Kevin. 2007. "New Dimensions of Political Cleavage." In *Oxford Handbook of Political Behaviour*, edited by Russell J. Dalton and Hans-Dieter Klingemann, 538–556. New York: Oxford University Press.

Dragojevic, Marko, Christofer Berglund, and Timothy K. Blauvelt. 2018. "Figuring out Who's Who: The Role of Social Categorization in the Language Attitudes Process." *Journal of Language and Social Psychology* 37: 28–50.

Enikolopov, Ruben, Vasily Korovkin, Maria Petrova, Konstantin Sonin, and Alexei Zakharov. 2013. "Field Experiment Estimate of Electoral Fraud in Russian Parliamentary Elections." *Proceedings of the National Academy of Sciences* 110: 448–452.

Evans, Geoffrey. 2006. "The Social Bases of Political Divisions in Post-communist Eastern Europe." *Annual Review of Sociology* 32: 245–270.

Evans, Geoffrey, and Stephen Whitefield. 1998. "The Structuring of Political Cleavages in Post-communist Societies: The Case of the Czech Republic and Slovakia." *Political Studies* 46: 115–139.

Fairbanks, Charles H., and Alexi Gugushvili. 2013. "A New Chance for Georgian Democracy." *Journal of Democracy* 24: 116–127.

Freedman, David A. 1999. "Ecological Inference and the Ecological Fallacy." *International Encyclopedia of the Social & Behavioral Sciences* 6: 4027–4030.

Freedom House. 2016. *Georgia Country Report. Freedom in the World 2016*. Accessed October 31, 2017. https://freedomhouse.org/report/freedom-world/2016/georgia

Gallina, Nicole. 2010. "Puzzles of State Transformation: The Cases of Armenia and Georgia." *Caucasian Review of International Affairs* 4: 20–34.

George, Julie A. 2008. "Minority Political Inclusion in Mikheil Saakashvili's Georgia." *Europe-Asia Studies* 60: 1151–1175.

George, Julie A. 2014. "Can Hybrid Regimes Foster Constituencies? Ethnic Minorities in Georgian Elections, 1992–2012." *Electoral Studies* 35: 328–345.

Golosov, Grigorii V. 2016a. "Factors of Party System Nationalization." *International Political Science Review* 37: 246–260.

Golosov, Grigorii V. 2016b. "Party Nationalization and the Translation of Votes into Seats under Single-member Plurality Electoral Rules." *Party Politics* 24: 118–128. doi: 10.1177/1354068816642808.

Gotua, Giorgi. 2011. "Different Governments in Tbilisi, Same People in Regions: Local Elites in the Years of Independence." In *Changing Identities: Armenia, Azerbaijan, Georgia*, by The Heinrich Boell Foundation South Caucasus Regional Office, 202–222. Tbilisi: Cezani.

Gugushvili, Dimitri. 2016. "Lessons from Georgia's Neoliberal Experiment: A Rising Tide Does Not Necessarily Lift All Boats." *Communist and Post-Communist Studies* 50: 1–14.

Harbers, Imke. 2010. "Decentralization and the Development of Nationalized Party Systems in New Democracies: Evidence from Latin America." *Comparative Political Studies* 43: 606–627.

Harbers, Imke. 2016. "Spatial Effects and Party Nationalization: The Geography of Partisan Support in Mexico." *Electoral Studies* 47: 55–66.

Harding, Luke. 2009. "Thousands Gather for Street Protests against Georgian President Mikheil Saakashvili." *The Guardian*, April 9, 2009. https://www.theguardian.com/world/2009/apr/09/georgia-protests-mikheil-saakashvili.

Jawad, Pamela. 2005. *Democratic Consolidation in Georgia after the 'Rose Revolution'?* Peace Research Institute, PRIF Reports No. 73. Frankfurt: PRIF.

Johnston, Ron. 2009. "Electoral Geography." In *Dictionary of Human Geography*, edited by Derek Gregory, Ron Johnston, Geraldine Pratt, Michael Watts, and Sarah Whatmore, 187–188. Oxford: Wiley.

Johnston, Ron, Kelvyn Jones, and David Manley. 2016. "The Growing Spatial Polarization of Presidential Voting in the United States, 1992–2012: Myth or Reality?" *PS: Political Science & Politics* 49: 766–770.

Johnston, Ron, David Manley, and Kelvyn Jones. 2016. "Spatial Polarization of Presidential Voting in the United States, 1992–2012: The 'Big Sort' Revisited." *Annals of the American Association of Geographers* 106: 1047–1062.

Johnston, Ron, Charles Pattie, Kelvyn Jones, and David Manley. 2017. "Was the 2016 United States' Presidential Contest a Deviating Election? Continuity and Change in the Electoral Map – Or 'plus Ça Change, plus Ç'est La Mème Géographie.'" *Journal of Elections, Public Opinion and Parties* 27: 1–20.

Jones, Stephen. 2000. "Democracy from below? Interest Groups in Georgian Society." *Slavic Review* 59: 42–73.

Jones, Stephen. 2013. *Georgia: A Political History since Independence*. London: IB Tauris.

Jones, Mark P., and Scott Mainwaring. 2003. "The Nationalization of Parties and Party Systems an Empirical Measure and an Application to the Americas." *Party Politics* 9: 139–166.

Kitschelt, Herbert. 1995. "Formation of Party Cleavages in Post-communist Democracies Theoretical Propositions." *Party Politics* 1: 447–472.

Kolsto, Pal. 1996. "Nation-building in the Former USSR." *Journal of Democracy* 7: 118–132.

Kupatadze, Alexander. 2012. "Explaining Georgia's Anti-corruption Drive." *European Security* 21: 16–36.

Lanskoy, Miriam, and Giorgi Areshidze. 2008. "Georgia's Year of Turmoil." *Journal of Democracy* 19: 154–168.

Leib, Jonathan, Nicholas Quinton, and Barney Warf. 2011. "On the Shores of the 'Moribund Backwater'?: Trends in Electoral Geography Research since 1990." In *Revitalizing Electoral Geography*, edited by Barney Warf and Jonathan Leib, 9–27. Farnham: Ashgate.

Lipset, Seymour Martin. 2001. "Cleavages, Parties and Democracy." In *Party Systems and Voter Alignments Revisited*, edited by Lauri Karvonen and Stein Kuhnle, 3–9. London: Routledge.

Lipset, Seymour M., and Stein Rokkan. 1967. "Cleavage Structures, Party Systems, and Voter Alignments: An Introduction." In *Party Systems and Voter Alignments: Cross-national Perspectives*, edited by Seymour M. Lipset and Stein Rokkan, 1–64. New York: Free Press.

Maškarinec, Pavel. 2017. "A Spatial Analysis of Czech Parliamentary Elections, 2006–2013." *Europe-Asia Studies* 69: 426–457.

de Miguel, Carolina. 2011. "The Geography of Economic Inequality, Institutions and Party System Territorialization." SSRN Scholarly Paper No. 1915225. Rochester, NY: Social Science Research Network.

de Miguel, Carolina. 2016. "The Role of Electoral Geography in the Territorialization of Party Systems." *Electoral Studies* 47: 67–83.

Mitchell, Lincoln A. 2006. "Democracy in Georgia since the Rose Revolution." *Orbis* 50: 669–676.

Moran, P. A. P. 1950. "Notes on Continuous Stochastic Phenomena." *Biometrika* 37: 17–23.

Morgenstern, Scott, Stephen M. Swindle, and Andrea Castagnola. 2009. "Party Nationalization and Institutions." *The Journal of Politics* 71: 1322–1341.

Moser, Robert G. 1999. "Electoral Systems and the Number of Parties in Postcommunist States." *World Politics* 51: 359–384.

Nilsson, Niklas, and Svante E. Cornell. 2008. "Georgia's May 2008 Parliamentary Elections: Setting Sail in a Storm." Policy Paper. Central Asia-Caucasus Institute & Silk Road Studies Program – A Joint Transatlantic Research and Policy Center. Accessed February 13, 2018. http://www.css.ethz.ch/en/services/digital-library/publications/publication.html/55485

Nodia, Ghia, and Álvaro Pinto Scholtbach. 2006. *The Political Landscape of Georgia: Political Parties: Achievements, Challenges and Prospects*. Delft: Eburon.

O'Loughlin, John. 2001. "The Regional Factor in Contemporary Ukrainian Politics: Scale, Place, Space, or Bogus Effect?" *Post-Soviet Geography and Economics* 42: 1–33.

O'Loughlin, John. 2003. "Spatial Analysis in Political Geography." *A Companion to Political Geography* (Feb.): 30–46.

O'Loughlin, John, Vladimir Kolossov, and Olga Vendina. 1997. "The Electoral Geographies of a Polarizing City: Moscow, 1993–1996." *Post-Soviet Geography and Economics* 38: 567–600.

O'Loughlin, John, Michael Shin, and Paul Talbot. 1996. "Political Geographies and Cleavages in the Russian Parliamentary Elections." *Post-Soviet Geography and Economics* 37: 355–385.

Office of the State Minister of Georgia for Reconciliation and Civic Equality. 2017. "Program '1 + 4.'" Accessed February 13, 2018. http://smr.gov.ge/detailspage.aspx?ID=56

Ordeshook, Peter C., and Olga V. Shvetsova. 1994. "Ethnic Heterogeneity, District Magnitude, and the Number of Parties." *American Journal of Political Science* 38: 100.

OSCE ODIHR. 2008a. "Georgia, Extraordinary Presidential Election, 5 January 2008." OSCE/ODIHR Election Observation Mission Final Report. Warsaw. http://www.osce.org/odihr/elections/georgia/30959.

OSCE ODIHR. 2008b. "Georgia Parliamentary Elections 21 May 2008." OSCE/ODIHR Election Observation Mission Final Report. Warsaw. http://www.osce.org/odihr/elections/georgia/33301?download=true.

OSCE ODIHR. 2012. "Georgia, Parliamentary Elections, 1 October 2012." OSCE/ODIHR Election Observation Mission Final Report. Warsaw. http://www.osce.org/ka/odihr/elections/98585.

OSCE ODIHR. 2013. "Georgia, Presidential Election 27 October 2013." OSCE/ODIHR Election Observation Mission Final Report. Warsaw. http://www.osce.org/odihr/elections/110301.

Ozen, Ilhan Can, and Kerem Ozan Kalkan. 2016. "Spatial Analysis of Contemporary Turkish Elections: A Comprehensive Approach." *Turkish Studies* 18: 1–20.

Parsons, Talcott, Robert Freed Bales, and Edward Shils. 1953. *Working Papers in the Theory of Action*. New York: Free Press.

Pearce, Neil. 2000. "The Ecological Fallacy Strikes Back." *Journal of Epidemiology and Community Health* 54: 326–327.

Pelkmans, Mathijs. 2002. "Religion, Nation and State in Georgia: Christian Expansion in Muslim Ajaria." *Journal of Muslim Minority Affairs* 22: 249–273.

Pelkmans, Mathijs. 2005. "Baptized Georgian: Religious Conversion to Christianity in Autonomous Ajaria." Accessed February 13, 2018. http://eprints.lse.ac.uk/28072/

Perepechko, Alexander S., Vladimir A. Kolossov, and Craig ZumBrunnen. 2007. "Remeasuring and Rethinking Social Cleavages in Russia: Continuity and Changes in Electoral Geography 1917–1995." *Political Geography* 26: 179–208.

Rekhviashvili, Lela, and Abel Polese. 2017. "Liberalism and Shadow Interventionism in Post-Revolutionary Georgia (2003–2012)." *Caucasus Survey* 5: 27–50.

Roudakova, Natalia. 2017. *Losing Pravda: Ethics and the Press in Post-Truth Russia*. Cambridge: Cambridge University Press.

Scala, Dante J., and Kenneth M. Johnson. 2017. "Political Polarization along the Rural-Urban Continuum? The Geography of the Presidential Vote, 2000–2016." *The Annals of the American Academy of Political and Social Science* 672: 162–184.

Schueth, Sam. 2011. "Assembling International Competitiveness: The Republic of Georgia, USAID, and the Doing Business Project." *Economic Geography* 87: 51–77.

Seligson, Mitchell A. 2002. "The Renaissance of Political Culture or the Renaissance of the Ecological Fallacy?" *Comparative Politics* 34: 273–292.

Shin, Michael. 2015. "Electoral Geography in the Twenty-First Century." In *The Wiley Blackwell Companion to Political Geography*, edited by John A. Agnew, 279–296. Oxford: Wiley-Blackwell.

Shubladze, Rati, and Tsisana Khundadze. 2017. "Balancing the Three Pillars of Stability in Armenia and Georgia." *Caucasus Survey* 5: 301–322.

Sichinava, David. 2015. "Cleavage Theory and the Electoral Geographies of Georgia." In *Security, Democracy and Development in the Southern Caucasus and the Black Sea Region*, edited by Ghia Nodia and Christoph H. Stefes, 27–44. Bern: Peter Lang.

Simón, Pablo. 2013. "The Combined Impact of Decentralisation and Personalism on the Nationalisation of Party Systems." *Political Studies* 61 (Suppl.): 24–44.

Siroky, David S., and David Aprasidze. 2011. "Guns, Roses and Democratization: Huntington's Secret Admirer in the Caucasus." *Democratization* 18: 1227–1245.

Tapiador, Francisco J., and Josu Mezo. 2009. "Vote Evolution in Spain, 1977–2007: A Spatial Analysis at the Municipal Scale." *Political Geography* 28: 319–328.

Tavits, Margit. 2005. "The Development of Stable Party Support: Electoral Dynamics in Post-Communist Europe." *American Journal of Political Science* 49: 283–298.

Taylor, Peter J., and Ronald John Johnston. 1979. *Geography of Elections*. London: Croom Helm.

Tiemann, Guido. 2012. "The Nationalization of Political Parties and Party Systems in Post-communist Eastern Europe." *Communist and Post-Communist Studies* 45: 77–89.

Van Hamme, Gilles, Alia Gana, and Maher Ben Rebbah. 2014. "Social and Socio-Territorial Electoral Base of Political Parties in Post-Revolutionary Tunisia." *The Journal of North African Studies* 19: 751–769.

Wahman, Michael. 2017. "Nationalized Incumbents and Regional Challengers: Opposition- and Incumbent-Party Nationalization in Africa." *Party Politics* 23: 309–322.

Wertsch, James V. 2006. "Georgia after the Rose Revolution." *The Caucasus & Globalization* 1: 54–66.

West, Jefferson. 2005. "Regional Cleavages in Turkish Politics: An Electoral Geography of the 1999 and 2002 National Elections." *Political Geography* 24: 499–523.

Wheatley, Jonathan, and Christoph Zürcher. 2008. "On the Origin and Consolidation of Hybrid Regimes: The State of Democracy in the Caucasus." *Taiwan Journal of Democracy* 4: 1–31.

Whitefield, Stephen. 2002. "Political Cleavages and Post-communist Politics." *Annual Review of Political Science* 5: 181–200.

Whitefield, Stephen, and Geoffrey Evans. 1999. "Class, Markets and Partisanship in Post-Soviet Russia: 1993–96." *Electoral Studies* 18: 155–178.

Zollinger, Daniel, and Daniel Bochsler. 2012. "Minority Representation in a Semi-democratic Regime: The Georgian Case." *Democratization* 19: 611–641.

The political geographies of religious sites in Moscow's neighborhoods

Meagan Todd

ABSTRACT

This paper addresses the spatial politics of Russia's increased religiosity in Moscow. It analyzes the rights of minority Muslim communities within the context of increased political support for expressions of Russian Orthodoxy in Moscow's public space. Moscow's Russian Orthodox and Muslim religious leaders claim that their communities have a lack of religious infrastructure, with one church per 35,000 residents and one mosque per three million residents, respectively. The Russian Orthodox Church has been more successful than Muslim organizations at expanding their presence in Moscow's neighborhoods. Drawing on ethnographic fieldwork, religious spaces are examined as sites of dissent as well as participatory, active citizenship at three different sites in Moscow. Protests over Russian Orthodox Church construction in one neighborhood are contrasted with the protests over mosque construction in two neighborhoods. This paper provides insights into how civil society and religious groups have increased their public presence in Moscow and shows the unequal access that different groups have to public space in that city.

Introduction

More Russians are identifying as religious since the collapse of the Soviet Union in 1991 and its formal state policy of atheism. The number of self-identified adherents to Russian Orthodoxy, which is Russia's largest religion, has grown from 37% in 1991 to 71% in 2015 (Pew Research Center 2017). Meanwhile, the number of adherents to Islam, which is Russia's second-largest religion, has grown to 10% in 2017, and the number of people who do not identify with any religion includes 15% of the population (2017). This paper addresses the spatial dynamics of Russia's increased religiosity in Moscow by analyzing the differing challenges and experiences that Muslim and Russian Orthodox communities face in constructing new places of worship. Religiosity consists of multidimensional measures, such as the beliefs in the tenets of religion, the practice of religious rituals, attendance at religious services, and self-identification as a member of a religious group (Cornwall et al. 1986). In the Soviet era, religion became a private practice, while public places of religion were monitored, transformed for other uses, or destroyed. Consequently, one enduring legacy of Soviet city planning and atheism is a paucity of churches, mosques, and other forms of religious infrastructure located near densely populated apartment blocks. The growth of religious adherents has created a demand for new places of worship in these formerly atheist neighborhood spaces.

Moscow's Russian Orthodox and Muslim religious leaders claim that their communities have a shortage of religious infrastructure. But with one church per 35,000 residents and one mosque per three million residents, respectively, Muslims are by orders of magnitude more underserved. Churches and mosques both provide places for rituals and shared practice. In Islam, the recitation and display of sacred words (usually in Arabic) designate places as sacred (Metcalf 1996). Unlike churches, mosques themselves are not considered sacred buildings due to the Muslim prohibition against *shirk*, or polytheism and idolatry. Still, mosques provide a place for both shared ritual and identity formation (Metcalf 1996; McLoughlin 2005).

Moscow is a majority Russian city, but its population consists of many ethnic minorities associated with Islam. According to the 2010 Russian census, Moscow's population was 11.5 million people, composed 91.65% of ethnic Russians, 1.42% of ethnic Ukrainians, and 1.38% of ethnic Tatars (a traditionally Muslim people), with .9% coming from other ethnicities associated historically with Islam (Rosstat 2010). Rosstat estimated this figure to have increased to 12.3 million in 2016, which includes temporary workers but not undocumented workers (Rosstat 2016b). Depending on seasonal fluctuation, there is an additional estimated one to three million undocumented workers living in Moscow (Light 2016), most of them undocumented male labor migrants from Muslim regions; these numbers include external migration from Central Asian republics and internal migration from the North Caucasus during the First and Second Chechen wars from 1994 to 1996 and from 1999 to 2009 (Ioffe and Zayonchkovskaya 2011). In 2009, the largest official migrant population in Moscow was from the North Caucasus at 26.4%, while the second largest was from countries in Central Asia, at 20.9% (Light 2016). In sum, Moscow's Muslim population is composed of many different ethnic groups with varying traditions of Muslim practice, including an "indigenous" (pre-Soviet era) Tatar population of 168,000, as well as a large proportion of registered and undocumented labor migrants (Ioffe and Zayonchkovskaya 2011). Researchers on Islam in Moscow estimate that due to migration from these regions, the city has 1.5–2 million Muslims (March 2010; Malashenko 2014).

Despite its sizeable Muslim minority population, Moscow has only four mosques, and plans to construct new ones have been fiercely opposed since the 2010 announcement to construct a mosque in Moscow's Tekstilshchiki District. Moscow's mayor since October 2010, Sergey Sobyanin, has endorsed a plan to construct over 200 Russian Orthodox churches while at the same time opposing plans to build new mosques, explicitly citing the Muslim population of Moscow as transient and illegal (Venediktov 2013). Moscow's Muslim leaders and imams have sought permission to build new mosques since the collapse of the Soviet Union, but due to protests, plans for new mosque construction in the city have come to fruition only twice since the end of state-based atheism and the collapse of the Soviet Union in 1991.

Mosque construction plans are one proxy for evaluating how Muslims are able to make claims to public space in Russia (McLoughlin 2005). Russia's 1997 Law on Freedom of Conscience and Religious Associations stipulates that religious congregations must be officially registered with the state to receive legal rights and recognition and is meant to curtail foreign missionary activity. This law created a two-class system of religions in Russia between religious organizations and religious groups. Religious organizations were defined by having at least a 15-year presence in Russia and include Russia's four traditional religions; Russian Orthodoxy, Islam, Buddhism, and Judaism. Religious groups meanwhile have been present less than 15 years and include foreign religious groups, such as Jehovah's Witnesses, Mormons, and other Protestant groups. Fully registered religious organizations enjoy a host of rights, such as the right to produce and distribute religious materials, the right to found educational institutions, and the right to found mass media (Fagan 2012a). Religious organizations also have the right to own property and the right to use land and property granted to them by the state or other public entities for their needs (Russian Federation Federal Law 1997). Religious groups, meanwhile, have the rights to conduct rituals and teach religion on property that they own or lease (Fagan 2012a). The difference between religious organizations and religious groups was heightened by the 2016 Yarovaya anti-extremism laws, which bans proselytization and prayer meetings outside of religious buildings and authorizes the arrest for violating "generally accepted norms of social behavior." The Yarovaya laws, intended to protect Russians from foreign influences, are widely criticized by human rights agencies as violating international norms of freedoms of conscience and religion and curtailing the rights of smaller denominations in Russia (Flikke 2016).

Despite the right to obtain or use property promised to religious organizations, Muslim religious organizations face challenges in its implementation. According to data aggregated at the national level, Russia has experienced a mosque construction boom. There were 870 registered mosques in 1991, with most of them located in the two historically Muslim regions of Russia: the North Caucasus, including the republics of Dagestan, with 240 mosques, and Chechnya-Ingushetia, with 162, and the Volga-Urals region, including the republics of Tatarstan, with 91 mosques, and Bashkortostan, with 65 (Halbach 2001). According to the chairman of two Russian Muslim organizations, the Russian Council of Muftis and the Spiritual Directorate of Muslims in the European part of Russia, Grand Mufti Ravil Gainutdin, 7000 new mosques have been built in Russia since Putin became president in 2000 (*RIA Novosti* 2014). This mosque construction is clustered in the historically Muslim regions in the North Caucasus and Volga-Ural regions; for example, in Tatarstan there are over 1300 mosques (Karimova 2017). At the national level, the increase in mosque construction is often cited by the Russian state to show its support for its Muslim minority populations.

In ethnically Russian majority regions, however, there is a pattern of opposition to new mosque construction (Fagan 2012a). For example, there have been controversies over mosque construction in administrative centers such as Astrakhan (Holland and Todd 2015), Vladimir, and Stavropol (Dzusati 2014). This opposition occurs not only in smaller cities, but also in Russia's largest and more cosmopolitan cities. Sochi's Muslim community has asked for and been denied permits to build a mosque for over 15 years, despite its international prominence as host to the 2014 Winter Olympics (Fagan 2012b), and Saint Petersburg has only two mosques to serve its Muslim population of 800,000 out of 5 million residents (Yalovkina 2016).

Despite the national-level data showing an exponential increase in mosque construction in the Russian Federation, Moscow fits the pattern of ethnically majority Russian regions that have seen conflict over new mosque construction. Two mosques have been built in Moscow since 1991, both in the 1990s as part of multi-confessional religious complexes housing Russian Orthodox Churches, Buddhist temples, and Jewish synagogues. At the time, both of these mosques faced protests from Russian nationals over their construction (Smith 2002), but they were ineffective in deterring the completion of the projects. Since then, no new mosques have been built and new mosque proposals face opposition. On one hand, the number of mosques has doubled in the city since 1991; however, the increase from 2 to 4 structures does not meet the needs of its estimated Muslim population of 1.5–2 million adherents.

The politics of religious construction serve as a lens to study how minority faith communities are embedded and subjected to networks of state and other forms of power (Kong and Woods 2016). Examining neighborhood and religious politics surrounding Muslim communities and religious construction projects, I address the following specific questions: What is the nature of the various discourses that surround the proposed construction of and opposition to new mosques in Moscow? Who are the opponents and advocates, and how are they organized and mobilized? How are these discourses promoted and received by interest groups including congregations, neighborhood associations, human rights organizations, and local government officials, concomitant to the evolution of Moscow's religious landscape?

My paper analyzes what debates about the place of Islam (i.e. Muslim life) in Moscow's public space suggest about the politics of religion and secularity in contemporary Russian democracy. Moscow is an iconic place and the center of Russian political culture. Political leaders have constructed their ideologies into Moscow's built environment: in the Russian Empire, Russian Orthodox churches in the Kremlin and Saint Basil's Cathedral in Red Square were built by tsars to display piety and celebrate military victories (Cracraft and Rowland 2003), but the city's seemingly endless apartment blocks were designed by planners to serve as the everyday setting of a classless, undifferentiated Soviet population (Collier 2011). This Soviet bloc landscape was deliberately atheist and devoid of religious buildings. Now, Moscow's urban landscape is again transforming. In 2017, Sobyanin proposed to replace outdated low-rise Soviet-era apartment blocks, called *khrushovki*, with high-rises (Associated Press in Moscow 2017). Developers have constructed IKEAs, malls, and luxury high-rise apartments along the city's edges. Amidst this changing urban geography, I examine the construction of new churches and mosques. Iconic due to its reputation as a political center, Moscow is also a representative case study of other ethnically Russian majority cities where Muslims face local opposition against the construction of new mosques.

Geographies of protest movements in Russia

Religious construction projects in general and mosques in particular generate new forms of local activism in Moscow. My paper analyzes local activism about religious construction projects in Russia's "managed" democracy. I use the term "managed" to describe the centrist nature of Russia's democracy under Putin. Managed democracy refers to how the characteristics of a liberal democracy are used in Russian politics to gather power. Elections regularly occur, but are neither free nor fair due to a lack of competition between political parties and the domination of the majority party over state-controlled media (Wilson 2005). Russia's managed democracy is often characterized by a lack of political activism and civic community, due to dismay about governmental corruption (Fish 2005; Lussier 2011). However, in Russia, civil society activism and political protest are still visible. Notably, protests in Moscow since Putin's 2011 bid for and re-election as Russian president resulted in mass demonstrations throughout Moscow and in regional capitals. These demonstrations, led by the middle class, about political fraud and corruption illustrate that many urban citizens are frustrated with managed democracy. They show that civil society in Russia is becoming more active and experienced in making rights-based demands through demonstrative protests. This can be witnessed in the ongoing protests throughout the country against Putin's re-election for his fourth term in 2018.

Robertson (2013) believes that the Moscow-based 2011 electoral fraud protests in Russia are a continuation of protest trends in Russia, rather than a landmark and singular event. He argues that in Russia, the nature, location, and meanings of protest have shifted considerably since the collapse of the Soviet Union. In the 1990s protests occurred in provinces and were about poverty and unpaid wages. Protest actions included industrial strikes, hunger strikes, and blockades. However, in the 2000s the majority of protests occurring in Moscow, including marches and demonstrations, were about environmental and city development issues, civil rights, and social justice. This is corroborated in Argenbright's (2016) research on civil society in Moscow, which analyzes the emergence of a "not-in-my-backyard"

(NIMBY) type of mindset among Muscovites as they protest developments in their neighborhoods. Beyond protests, oppositional tactics to development include more participatory tactics such as the creation of citizen's watch groups that challenge planning decisions, the mobilization of neighbors to block construction projects, and coalition-building among these neighborhood watch groups (Argenbright 2016).

This article adds to the larger literature on the changing nature of protest and civil society groups in Russia. Drawing on qualitative data, I examine how religious spaces become sites of dissent and citizenship at three different sites in Moscow. First, I overview protests over mosque construction in Tekstilshchiki and Mitino, two of Moscow's *spalnie rayoni*, or sleeping districts, which were designed by Soviet city planners to provide for the activities of daily living of Soviet people. They were built starting in the 1950s to address the housing crisis caused by increased urbanization in the Soviet Union and consisted of prefabricated apartment buildings, *dvori*, or courtyards for play and exercise, and schools (Harris 2013). They were often named after the outlying villages surrounding Moscow, such as Yasnevo or Belyayevo, in which they were erected (Snopek 2013). These blocks were a main component of Soviet planning, which sought to manage the collective life of the Communist workforce (Collier 2011). I next introduce "200 Churches," the project to build 200 Russian Orthodox churches in suburban Moscow. I investigate how the protest over church construction in one neighborhood, Babushkinsky, contrasts with the protests over mosque construction in two other neighborhoods, Tekstilshchiki, and Mitino. I chose these three locations because they are the most prominent cases that occurred in my database of Russian media sources prior to fieldwork in 2013 and 2015 (see below). This analysis provides insight into the types of conflicts that the public presence of religion engenders in Moscow.

The immediate and long-term outcomes of protest movements show how normative publics and hegemonic norms are formed. Normative publics refer to those groups whose practices, values, and esthetics receive support or reflect the ideals of the state (Mitchell 2012). Geographers have argued that public spaces are sites of contestation: they are spaces where people can make claims against the state and hegemonic normative public practices (Staeheli and Mitchell 2007). Not all groups have equal access to public space and the ability to claim representation. Historically, public spaces like the agora or the park have functioned as zones of inclusion and exclusion. This means that these "open" spaces are normally accessible to the public, but through bans and restrictions serve as a means of disenfranchisement to some groups. One example includes the prohibition of people experiencing homelessness sleeping in public parks in the United States (Mitchell 2012). This paper examines what the access of different religious and civic groups to public space says about the formation of publics in Russia's managed democracy.

My paper is informed both by ethnographic research and by analysis of a database of key print and digital media documents. My qualitative data consists of 29 in-depth interviews with key informants, 95 short interviews, and 3 group interviews (see the Appendix 1 for a partial list), as well as participation-observation field notes taken over a period of 10 months in 2013 and 2015 in Moscow. Participation-observation occurred at three sites of protests over religious construction projects in Moscow's neighborhoods: Teskstilshchiki, Mitino, and Babushkinsky. To supplement my ethnographic research, I built a database or "clippings file" of news articles, relevant media, and government documents on topics of mosque and church construction and its opposition from 2000 (the beginning of Putin's presidency) to the present. This database consists of 469 items, including news articles from national-level newspapers such as *Izvestiya*, Moscow newspapers like *Bolshoi Gorod* and *Moskovskie Novosti*, transcripts from the radio station, Echo of Moscow, press releases from the Russian Council of Muftis, and documents from local organizations such as *Moj Dvor*. My data collection captures the Russian managed democracy in action and includes trial proceedings, summaries, and reports from local courts, as well as letters from local politicians. It also contains interviews with some neighborhood government officials on the issue of mosques. My field notes and street interviews from participant observation provided me with insight into neighborhood dynamics of both areas of proposed mosque construction and their neighborhood environs.

NIMBYism in Moscow's suburbs

Tekstilshchiki

Mosque construction serves as a touchstone for debates over Muscovite identity in the suburbs. Two notable cases of proposed and failed mosque projects occurred in the sleeping districts and involved high levels of citizen activism and protest activity. The first case occurred in Tekstilshchiki, a sleeping district and municipality in the Southeastern Administrative District of Moscow. Its name refers to its historical textile industries. In the mid-nineteenth century, the area was home to a wool factory of a wealthy merchant. More textile industries developed in the area throughout the 1920s and 1930s, including a factory for fabric printing and dyeing, a calico factory, and a wool-processing mill. In the 1960s, the region was converted into a mass housing development and incorporated into Moscow. It is densely populated, with 104,516 residents (Rosstat 2016a). The rents in the area are lower than in more prestigious neighborhoods, and there is less green space than in other Moscow neighborhoods. The neighborhood is historically working class and a classic example of a Soviet housing bloc (Akhmetzyanova 2015).

In September 2009, the Prefect of Moscow's Southeastern Administrative District met with the head mufti of the Russian Council of Muftis, Ravil Gaynutdin, and the Rector of Moscow Islamic University, Marat Murtazin, to discuss the construction of a mosque on Volga Boulevard in Tekstilshchiki (Sedakov and Djemal 2010). The Russian Council of Muftis received official permission from the Moscow mayor's office to build a mosque at the corner of Volga Boulevard and Saratov Street (Figure 1). This site was selected by Moscow city planners and was on land

Figure 1. The proposed site of mosque construction on Volga Boulevard, an undeveloped lot with above-ground pipelines. Fall 2013. Photo by author.

owned by the city. In April 2010, the site was zoned by the Moscow Department of Land Resources for religious building construction, and the cadastral passport to build the mosque was issued in July 2010 (*Interfax Religion* 2010). The proposed construction site was what Muscovites call *pustin*, a term that refers to empty space, such as deserts, forests, or, in cities, vacant lots. Volga Boulevard has a lack of development, and contains grass, trees, old construction materials, and above-ground pipelines (Figure 2). Makeshift sidewalks and small concrete bridges go over pipelines that transect the space. Large apartment buildings surround Volga Boulevard, older 1960s 12-story buildings to one side and post-Soviet 23-story high-rise buildings to the other.

Notably, another national Muslim organization, the Central Spiritual Board of Muslims of Russia, had applied for over eight years to build a mosque in the same location and had continuously had their application rejected for claims of mishandled paperwork (*Portal-Credo.ru* 2010). The success of the Russian Council of Muftis in receiving permission to build their mosque shows the close political relationship of Gainutdin, the leader of the Council of Muftis, to the Moscow city government. However, the proposed mosque project faced opposition from nationalists, an anti-immigration group, and an environmentalist group, *Moj Dvor* (My Courtyard). The contestation over the site shows how even *pustin* matters in neighborhood identity and the differing access of different groups to neighborhood spaces.

Figure 2. *Pustin*, or undeveloped land, in Tekstilshchiki. Fall 2013. Photo by author.

Mosque construction debates in Europe frequently occur in cities with large migrant populations. Debates about mosque construction often center on two issues of democracy: Muslim group representation in public space and the ability of locals to participate and influence urban planning and municipal decisions (Stoop 2012). In Europe, these mosque construction debates are often co-opted by nationalists who argue that mosque construction will bring the "clash of civilizations" to neighborhood settings, as seen in Stoop's (2012) analysis of mosque construction in the German cities of Cologne and Duisburg and Astor's (2012) analysis of mosque construction in Catalonia. The co-option of mosque construction debates by nationalists results in the framing of Muslim and neighborhood rights as in opposition to each other (Gale 2004; Cesari 2005; Stoop 2012). Neighborhood rights often coalesce under "Not in my backyard" (NIMBY) demands, which portray mosques as disruptive to neighborhood identity and to the everyday lives of locals.

As in the European case studies, discourses of NIMBY-based environmental activism and anti-Muslim nationalism frame the campaigns against mosque construction in Tekstilshchiki. Although the proposed site was an undeveloped lot, citizen action blocked the proposed construction of the mosque. The Russian All-National Union and the Movement Against Illegal Immigration joined the campaign against the mosque construction and asked the local neighborhood authorities for political support (Sibireva and Verkhovsky 2011). But, an environmental social movement called *Moj Dvor* (My Courtyard) organized most activism against the mosque. The name of the group is indicative of their NIMBY platform: a *dvor* is the shared outdoor space for those living in Russia's apartment buildings. Courtyards are sites for social interaction and community building. They often have picnic tables, benches, exercise equipment, and playgrounds. *Moj Dvor* relied on discourses of concerns for environment couched in concerns for public health and Moscow's green space to protest mosque construction. According to the group's leader, *Moj Dvor* was in principle not against a mosque, but rather pro-park (Interview A, May 2013). The group argued that the chosen site was one of the few green spaces in the neighborhood, and hence a poor location for a mosque. One of *Moj Dvor's* protest actions against mosque construction involved gathering locals to plant trees in the *pustin,* signaling their desire to transform the land into a park. They planted 23 maples, ash, and lindens, as well as posted a sign stating, "There will be a park." (Lebedeva 2010).

Moj Dvor's tactics of using NIMBYism were especially effective in Moscow. Although Russia's low levels of NGO activity have been viewed as signs of a weak civil society, research shows that place-based opposition to redevelopment is a growing form of political participation in Moscow (Argenbright 2016). Residents are increasingly opposing infill redevelopment and other projects which they view as disruptive to their neighborhoods. Within this environment of increased urban activism by citizens, *Moj Dvor* was successful in galvanizing local support. According to *Moj Dvor* 9616 locals signed their petition against the mosque, addressed to Russia's then-President Dmitry Medvedev. In an interview, *Moj Dvor* activists stated that the petition included every second or third resident living in the apartment buildings surrounding the proposed construction site. The group stressed the multi-ethnic nature of the signees, stating they included Tatars, an ethnic group historically associated with Islam (Interview A, May 2013).

From interviews with residents and analyses of local newspapers, two NIMBY explanations rooted in a sense of ownership emerged for why locals were drawn to *Moj Dvor's* cause. First, *Moj Dvor* rallied locals around a shared sense of ownership over the *pustin.* Opponents to the mosque claimed that local officials had promised that more urgent infrastructure would be built in the area, such a park, hospital, or a sports complex. For example, a local resident expressed her frustration with the decision of the planning office, stating,

> I care and respect other people's rights. But, this spot was poorly chosen. We have been trying to get a hospital and school built in the neighborhood for years. We need this space for other things to benefit more people, not for a mosque which will only benefit some. (Interview C, March 2013)

This statement shows how the debate over mosque construction in Tekstilshchiki was connected to the increased sense of ownership and growth of participatory democracy at the neighborhood level.

The second rationale for increased residential activism is based in the growth of privatized ownership. In the Soviet era, the state- controlled housing, and Soviet citizens were generally assigned housing according to their residency, as

POLITICAL GEOGRAPHIES OF THE POST-SOVIET UNION

demarcated on their *propiska,* or domestic passport. According to a *Moj Dvor* representative, the privatization of residential property ownership explains why some Muscovites are likely to participate in local movements.

> People from the older buildings got their apartments in the 1960–1970s, then they died and passed on their apartments. It means that people who live in them right now made no effort to get their flats. The people who live in the new buildings (the high rises) worked and had to display initiative on order to purchase an apartment. And those people were the most active. Therefore, a person who works and earns money is more responsible for his future, including city planning. (Interview A, May 2013)

Because the new apartments were purchased by owners in the post-Soviet capitalist economy rather than parceled out by the state, this interviewee proposed that those owners were more likely to participate in citizen activism. As part of their platform against the mosque, *Moj Dvor* contends that since the developers of the 23-floor buildings adjacent to the green space of Volga Boulevard did not show a mosque on the plan for the neighborhood, new property owners were deceived. The group states that a new mosque would hurt property values (Nash Manifest (Our Manifest) 2010).

The issue of property rights raises the question of why the proposed mosque was not seen as a valuable community addition, and why *Moj Dvor* combined their NIMBY platform with anti-Muslim propaganda. On their web page and in social media they circulated mass media stories highlighting issues of overcrowding at the four other mosques in Moscow, violence and police arrests of Muslims. *Moj Dvor* also shared articles on the intention of Saudi Arabia to sponsor more mosques in the city (mecheti.net). One scare tactic they used was to highlight the intention of Saudi Arabia to sponsor mosque construction and to associate Saudi mosque construction with Wahhabism. The issue of Saudi mosques being constructed in Moscow is especially controversial: the Russian state and media treat Wahhabism as a foreign violent jihadi movement responsible for the Salafist separatist movements in the North Caucasus and Second Chechen War. This sensationalist account elides Wahhabism with Salafism, instead of treating Wahhabism as a historically fundamentalist conservative movement and state religion of Saudi Arabia and Salafism as a modern conservative reform movement in Islam (Knysh 2007). *Moj Dvor* actively works to make Moscow's Muslim community look vulnerable to foreign radicalism. According to a representative of the Russian Council of Muftis, such news articles are evidence of the intent of the state to use government-controlled mass media to stir up Islamophobia (Interview B, July 2013). *Moj Dvor* merged locals' environmental concerns with negative representations of Muslims to further their agenda. By associating the potential mosque with representations of Wahhabism, arrests, and overcrowding, the environmental group portrayed a clash of civilizations emerging in the sleeping districts.

According to human rights agencies such as the SOVA center, a non-profit organization that conducts research on social issues in Russia, *Moj Dvor* served as a xenophobic front led by ultra-right former neo-Nazis (Sibireva and Verkhovsky

2011). Moscow's muftis agree with human rights agencies that the controversy caused by the mosque construction was not rooted in NIMBY concerns over green space but rather motivated by identity politics and latent Islamophobia. In a sermon at Cathedral Mosque, the head of the Council of Muftis, Ravil Gaynutdin, stated,

> In the subconscious of the modern citizen of the titular nation, in Russia, there is fear – fear one morning of waking up in a Muslim country, where after the word "Muslim" stands for the lack of knowledge about Islam as a monotheistic religion. (Shishlin 2011)

He believes that the media carried out a smear campaign against the mosque through exacerbating Islamophobic sentiment.

The leader of the *Moj Dvor* movement denied that his movement was motivated by xenophobia, stating,

> People tried to label us as xenophobes. "Phobos" means fear. We don't have fear, we have common sense and we are trying to show it to other people. And our opponents can scream whatever they want. I don't think we are xenophobes. (Interview A, May 2013)*Moj Dvor's* appeal to common sense shows how NIMBY discourses intersect with anti-Muslim rhetoric – by stating that a mosque is incompatible with daily activities of communities, the conversation of how mosques can enhance life in Moscow's suburbs is foreclosed. "Common sense" rhetoric pits local rights against Muslim rights in regard to access to public space.

The mosque's supporters were also active; a Tatar leader, Gayz Yambayev, led a picket with a Tatar organization, the Tatar Cultural National Autonomy (RTNKA) (Jalilov 2010). The supporters of the mosque refuted NIMBY arguments against the mosque. They attested that a mosque would enhance property values and the sense of community in the neighborhood, rather than be detrimental. Representatives from the RTNKA also stated that *Moj Dvor* mislead their supporters against the mosque, arguing that the chosen site was not green space or an appropriate location for major development. *Moj Dvor's* claim of promised infrastructure construction from the local municipal government was an exaggeration, as examinations of previous applications for construction in the area shows that local officials thought that there were too many communications and gas pipelines in the *pustin* to allow for major construction projects (Sedakov and Djemal 2010).

Instead, proponents of the mosque argue that the success of *Moj Dvor* was due to widespread xenophobic feelings and a lack of support for Muslims from the Moscow Government (Interview E, October 2013). They believed local officials heeded complaints because it allowed them to block the constitutional right of Muslims to practice their religion while making it seem like a democratic decision. Muslim proponents of the mosque state that protests against construction projects are nothing new, and many planning projects have been carried out despite protests, including those for shopping centers and in churches (Interview D, July 2013).

Moj Dvor and local residential opposition movements were successful, and the plan for the mosque project was canceled on the 23rd of November 2010. Vladimir Zotov, a representative of the Southeastern Administrative District, stated that the

Figure 3. A dog park constructed in Tekstilshchiki in 2013 at the site of the proposed mosque. November 2013. Photo by author.

assigned place of construction in Tekstilshchiki was a mistake on the part of the Moscow Department of Architecture. In a press release he stated, "A new mosque will be built, but not in this place" (Shishlin 2011). He promised that instead of a mosque, the vacant site would become either a park or something else the community needed, such as a kindergarten. (2011).

Over time, Zotov proved to be right about the park. Following the citizen activism in Tekstilshchiki, I visited Volga Boulevard in 2013 to see what had developed in the site of proposed construction. I noticed that nothing had yet been built on the site, and that the area was still *pustin*. However, just because it was technically undeveloped did not mean it went unused. The crumbling paths were busy with people walking their dogs and with children. There were a few shady trees along the boulevard, under which I saw many picnics as well as men drinking alone or in groups. While visiting the site one evening in spring 2013, I noticed a sign about a protest against further proposed construction in the area. Despite local opposition, the site on Volga Boulevard once designated for mosque construction became a dog park in fall 2013 (Figure 3). Upon a return visit in summer 2015, playground equipment had been installed in the space, marking its transition from *pustin* to park.

Figure 4. An apartment building located near the site of proposed mosque construction in Mitino. November 2013. Photo by author.

Mitino

In 2012, another proposed plan to build a mosque in Moscow's neighborhoods was canceled due to activism. On 6 September 2012, the Moscow Urban Planning commission announced plans to construct a mosque and Muslim cultural center in Mitino, sponsored by the United Islamic Congress of Russia, headed by Shavkat Avyasov. Mitino is a suburban Moscow district on the fringes of the city, located in northwest Moscow. It is one of the most prestigious high-density districts in the city, composed of 13 neighborhoods (Vendina 1997; Nozdrina 2006; Akhmetzyanova 2015). Mitino had a population of 188,342 in 2016 (Rosstat 2016b) and is mainly residential, consisting of 1990s-era high-rises. The residents tend to be wealthier than those in Tekstilshchiki (Nozdrina 2006). The site for proposed mosque construction within Mitino was Baryshiha Street near apartment building 57, a 17-story high-rise. This site was located in an undeveloped, wooded area at the fringe of the city limits (Figure 4), and plans were approved for a building up to 35 meters high (or about 10 stories or 100 feet).

Once the mosque construction plan was publicized, opponents quickly mobilized and organized an unauthorized rally. In Moscow, locals must obtain permits to carry out protests; however, this protest occurred despite the lack of permits. More than 1000 people attended the rally held on 19 September 2012. The size of the protest was notable, as the estimated attendance of 1000 or more that gathered to protest the mosque was much larger in comparison to other gatherings that occurred in the neighborhood (Malashenko 2014). This level of popular dissent was unprecedented, and according to human rights organizations, a fabrication by nationalists. Human rights agencies believed that the protest attendees were paid to protest and bussed in from other districts by xenophobic nationalist groups, rather than a locally organized response (Interview G, July 2013). The protest was effective, as the day after, on 20 September 2012, the Moscow Urban Planning and Land Commission canceled the project.

According to a participant, the social movement in Tekstilshchiki inspired the protest in Mitino. He stated:

> The events in Tekstilshchiki showed citizens that they could resist, struggle, and win, and they also showed the authority that citizens didn't want to live with guest workers and share their land with them. After Tekstilshchiki the same situation happened in Mitino when the construction of a new mosque in their neighborhood was announced. And a great meeting of 2000–3000 people showed they didn't want any mosque to be there (Interview C, May 2013).

Unlike the anti-mosque movement in Tekstilshchiki, however, protestors in Mitino did not have to go to local meetings of the district government or collect signatures to achieve their desired result – Mitino protestors' demands were quickly met. Although the protests in Tekstilshchiki and Mitino were not led by the same organizations, they are both examples of increased participatory citizenship in Moscow.

As in Tekstilshchiki, the potential mosque in Mitino was portrayed as incompatible to local lifestyles. As in the protests against mosque construction in Tekstilshchiki, local rights and NIMBY concerns rooted in ownership and the disruption of daily life were voiced in the Mitino protest. Anti-mosque protestors distributed leaflets throughout the neighborhoods in Mitino to draw attention to the proposed mosque, stating, "Have residents been asked about the Construction of a Mosque in Mitino?" The mosque and Islamic center's allocated height of 35 meters alarmed protestors, as they said that they were concerned that the mosque and cultural center would be massive and out of line with neighborhood characteristics. Residents expressed frustration at the decision to construct a mosque instead of other infrastructure projects, such as skating rinks, swimming pools, or a woman's health clinic, showing that they felt disconnected with planning decisions and that in turn, they had a right to question them and be part of the planning process.

In concert with these local concerns over how a mosque would detract from the quality of life in Mitino, a city official stated that the main reasons the project was shelved is because increased traffic caused by the mosque might obstruct the nearby fire station. He believed that the chosen lot was ill-suited, from a logical perspective, for a mosque (Interview F, November 2013). He explained that mosques were overcrowded, and that if there was an emergency, the crowds could impede firemen. Thus, the mosque was portrayed as a safety threat instead of as an enhancement to neighborhood infrastructure. Rather than seeing the value in providing worship space to Muslims, any potential overcrowding was viewed negatively.

Differing from *Moj Dvor*'s couching of any Islamophobic sentiment into environmental or NIMBY concerns, protestors in Mitino voiced anti-immigrant rhetoric in their anti-mosque campaign. Four protestors were arrested for voicing nationalist sentiment, only one of which was a local. Attendees directly linked the issue of mosque construction with the concern over drawing immigrants into the district that would create ethnic enclaves, live in overcrowded conditions, and lower property values (Vinogradov 2012). As property values and rents in Moscow are high, labor migrants in Moscow often share apartments. Homeowner's associations look out for high rates of utility usage in units to find out if their apartments are hosting illegal labor immigrants, and vigilante nationalist groups often look for densely occupied apartments to raid and report to police (Krainova 2013).

As in the Tekstilshchiki plan, proponents of mosque construction included Muslim organizations, such as the Russian Council of Muftis, and human rights organizations, such as the SOVA Center for Information and Analysis. Muslim leaders argued that concerns of increased Muslim presence in the neighborhood were unjustly linked with terrorism and that Moscow's mayor should intervene on behalf of the Muslim minority to support mosque construction (Fagan 2012a). However, since the cancelation of the mosque plan in Mitino, Moscow's mayor, Sobyanin, has spoken strongly against the construction of new mosques in several public forums. In an interview with Echo of Moscow, an Internet radio outlet devoted to discussing political and social issues, Sobyanin stated that his stance against mosque construction is based on Muscovite opinion:

> The Muslim religious holidays are mostly attended by worshipers who arrive from the Moscow Region and other regions of the country. From 60 to 70% of them are outsiders. We cannot provide for all those who need something. Muscovites are becoming irritated by people who speak a different language, have different customs, and display aggressive behavior. This is not a purely ethnic issue, but it is connected with some ethnic characteristics (Venediktov 2013).

Sobyanin's adoption of the view of mosques as immigrant spaces was timely and politically motivated; indeed, 2013 was a significant year for Sobyanin. He had been appointed Moscow mayor in 2010 after then-Russian President Medvedev fired former Mayor Yuriy Luzhkov by presidential decree for loss of trust. The firing of Luzhkov and appointment of Sobyanin is illustrative of managed democracy in national Russian politics, as the role of Moscow mayoral is an elected position. In summer 2013, Sobyanin stepped down from his mayoral appointment in order to allow a popular election for mayor. The goals of the fall 2013 election were to increase political legitimacy for Sobyanin and the United Russia party, which is aligned to Putin, while at the same time to allow some political action for the opposition. Many political parties challenged Sobyanin and United Russia in the election, most notably opposition leader Alexei Navalny, who ran on an anti-corruption platform (Orttung and Waller 2013). According to official results, Sobyanin won the popular election, earning 51.37% of the vote, while Navalny earned 27.24%. The turnout rate was 32%. Sobyanin is regarded by most Muscovites as having a positive influence on the city and allowing for more democratic participation at the local level as compared to his predecessor, Yuri Luzhkov. (Levada Center 2013).

The year 2013 was also the height of anti-immigrant sentiment in Moscow, with residents citing migrants from the southern republics of the former USSR and North Caucasus as the main issue facing the city (Levada Center 2013). Ethnic tensions were illuminated in the 2013 anti-immigrant riots and migration raids in another sleeping district, Biryulevo, in southeastern Moscow after a 25-year old Azeri immigrant murdered a Russian. As evidenced in his interview with Echo of Moscow (Venediktov 2013), Sobyanin painted opposition to new mosques as democratic, as the will of Muscovites to protect their limited territory from outsiders who do not belong. He also drew from the discourses used in anti-mosque protests to express the lack of mosques as a problem of shortage – as if Moscow was unable to balance the needs of minorities with the needs of majorities. Muscovite Muslims are purported to have enough space to accommodate their needs, even though they too experience overcrowding and voice a desire for a new mosque and easier access (Group Interview, November 2017). According to the logic of Sobyanin, as the Muslim population is not permanent, neither is the spatial solution. Because a majority of Moscow's Muslims were considered to be transient economic migrants, city officials cited no need for new mosques.

No new construction plans have been announced since the cancelation of the mosque plan in Mitino in 2012. Although Moscow's central mosque was expanded in a renovation project and reopened in 2015, this expansion did not meet the needs of the growing number of Muslims in the city. Imams and Muslim leaders have unsuccessfully advocated for new mosque construction projects in the city's neighborhoods (Interview E, October 2013). They attest that mosques can provide a place for immigrants to learn about traditional Russian Islam, and several mosques offer Russia language lessons (Interview H, September 2013; Interview P, October 2013). The shortage of mosques has not prevented other types of Muslim spaces from forming in the city. Moscow's Muslim communities have been active in creating other spaces of belonging, from community centers, educational organizations, and charities, to new halal businesses.

"200 Churches" program: building Russian orthodox churches in Moscow's neighborhoods

Whereas Moscow's Muslim community has been unsuccessful at advocating for and constructing new mosques in Moscow's neighborhoods, the Russian Orthodox Church has been comparatively successful. In 2010, the Russian Orthodox Church initiated a plan to build 200 Orthodox churches throughout Moscow, entitled "200 Churches" (although they have contemplated changing the name in order to expand the number of new churches; Novosti stroitel'stva (Construction News), n.d.). The suburbanization and expansion of "200 Churches" is Moscow's governmental spatial strategy to incorporate Russian Orthodoxy into the everyday lives of its inhabitants (Sidorov 2015). Because the Soviet Union had a religious policy of atheism, religious buildings were not part of the Soviet Union's urban design. Post-Soviet Moscow therefore inherited a religious infrastructure where churches were clustered in the city center and sparse in the outlying *spalnie rayoni,* where most people live (2015). The "200 Churches" program intends to construct churches in the *spalnie rayoni*, so that, according to priests, "Grandmothers can take their grandchildren to church every day, without having to ride the metro" (Interview J, November 2013). Although churches are overcrowded on Easter, they are not overcrowded on other holidays or ordinary Sundays. Thus, the program is not so much in answer to a demand as it is to aspirations to increase spirituality of ordinary Muscovites (Interview K, August 2015).

The first church built for the project was symbolic, located in front of the Dubrovka theater – the site of the 2002 siege of 912 hostages by radical terrorists from the Caucasus that resulted in 130 deaths (Sidorov 2015). Most of the churches, however, are not symbolic and are meant to integrate into the everyday landscape of Moscow. Often, these are temporary wooden structures located on empty parcels in the *spalnie rayoni*: 41 churches have been built, while 40 are under construction (Novosti stroitel'stva (Construction News), n.d.). While some structures consist of a small building to provide room for service, others include small *lavka*, or shops selling icons, candles, and religious books. In some cases, the wooden structures are being replaced or complemented by larger, more permanent structures, as seen in Figure 5. The church, located in Moscow's southwest district Novye Cheryomushki on the left in the figure is being complemented with a larger stone structure, on the right.

In an address to the supervisory board and working group of the Foundation to Support Church Construction in Moscow, Mayor Sobyanin explained the challenges of finding proper locations for church construction:

> Your Holiness, colleagues, the city has been implementing the church construction program for 3.5 years – and, frankly, the achievements during this period surpass our expectations because, to find 200 sites in Moscow, a city whose construction density exceeds that of all other similar European cities, is not a trivial task, to put it mildly. It is twice as hard to find a location for constructing Orthodox churches that would not be criticized by city residents. Yet, together with you and municipal deputies, vast work has been done and most sites were selected. As regards a large number of sites, as you have mentioned, alternative options were proposed and they were accepted by residents. We chose the locations that neighborhood residents agreed upon, with churches blending into the local urban landscapes and complementing the city both in architectural and beautification terms (Press Service of the Patriarch of Moscow and All of Russia 2014).

Sobyanin's statement here shows similarities and differences in his publicized views against mosque construction. Sobyanin recognizes the hard work of finding places

Figure 5. A concrete church is erected next to a wooden temple built as part of the "200 Churches" program in southwest Moscow. Photo by author.

to construct religious buildings in Moscow's densely populated neighborhoods, but he acknowledges how churches can become part of these neighborhoods. The introduction of Russian Orthodox churches into Moscow's *spalnie rayoni* has been met with some pushback from residents, but has also resulted in compromise. In 35 out of 70 NIMBY-like cases against new church construction, locals have met with success in requesting the relocation of construction sites (Sidorov 2015). The churches are intended to be quotidian aspects of the urban fabric, complementing existing neighborhood esthetics. Mosques, however, were portrayed as foreign spaces.

One project in "200 Churches" resulted in overt citizen action and extended protest, thereby interrupting Sobyanin's narrative of harmony between municipal authorities and local communities. The proposed construction of a Russian Orthodox Church devoted to Our Lady of Kazan in Moscow's northern Babushkinsky District elicited protest movements and citizen action. The planned location in Torfyanka Park was met with resistance, as locals felt uninformed and unhappy with the site selection. On 30 April 2013, Sobyanin signed a decree to build the church complex in Torfyanka Park as part of the "200 Churches" program. The project design published on the webpage of "200 Churches" depicted the proposed church as having space for 500 worshippers. The local residents appealed to the Babushkinksy District court, stating that the public hearing for the project had

Figure 6. The people's picnic: The makeshift campsite erected by locals to block church construction in Torfyanka Park. August 2015. Photo by author.

not been held legally, as it was not published in print media and was a violation of their rights to be heard. Their appeal to the district court over the results of the public hearing was denied in February 2015, and in April 2015, the Moscow City Court sent the case back to the local court.

On 18 June 2015, residents physically halted the start of construction efforts. To oppose the construction, residents set up camp in the proposed site to block construction equipment (Figure 6). One local described the action stating, "One day we noticed construction trucks in the area and grew concerned. We banded together and locked our arms together, forming a human chain" (Interview M, August 2015). As the construction began on a weekday morning, this human chain consisted of grandmothers, pensioners, and stay-at-home mothers. Other locals joined in efforts to construct an on-site protest camp.

"We surprised ourselves. Who knew that we could stop them?" stated an elderly female resident as she recollected the events of that day (Interview N). "We set up a camp and collected signatures to protest the construction. There is always someone at the camp now ... it has been a good place to get to know our neighbors." The camp was makeshift, consisting of umbrellas and tables gathered by residents. They displayed banners with the name of their social movement, "For Torfyanka Park." The protest camp lasted for a period of four months, starting in June 2015. The local movement received political support from two opposition political parties – the Communist Party and Yabloko, the social liberal democratic party.

One interviewee explained, "We are not against churches, but we are against churches in this park. It is a form of NIMBYism, is that the term? We do not need a church here. It can go elsewhere" (Interview N, August 2015). Babushkinsky consists of smaller *khrushovki*, Soviet-era apartment buildings with between 5 and 9 floors, instead of the mega-high rise apartments that are a hallmark of late and post-Soviet construction projects. Because of this, locals consider the neighborhood cozier

Figure 7. Camp devoted to constructing the church in Torfyanka Park. August 2015. Photo by author.

and more comfortable than others in Moscow. Many flats had been in families for two generations.

The protest camp sat next to an area cordoned off with fencing, with signs displaying the official city plan to construct a new church. Within this fence stood another protest camp (Figure 7). These protestors were from the Orthodox movement "*Sorok Sorokov*." This symbolic name refers to the Russian Orthodox word for parish, and it originates from a Russian idiom: "There are *sorok sorokov* (a multitude) of churches in Moscow" (Gumerov 2009). This idiom refers to the idea of Moscow as a Russian Orthodox center. The movement *Sorok Sorokov* is composed of Orthodox activists who support and sponsor the construction of Russian Orthodox churches in Moscow, many of whom are also nationalists (Interview O, August 2015). The *Sorok Sorokov* protestors encamped in Babushkinsky at the time of my visit included five members: a pensioner and her son, two non-local Muscovites, and a Cossack. (Cossacks are a Russian Orthodox, ethno-cultural group who historically served the tsar). Modern Cossacks function as a group of militant patriotic morality police (Markowitz and Peshkova 2016). They defend conservative Russian Orthodox values and target enemies of the current regime such as anti-corruption protest leader Alexei Navalny and Pussy Riot. Pussy Riot is a feminist performance group of protest artists based in Moscow, and one of their performance pieces was as a punk band performing in the Cathedral of Christ the Savior Church in Moscow in 2012. Their performance piece at the Cathedral of Christ the Savior Church was called "Punk Prayer–Mother of God, Chase Putin Away!" The song drew attention to the growing relationship between Putin and the Russian Orthodox Church, asking Mary, Mother of God, to banish Putin from the church. Three members of Pussy Riot were arrested for hooliganism during the live performance and tried for "hooliganism motivated by religious hatred." Two, Mariya Alyokhina and Nadezhda Tolokonnikova, were found guilty and sentenced to two years in Russian prison. One effect of their trial was the subsequent 2013 criminalization of insulting religious feelings in Russia, including the public desecration of all religious books or objects.

The Cossack protestor at Torfyanka Park explained that his duty was to protect the church, even though he was not from the Moscow area. He stated that he was paid daily to stay at the site, but did not indicate who paid him. *Sorok Sorokov* and the proponents of the church felt that they were providing a service to the locals. The pensioner stated, "Moscow is a Christian city and needs more churches. This is horrible. A church will be built here." Another protester stated, "Yes, another location would be suitable. One church can be built here, and another in the new location as well. Even better." (Group interview, August 2015).

The different protests movements involved in the debate over church construction had negative views of each other. "For Torfyanka Park!" was viewed by *Sorok Sorokov* as disruptive, and they viewed "For Torfyanka Park!" as a foreign-funded movement. Although the "For Torfyanka Park!" movement did not engage in any anti-Russian Orthodox sentiment in their protest movement, *Sorok Sorokov* disagrees. In an interview with *The Moscow Times*, a representative from *Sorok Sorokov* stated that foreign foundations seeking to undermine the influence of the Russian Orthodox Church finance protest movements against church construction in Moscow (Litvinova 2015). Ordinary public spaces are at the forefront of identity politics. For the *Sorok Sorokov* protestors, Moscow neighborhood church projects are tied to questions of national identity and subject to foreign intervention.

Locals were able to block the proposed construction, and the church plan was eventually relocated to a different space. In 2015, the Babushkinsky municipal court appealed the city's decision to build a church in the park and held an online vote of local residents to choose the new location. The new church location is near a train station and more accessible by public transportation, although *Sorok Sorokov* believes the former location was more accessible to locals especially the elderly (*RIA Novosti* 2016; TASS 2016).

The outcomes of the protest against the church for different groups are significant indicators of the range of citizen activism permitted in Russia's managed democracy. Despite their success in relocating the church, the activists against church construction have experienced harassing reprisals from the state and the Russian Orthodox Church. In November 2016, the Moscow police raided two of the activists' apartments in response to tips that they were radicals and had violated criminal laws against insulting religious feelings. These laws were passed as a result of the Pussy Riot show trials and seen as an example of the installation of social conservatism into Russian law (Gorbunova and Ovsyannikova 2016). The activists were not charged with the crimes, but their homes were raided and computers and phones searched. The treatment of these activists shows the contours of Russia's managed democracy: Although the activists were engaged in legal forms of protest, they still were treated like criminals.

Discussion

Neighborhoods are often contentious sites in Russia's managed democracy, and local politics are tied to questions of civil society and identity politics. In neighborhoods, the interactions between activists, locals, municipal governments, and religious leaders create priorities, values, and norms. Religious construction projects in the neighborhoods have generated new forms of urban activism. In regard to civil society development, these three case studies of protests against religious construction projects in Moscow supports research showing that although NGO activity in Moscow is weak, neighborhood associations and local social movements are growing forms of activism (Argenbright 2016). Even though current Russian laws curtail the activities of foreign-sponsored NGOs, drawing criticism in international freedom reports, these laws have not stopped the emergence of local forms of participation in Russia's authoritarian democracy (Crotty, Hall, and Ljubownikow 2014). In response to the loss of Soviet public spaces and infill development, NIMBYism and local activism are new forms of participatory democracy in Russia (Argenbright 2016). These local social movements against religious construction are notably successful forms of opposition within the contexts of Russia's managed civil society. Whereas larger protests make anti-regime demands against corruption, they have not been successful at altering policies, with the caveat that achieving public representation for oppositional viewpoints is an achievement in its own right. The movements against church and mosque construction were smaller than the Navalny-led anticorruption protests in 2012 or 2017, but they have been effective in fulfilling their agendas. Protests against mosques were motivated by and incorporated xenophobic and Islamphobic elements, while the protest against the church maintained a NIMBY anti-development position.

In regard to identity politics, these case studies show the spatial politics of Russia's religious revival. Within Moscow, debates over creating new religious sites in neighborhood public spaces have clarified the strong role of the Russian Orthodox Church in cultural politics. Although all three protest movements discussed in this article were successful in canceling plans against church and mosque construction, Russian Orthodox church construction continues in Moscow's neighborhoods, whereas new mosque construction has not occurred. These case studies show that protests against mosques were quicker in stopping construction than those against church construction. In the case of Mitino, anti-mosque protestors were successful almost overnight. This experience contrasts to that of the protest movement in Torfyanka Park, where citizens protested against church construction for over a month before the municipal government met their demands. Reprisals from the state against anti-church protesters show that the Moscow Government valorizes Russian Orthodox churches. Moscow's promotion of Russian Orthodox church construction is indicative of the rise of moral politics in Russia, wherein the Russian Government has securitized the Russian Orthodox Church and turned religious values into national security and identity issues (Sharafutdinova 2014; Østbø 2017). Mosques are meanwhile excluded from the everyday public spaces of Moscow's neighborhoods through their xenophobic portrayal as disruptive and disorderly immigrant spaces. The proliferation of Russian Orthodox churches in Moscow's neighborhoods illustrates the everyday spaces of authoritarianism in Russia's managed democracy, which protects the rights of the influential Russian Orthodox Church while curtailing those of Moscow's minority Muslim communities.

Disclosure statement

No potential conflict of interest was reported by the author.

Funding

This work was supported by Social Science Research Council [Eurasia Program Dissertation Award]; U.S. Department of Education [Fulbright-Hays Doctoral Dissertation Research Abroad Fellowship]; and the Department of Geography at the University of Colorado, Boulder [Jennifer Dinaburg Memorial Research Fellowship].

References

Akhmetzyanova, Leyla. 2015. "Modeling Income-Based Residential Segregation in Moscow, Russian Federation." Unpublished thesis, Umeå University.

Argenbright, Robert. 2016. *Moscow under Construction: City Building, Place-Based Protest, and Civil Society*. London: Lexington Books.

Associated Press in Moscow. 2017. "Protests in Moscow at Plan to Tear down Soviet-Era Housing in Affluent Areas." *The Guardian*, May 14. https://www.theguardian.com/world/2017/may/14/people-protest-moscow-plans-to-tear-down-housing.

Astor, Avi. 2012. "Memory, Community, and Opposition to Mosques: The Case of Badalona." *Theory and Society* 41: 325–349.

Cesari, Jocelyne. 2005. "Mosque Conflicts in European Cities: Introduction." *Journal of Ethnic and Migration Studies* 31: 1015–1024.

Collier, Stephen J. 2011. *Post-Soviet Social: Neoliberalism, Social Modernity, Biopolitics*. Princeton, NJ: Princeton University Press.

Cornwall, Marie, Stan L. Albrecht, Perry H. Cunningham, and Brian L. Pitcher. 1986. "The Dimensions of Religiosity: A Conceptual Model with an Empirical Test." *Review of Religious Research* 27: 226–244.

Cracraft, James, and Daniel Bruce Rowland, eds. 2003. *Architectures of Russian Identity: 1500 to the Present*. Ithaca, NY: Cornell University Press.

Crotty, Jo, Sarah Marie Hall, and Sergej Ljubownikow. 2014. "Post-Soviet Civil Society Development in the Russian Federation: The Impact of the NGO Law." *Europe-Asia Studies* 66: 1253–1269.

Dzusati, Valery. 2014. "Mosque Construction in Stavropol Sparks Debate over Role of Islam in Region." *Eurasia Daily Monitor* 11, June 18. https://jamestown.org/program/mosque-construction-in-stavropol-sparks-debate-over-role-of-islam-in-region-2/.

Fagan, Geraldine. 2012a. *Believing in Russia: Religious Policy after Communism*. London: Routledge.

Fagan, Geraldine. 2012b. "Russia: No More Mosques outside "Muslim Areas'?." *Forum 18 News Service*, October 3. https://wwrn.org/articles/38232/.

Fish, M. Steven. 2005. *Democracy Derailed in Russia: The Failure of Open Politics*. Cambridge: Cambridge University Press.

Flikke, Geir. 2016. "Resurgent Authoritarianism: The Case of Russia's New NGO Legislation." *Post-Soviet Affairs* 32: 103–131.

Gale, Richard. 2004. "The Multicultural City and the Politics of Religious Architecture: Urban Planning, Mosques and Meaning-Making in Birmingham, UK." *Built Environment* 30: 30–44.

Gorbunova, Yulia, and Anastasia Ovsyannikova. 2016. "In Russia, Thou Shalt Not Disagree with the Orthodox Church." *Human Rights Watch*, November 18. https://www.hrw.org/news/2016/11/18/russia-thou-shalt-not-disagree-orthodox-church.

Gumerov, Iov. 2009. "Chto oznachayet vyrazheniye «Sorok Sorokov »? [What Does the Phrase 'Soros Sorokov' Mean?]." Accessed March 27, 2017. http://www.pravoslavie.ru/33084.html.

Halbach, Uwe. 2001. "Islam in the North Caucasus." *Archives de Sciences Sociales Des Religions* 46: 93–110.

Harris, Steven. 2013. *Communism on Tomorrow Street: Mass Housing and Everyday Life after Stalin*. Washington, DC: Woodrow Wilson Center Press.

Holland, Edward C., and Meagan Todd. 2015. "Islam and Buddhism in the Changing Post-Soviet Religious Landscape." In *The Changing World Religion Map*, edited by S. D. Brunn, 1515–1530. New York: Springer.

Interfax Religion. 2010. "Prefektura YUVAO predlagayet meru Moskvy postroit' mechet' na drugom zemel'nom uchastke [Prefecture of the southeastern district asks the mayor of Moscow to build a mosque on another land plot]." October 21. http://www.interfax-religion.ru/print.php?act=news&id=37889.

Ioffe, Grigory, and Zhanna Zayonchkovskaya. 2011. "Spatial Shifts in the Population of Moscow Region." *Eurasian Geography and Economics* 52: 543–566.

Jalilov, Rustam. 2010. "Zhiteli Tekstil'shchikov zystupili v zashchitu mecheti [Residents in Tekstilshchiki spoke in defense of the mosque]." *Islam News*, October 26. https://www.islamnews.ru/news-27451.html.

Karimova, Liliya. 2017. "Russia's Muslims are as Diverse as Their Experiences." *The Russia File*. Washington, DC: The Kennan Institute. http://www.kennan-russiafile.org/2017/05/25/russias-muslims-are-as-diverse-as-their-experiences/.

Knysh, Alexander. 2007. "Contextualizing the Salafi – Sufi Conflict (from the Northern Caucasus to Hadramawt)." *Middle Eastern Studies* 43: 503–530.

Kong, Lily, and Orlando Woods. 2016. *Religion and Space: Competition, Conflict and Violence in the Contemporary World*. London: Bloomsbury.

Krainova, Natalya. 2013. "Police to Raid Migrants' Apartments Every Friday." *The Moscow times*, October 21. https://themoscowtimes.com/news/police-to-raid-migrants-apartments-every-friday-28764.

Lebedeva, Ekaterina. 2010. "Tekstil'shchikakh, gde dolzhny vozvesti mechet', zasadili derev'yami [The empty space in Tekstilshchiki, where the mosque is to be built, was planted with trees]." *Komsomol'skaya Pravda*, September 28. http://www.interfax-religion.ru/?act=print&div=12010.

Levada Center. 2013. "Moskva nakanune vyborov mera: Polnoe issledovanie [Moscow on the eve of mayoral elections: A full study]", July 17. https://www.levada.ru/2013/07/17/moskva-nakanune-vyborov-mera-polnoe-issledovanie/.

Light, Matthew. 2016. *Fragile Migration Rights: Freedom of Movement in Post-Soviet Russia*. London: Routledge.

Litvinova, Daria. 2015. "Protesters against Church Construction Gain Rare Victory." *The Moscow times*, August 3. https://themoscowtimes.com/articles/protesters-against-church-construction-gain-rare-victory-48718.

Lussier, Danielle. 2011. "Contacting and Complaining: Political Participation and the Failure of Democracy in Russia." *Post-Soviet Affairs* 27: 289–325.

Malashenko, Alexey. 2014. "Islam in Russia: Changes in the Kremlin's Rhetoric." *Russia in Global Affairs*, September 23. http://eng.globalaffairs.ru/number/Islam-in-Russia-17002.

March, Luke. 2010. "Modern Moscow: Muslim Moscow?" In *Russia and Islam: State, Society and Radicalism*, edited by Roland Dannreuther and Luke March, 84–102. London: Routledge.

Markowitz, Lawrence P., and Vera Peshkova. 2016. "Anti-Immigrant Mobilization in Russia's Regions: Local Movements and Framing Processes." *Post-Soviet Affairs* 32: 272–298.

McLoughlin, Seán. 2005. "Mosques and the Public Space: Conflict and Cooperation in Bradford." *Journal of Ethnic and Migration Studies* 31: 1045–1066.

Metcalf, Barbara Daly. 1996. *Making Muslim Space in North America and Europe*. Berkeley: University of California Press.

Mitchell, Don. 2012. *Right to the City: Social Justice and the Fight for Public Space*. New York: Guilford Press.

Nash Manifest (Our Manifest). 2010. *Verum Index*, September 23. https://web.archive.org/web/20141222123004/http://www.verum-index.com/.

Novosti stroitel'stva (Construction News). n.d. *Program to Construct Russian Orthodox Churches in Moscow*. Accessed March 26, 2017. http://www.200hramov.ru

Nozdrina, N. N. 2006. "The Housing Market of Moscow: Development and Area Differentiation." *Studies on Russian Economic Development* 17: 587–595.

Orttung, Robert, and Julian Waller 2013. "Navalny and the Moscow Mayoral Election." *Russian Analytical Digest* 136. https://www.research-collection.ethz.ch/bitstream/handle/20.500.11850/73397/eth-7419-01.pdf?sequence=1&isAllowed=y.

Østbø, Jardar. 2017. "Securitizing 'Spiritual-Moral Values' in Russia." *Post-Soviet Affairs* 33: 200–216.

Pew Research Center. 2017. "Religious Belief and National Belonging in Eastern and Central Europe." May 10. http://www.pewforum.org/2017/05/10/religious-belief-and-national-belonging-in-central-and-eastern-europe/.

Portal-Credo.ru. 2010. Vlasti Moskvy otkazali TSDUM v stroitel'stve mecheti v Tekstil'shchikakh, odnako razreshili eto Sovetu Muftiyev RF [The authorities of Moscow denied TsDUM in the construction of a mosque in Tekstilshchik, but they allowed it to the Council of Muftis of the Russian Federation]. September 23. http://www.portal-credo.ru/site/print.php?act=news&id=79831.

Press Service of the Patriarch of Moscow and All of Russia. 2014.Predstoyatel' Russkoy tserkvi vozglavil chetvertoye zasedaniye popechitel'skogo soveta fonda podderzhki stroitel'stva khramov g. Moskvy [The patriarch of the Russian Church headed the fourth meeting of the Board of Trustees of the Fund for Support of the Construction of the Temples of Moscow]. *Official Website of the Moscow Patriarchate of the Russian Orthodox Church*, April 1. http://www.patriarchia.ru/db/text/3615968.html.

RIA Novosti. 2014. "SMR: Chislo mechetey v RF prevysilo 7 tysyach - eto bespretsedentnyy rost [SMR: the number of mosques in Russia has exceeded 7 thousand–this is an unprecedented growth]", November 17. https://ria.ru/religion/20141117/1033687343.html.

RIA Novosti. 2016. "Vosem'mesyatsev protivostoyaniya na Torfyanke – kto prav? [Eight months of protest in Torfyanka–who is right?]", February 12. https://ria.ru/religion/20160212/1373501203.html.

Robertson, Graeme. 2013. "Protesting Putinism." *Problems of Post-Communism* 60: 11–23.

Rosstat. 2010. "Russian Census 2010. Population Structure by Municipalities." *Federal State Statistics Service*. Accessed March 18, 2017. http://www.gks.ru/free_doc/new_site/perepis2010/croc/perepis_itogi1612.htm

Rosstat. 2016a. "The Population of the Russian Federation by Municipality." *Federal State Statistics Service*. Accessed March 14, 2018. http://www.gks.ru/wps/wcm/connect/rosstat_main/rosstat/ru/statistics/publications/catalog/afc8ea004d56a39ab251f2bafc3a6fce

Rosstat. 2016b. *Russia in Figures 2016 Statistical Handbook*. *Federal State Statistics Service*. Accessed March 14, 2018. http://www.gks.ru/free_doc/doc_2016/rusfig/rus16e.pdf

Russian Federation Federal Law: 'On Freedom of Conscience and on Religious Associations.' 1997. *Journal of Church and State* 39: 873–889.

Sedakov, Pavel, and Orxan Djemal. 2010. "Damoklova mechet [Damocles' Mosque]." *Russian Newsweek*, September 21. http://www.interfax-religion.ru/print.php?act=print_media&id=11959.

Sharafutdinova, Gulnaz. 2014. "The Pussy Riot Affair and Putin's Démarche from Sovereign Democracy to Sovereign Morality." *Nationalities Papers* 42: 615–621.

Shishlin, Vladimir. 2011. "Mecheti v Tekstil'shchikakh ne budet [There will not be a mosque in Tekstilshchiki]." *Interfax Religion*, February 16. http://www.interfax.ru/moscow/177760.

Sibireva, Olga, and Alexander Verkhovsky. 2011. "Freedom of Conscience in Russia in 2010: Restrictions and Challenges." *SOVA Center for Information and Analysis*, April 21. http://www.sova-center.ru/en/religion/publications/2011/04/d21460.

Sidorov, Dmitrii. 2015. "Changing Russian Orthodox Landscapes in Post-Soviet Moscow." In *The Changing World Religion Map*, edited by S. D. Brunn, 2453–2473. New York: Springer.

Smith, Kathleen. 2002. *Mythmaking in the New Russia: Politics and Memory during the Yeltsin Era*. Ithaca, NY: Cornell University Press.

Snopek, Kuba. 2013. *Belyayevo Forever: Preserving the Generic*. Moscow: Strelka Press.

Staeheli, Lynn, and Donald Mitchell. 2007. *The People's Property: Power, Politics, and the Public*. London: Routledge.

Stoop, David Christopher. 2012. "Mosque Debates in Germany: Between Democratic Participation and Social Exclusion." *Annales UMCS, Sectio K*. 19: 35–49.

TASS. 2016. "Dvizheniye "Sorok Sorokov" prekratilo uchastiye v aktsiyakh na meste stroyki khrama v Torfyanke." [Movement "Forty-Forty" stopped participation in the actions on the site of construction of the temple in Torfianka], July 27. http://tass.ru/obschestvo/3491066.

Vendina, Olga I. 1997. "Transformation Processes in Moscow and Intra-Urban Stratification of Population." *GeoJournal* 42: 349–363.

Venediktov, Alexsei. 2013. "Ishchem vykhod… [Looking Out…]." *Echo MSK OnlineTV*, February 28. http://echomsk.onlinetv.ru/record/18253.html.

Vinogradov, Dmitriy. 2012. "Mecheti, ne pustili v Mitino, budet trudno nayti drugoye mesto [It will be difficult to find another place for the mosque which was not allowed in Mitino]." *RIA Novosti*, September 21. https://ria.ru/ocherki/20120921/755786014.html.

Wilson, Andrew. 2005. *Virtual Politics: Faking Democracy in the Post-Soviet World*. New Haven, CT: Yale University Press.

Yalovkina, Anna. 2016. "Neprozrachnyj Islam Petersburga [Unclear Islam of Saint Petersburg]." *Kavpolit*, June 1. http://kavpolit.com/articles/neprozrachnyj_islam_peterburga-25953/.

Appendix 1

Interviews

A	Male, Russian	Founder, Moj Dvor	May 2013
B	Male, Tatar	Employee, Russian Muftis Council	July 2013
C	Female, Russian	Resident, Tekstilshchiki District	March 2013
D	Male, Tatar	Resident, Tekstilshchiki District	July 2013
E	Male, Tatar	Imam, Moscow Islamic University	October 2013
F	Male, Russian	Employee, Mitino District Planning Department	November 2013
G	Male, Russian	Apartment Owner (Leases apartment to family), Mitino District	July 2013
H	Male, Tatar	Imam, Memorial Mosque	September 2013
I	Male, Tatar	Muslim Entrepreneur	August 2015
J	Male, Russian	Russian Orthodox Priest	November 2013
K	Male, Russian	Russian Orthodox Priest	August 2015
L	Male, Russian	Russian Orthodox Priest	August 2015
M	Female, Russian	Resident, Babushkinsky District	August 2015
N	Female, Russian	Resident, Babushkinsky District	August 2015
O	Male, Russian	Resident, Babushkinsky District	August 2015
P	Male, Tatar	Imam, Historical Mosque	October 2013

Index

Note: Page numbers in *italics* refer to figures
Page numbers in **bold** refer to tables
Page numbers followed by "n" refer to notes

Abaev, Vasily 143
Abdulatipov, Ramazan 157
Abkhazia 3, 133; cultural icons 136, 138, 146, *147*; national identity 136, 138, 146–147; political icons 146, *147*; and Russian World 18, *23*, 24, 29, *30*, 31, 32
activism: civil society, in Russia 227–228; local, and religious construction projects 227, 245; environmental, in Tekstilshchiki (Moscow) 231–232
Afghanistan 92; governmental spending, and violence rate 171; return migration of fighters from 160; and US Silk Road 68, 69, 71, 72, 73
Agassi, Andre 141
Agnew, John 203, 217
Akhalaia, Bacho 207
Ak Zhol (Kazakhstan) 55
Alam, Muzaffar 65–66
al-Khattab, Ibn 154
Alley of Classics (*Aleea Clasicilor*) 145
all-source analysis 184
Alyokhina, Mariya 244
Anderson, Benedict 148
Anselin, Luc 208, 212
Arab Spring 92
Ardzinba, Vladislav 146
Argenbright, Robert 227–228
Armed Conflict Location Event Data-set (ACLED) 153
Armenia: and Eurasian Economic Union 41, 48, 49, 84; and Nagorno-Karabakh 94, 139, 141; and Russia 53
Asian Development Bank 70
Asilderov, Rustam 159
Assad, Bashar 93
Association Agreements (European Union) 50, 104, 111, 117
Astor, Avi 231

Atambayev, Almazbek 93
Atlantic Council 69
authoritarian cooperation 2, 40–41, 57–58; adjective regionalism 43–44; commonalities 43; and domestic pressures 46; Eurasian economic integration 48, 55–57; post-Soviet regionalism 43; puzzle of 42–44; and regime identity 46–48, 55–57; and regime legitimacy 47; and regime security 44–48, 49–55; and regional integration 43; and regional organizations 43; *see also* Eurasian Economic Union (EAEU)
automotive industry *see* Ukraine, automotive industry of
AvtoInvestStroy (AIS Group) 111, 113
AvtoZaZ 106, 108, 109, 111, 112, 113, 114, 116
Avyasov, Shavkat 236
Azerbaijan 94, 104, 139
Aznavour, Charles 141

Babushkinsky (Moscow), protests over church construction in 228, 241–244, *242*, *243*; "For Torfyanka Park!" movement 244; Pussy Riot 243–244; *Sorok Sorokov*" movement 243, 244
Bagapsh, Sergey 146
Baker, Peter 196
Bakradze, Davit 207, 208
Barth, Fredrik 129
Basayev, Shamil 154, 155, 161
Belarus: and China 50; decline in oil prices 49; direct support from Russia 51; economic concession from Russia 91; economic development of 56; and Eurasian economic integration 90, 91; and Eurasian Economic Union 48, 49, 50, 51, 53, 56, 84; EU sanctions on 51; fifth column in 51;

internal tensions 54; national identity of 55; regime identity of 55–56; rent-seeking elites in 104; and Russia 52, 53–54, 90, 91; state ownership 55–56; and Ukrainian crisis 90, 91
Belaventsev, Oleg 158
Bhabha, Homi 133
Billig, Michael 133
Blake, Robert O., Jr., 73
Bogdan Corporation 106, 107, 108, 111, 113, 115
Bratersky, Maxim 2, 80
Breedlove, Phillip 192
Brezhnev, Leonid 144
Bruneau, Michel 131
Brzezinski, Zbigniew 74
Burns, William 73
Byutukavev, Aslan 159

Caramani, Daniele 202
Caucasus Emirate (CE) 155, 157, 159, 164, 175n2
Center for Strategic and International Studies 69
Central Asia 64, 74; Central Asia–South Asia axis 68; Eurasia 65, 67, 75–76; and Eurasian economic integration 83; investment of China in 66; Transport Corridor Europe-Caucasus-Asia (TRACECA) project 68, 76n5; see also Silk Road; US Silk Road
Central Asia-Caucasus Institute (CACI) 68, 69
Central Spiritual Board of Muslims of Russia 230
Champion, Tony 130
Chechen Republic of Ichkeria 155, 162, 175n2
Chechnya, violence in 152–153, 161, 174; action-reaction models 167, 167, 168, 169, 170; arrests 162–163; budgetary transfers from Russia 158; civilian casualties 163; conflict with Russia 154–155; decline in 164; First Chechen War 154–155; and forest cover 172; Second Chechen War 155, 161, 163
Child, Travers 170
China: and Eurasian economic integration 94–95; and Eurasian Economic Union 50; and Kazakhstan 90; and Silk Road 66, 72; and US Silk Road 72, 76
Chou, Tiffany 171
churches, in Moscow neighborhoods 4, 224–225, 228, 239–245; see also Moscow, religious sites in neighborhoods of
Churkin, Vitaly 182
citizenship: participatory 237; and Russian World 26
civilizational identity, and Russian World 12
civil society activism, in Russia 227–228
cleavages: electoral 203, 215–216; geographic 25–27; societal 204, 205, 209;

spatial 203; and UNM vote territorialization 212–215, 213, 214
Clem, Ralph S. 1, 3, 134, 181
Clinton, Hillary 67
cloisonnement 130
cold war 7, 8
collective consciousness, and political icons 139
Collective Security Treaty Organization (CSTO) 83, 93, 94
Commonwealth of Independent States (CIS) 82–83, 98n1, 114
Communism, and Russian World 12, 27
compatriots (sootechestvennik) 10–11, 15
Congress of Russian Communities (KRO) 9
construction projects, religious 226, 227, 245; see also Moscow, religious sites in neighborhoods of
containment strategy 74
Crimea 18; Russian annexation of 1, 7, 18, 19, 24, 33, 54, 89, 181; and Russian World 15, 22, 23–24, 28, 29, 29, 30, 31, 32; see also Ukraine
critical geopolitics 64–65, 75
Croatia 10
cultural icons, and national identity 129, 136, 137, 138, 148; Abkhazia 146, 147; Nagorno-Karabakh 139, 140, 141; South Ossetia 142, 143; Transdniestria 144–145, 144, 146
culture: legacies, of post-Soviet states 85, 88; and Russian World 10, 11, 15, 20, 23, 85, 86
Customs Union (CU) 83, 84, 85, 94, 104

Daewoo 108, 109, 110–111, 116
Dagestan, violence in 153, 154, 157, 160, 161, 164; action-reaction models 167, 167, 168, 169, 170; arrests 163; civilian casualties 163
Deep and Comprehensive Free Trade Agreement (DCFTA) 85, 104, 117, 118, 119, 121–122
de facto states, building identities in 3, 127–129, 133; Abkhazia 146–147, 147; age of respondents 140–141; attachments to Russia 138; cultural icons 136, 137, 138, 148; distinctiveness 131, 134, 138–139; local icons 132, 134, 136, 138; Montevideo Treaty (1933) 134; Nagorno-Karabakh 139–141, 140; political icons 135, 136, 138; research questions 133–134; South Ossetia 141–143, 142; Soviet figures 134, 136, 139, 141, 143, 144, 145, 147; Soviet legacy 138, 141, 145, 146; Transdniestria 143–146, 144; Tsarist figures 134, 136, 138, 139, 141, 143, 147; Western figures 138, 140, 147
de Miguel, Carolina 203, 205, 217
democracy: and authoritarian cooperation 40, 42; electoral 203; and foreign policy 46; and international organizations 45; managed 227, 229, 238, 244, 245; and mosque construction debates 231;

participatory 245; promotion 70, 206–207; and regional integration 41, 44
diaspora 9, 10–11, 141
Digital Globe 188, 192, 194
distinctiveness 134
Djusoev, Nafi 143
Dnipro (Ukraine) 21
Doga, Evgeny 146
Donbas conflict (Ukraine) 3, 7, 15, 181, 186–187; *see also* Russia–Ukraine conflict, storyline evidence of
Donetsk Oblast (Ukraine) 184–185, 186, 193
Donetsk People's Republic (DNR) 185, 186
dual-level game 46
Dugin, Alexander 67
Dzutsati, Valery 159, 174

Earnest, Josh 191
Eastern Partnership initiative (European Union) 51, 104
economic integration *see also* authoritarian cooperation; Russia, approach to regional economic integration; Ukraine, automotive industry of
economic liberalism 57
economic liberalization 2, 56, 117
Èech, Adam 190
Eggert, Konstantin 93
electoral authoritarian regimes 47
electoral geography *see* Republic of Georgia, territorialization of political parties
ethnic cleansing 129
ethnicity: ethnic minorities, in Georgia 214–216, 217; homogeneity, in Nagorno-Karabakh 138, 139; multi-ethnic character of Transdniestria 144; and nationalization of political parties 204–205; and regime identity 55, 56; and Russian World 9–10, 16, 20, 21, 26
ethnography 228–229
ethnonationalism 139; Kazakh 56; Russian 10, 89–91
ethno-symbolism 130, 131
Eurasia: neo-Eurasianism 67; and US Silk Road 65, 67, 75–76
Eurasian Development Bank (EDB) 51, 56
Eurasian Economic Commission 42, 52, 54
Eurasian Economic Community 83
Eurasian economic integration: and authoritarian cooperation 48, 55–57; and regime identity 55–57; *see also* authoritarian cooperation; Ukraine, automotive industry of
Eurasian Economic Union (EAEU) 1, 2, 16, 41, 42, 45, 48, 57, 94; and China 50; coordination problems between member states 52; Court 52; and economic modernization 50; and Eurasian economic integration 84, 88, 90–91, 94–95; and European Union 50–51, 97; external

security and internal tensions 54–55; and global financial crisis (2008) 49, 50; non-tariff barriers 52, 53; and regime security 49–55; reliance of smaller states with Russia 53–54; security-focused integration projects 51; as source of regime credit 51; Supreme Council 52; and Ukrainian crisis 88; weakness of institutions and enforcement mechanisms 52–53
Eurasian Stabilization and Development Fund (ESDF) 51
Eurocar (Evrocar) 113, 114
Europe: mosque construction debates in 231; and nationalization/territorialization of political parties 205; and US Silk Road 72–73
European Commission 114, 117, 119
European Neighborhood Policy (European Union) 104
European Parliament 13
European Union (EU) 1; Association Agreements 50, 104, 111, 117; Eastern Partnership 51; and Eurasian Economic Union 50–51, 97; interests in shared/contested neighborhood 103–104; and Kazakhstan 90; and Russia–Ukraine conflict 181, 187; sanctions 51; and Ukraine 85, 86, 87, 89
European Union (EU), and Ukrainian automotive industry 2–3, 105, 110–111, 114–116, 118–120, 121–122; asymmetric liberalization 110; Daewoo 110–111; DCFTA negotiations 119; exports 115; foreign investors 116; import tariffs 110, 117, 118; leverage on pro-Western governments 119; oligarchs 114; second-hand car market 110; semi-knock-down (SKD) assembling 114; state aid policy 114; trade liberalization 110, 111, 114, 117, 121; WTO accession negotiations 119; *see also* Russia, and Ukrainian automotive industry
Evans, Geoffrey 217
exploratory spatial data analysis 208

Fearon, James 170
fixed effects model 209, 215
foreign policy: in democratic states 46; and Eurasian Economic Union 52, 54, 56; and identity 47; Kazakh 56; Russian, and regional integration 83, 87; and Russian World 12; and state security 44, 45; of Ukraine 114; *see also* authoritarian cooperation
"For Torfyanka Park!" movement 244
Foundation for Efficient Politics 9
founding fathers 128, 144
Fung, Anthony Y. H. 130

Gabaraev, Ilya 143
Gachechiladze, Levan 212

Galeotti, Mark 190
Gamsakhurdia, Zviad 214
Gaynutdin, Ravil 225, 229, 230, 234
GAZ 114
General Motors (GM) 111
genius loci (protector of the place) 131
geopolitical culture 7–8
geopolitical frames 8, 33
George, Julie A. 214
Georgia *see* Republic of Georgia
Gerasimov, Valery 159
Gergiev, Valery 139, 143
Ghukasyan, Arkady 139
Girkin, Igor 15
global financial crisis (2008): and Eurasian economic integration 84; and Eurasian Economic Union 49, 50
Global Positioning System (GPS) 183
Gogol, Nikolai 131
Golosov, Grigorii V. 204
Gorbachev, Mikhail 10, 144
Gottmann, Jean 130–133, 149
"Grand Chessboard" theory 74
Granger causality analysis 166
ground-level digital imagery 184
GUAM (Georgia-Ukraine-Armenia-Moldova) project 68
Gulf Cooperation Council (GCC) 45
Guliya, Dmitry 139, 146
Gumilev, Lev 14, 67

Harbers, Imke 209
hard power, use by Russia 2, 81, 83–84, 89, 91, 92, 93–94, 95, 96, 97–98
Härtel, André 117
Hashimoto Ryutaro 65
Haushofer, Karl 74
Heartland 74–75
Heathershaw, John 75
Herszenhorn, David M. 196
Higgins, Eliot 192
Hinderstein, Corey 183
Holland, Edward C. 3, 152, 153
Hungary 10

icons 127–128; and birthplace 131; cultural *see* cultural icons, and national identity; in de facto states 133–147; *genius loci* (protector of the place) 132; local 132, 134, 136, 138; micro-nationalist studies 131; narratives 133; and national identity 130–133; policy of memory 133; political *see* political icons, and national identity; political units 130; *see also* de facto states, building identities in
identity politics 47, 234, 244, 245
immigration 15, 16, 88, 238, 239, 246
Independent Institute for Socio-Economic and Political Studies (IISEPS) 90
India: and Silk Road 65–66; and US Silk Road 68–69

Ingushetia, violence in 153, 154, 155, 160; arrests 163; decline in 164
intelligence analysis 184
International Crisis Group 160
International Security Assistance Force (ISAF) 71
Internet 183
Iran, and US Silk Road 72
Iraq 159–160, 168–169, 183
irredentism 89, 90, 92
ISIS 81, 92, 93, 153, 159
Islamist insurgency, in North Caucasus *see* North Caucasus, violence in (2010–2016)
Islamophobia 233, 234, 245
Ivanishvili, Bidzina 207, 213

Jamestown Foundation 68, 69, 73
Janda, Jakub 190
Japan, and Silk Road 65

Kabardino-Balkaria, violence in 153, 154, 155, 160, 163, 164
Kadyrov, Ramzan 3, 157, 162, 174, 175
Kanokov, Arsen 156
Kavkazskii Uzel (Caucasian Knot) 153, 157, 160, 163
Kazakhstan: ban on imports 52; and China 50; decline in oil prices 49; domestic opposition in 51; economic concession from Russia 91; and Eurasian economic integration 83, 90–91; and Eurasian Economic Union 48, 49, 50, 52, 54, 56, 84, 90–91; Eurasianism 67; internal tensions 55; regime identity of 56–57; and Russia 53, 54, 56, 90–91; and Russian annexation of Crimea 54, 90; threats to regime stability 92; and Ukrainian crisis 90, 91; and US Silk Road 71
Kebekov, Aliaskhab 157
Kennan, George 74
Kerry, John 69
Khachaturyan, Aram 139, 141
Khalimov, Gulmurod 93
Kharkiv (Ukraine) 20–21
Kherson (Ukraine) 21
Khetagurov, Kosta 142, 143
Khloponin, Alexander 156
KIA 106
Kirillov, Valery 15
Kitschelt, Herbert 25, 33
Kocharyan, Robert 139
Kokoity, Eduard 143
Kokov, Yuri 158
Kolomoisky, Igor 21
Kolosov, Vladimir 2, 3, 6, 127
Kolstø, Pål 129
Kortunov, Andrei 97–98
KraSZ 111
Kremenchuk-AutoGAZ 111
Krickovic, Andrej 2, 80

Kuchins, Andrew 69, 71, 72
Kuchma, Leonid 107, 108, 114, 119, 131
Kyrgyzstan: economic concession from Russia 91; and Eurasian Economic Union 41, 48, 49, 50, 84, 94; remittance from Russia to 88; and Russia 53, 91; threats to regime stability 93, 94; and Ukrainian crisis 91

Lagarde, Christine 86
Lagrange multiplier 173
Lake, David 81, 95, 96
Langbein, Julia 2, 103
Larkin, Sean 183
Laruelle, Marlene 2, 14, 64
Lavrov, Sergei 187
Law on Freedom of Conscience and Religious Associations (1997), Russia 225
Lenin, Vladimir 139, 144
Libman, Aleksandr 43
Lieven, Anatol 154
Linke, Andrew M. 161
Lipset, Seymour Martin 25, 33, 205, 209
local activism, and religious construction projects 227, 245; *see also* Moscow, religious sites in neighborhoods of
Lokshina, Tanya 157
LuAZ 111
Luhansk Oblast (Ukraine) 184–185, 186, 191
Luhansk People's Republic (LNR) 185, 186
Lukashenko, Aleksandr 51, 53–54, 55–56, 90, 91
Luzhkov, Yuriy 238

Ma, Eric K. W. 130
Mackinder, Sir Halford 74–75
Magomedov, Magomedsalam 157
Malaysia Airlines Flight 17 (MH17), downing of 3, 185, 196
managed democracy 227, 229, 238, 244, 245
Marcus, Jonathan 196
maritime trade 73
Maskhadov, Aslan 155, 161
mass media 8, 138, 139, 148, 233
Medvedev, Dmitry 143, 155, 156, 232, 238
Megoran, Nick 75
Melikov, Sergey 157–158
Memorial (human rights organization) 160, 163
Merabishvili, Vano 207
Merkel, Angela 97
Mezo, Josu 208
Miller, James 191
Millward, James 66
Milosevic, Slobodan 89
Mitino (Moscow), protests over mosque construction in 228, 236–239, *236*; anti-immigrant rhetoric 238; arguments of mosque's supporters 238; incompatibility with local lifestyles 237; overcrowding

237–238; property values 238; and Sobyanin 238–239; and Tekstilshchiki protests 237
Modern Silk Road (MSR) 69–70
Moj Dvor (My Courtyard) 230, 231–232, 233–234
Montevideo Treaty (1933) 134
Moran's I 208, 211
Moravcsik, Andrew 47
Morgan, Patrick M. 81, 95, 96
Moscow, religious sites in neighborhoods of 4, 223–227; demographics of Moscow 224; Law on Freedom of Conscience and Religious Associations (1997) 225; Mitino 228, 236–239, *236*; religious organizations and groups 225; Tekstilshchiki 229–235, *230, 231, 235*; "200 Churches" program 239–245, *241, 242, 243*; Yarovaya anti-extremism laws (2016) 225
mosques, in Moscow neighborhoods 224, 225–226, 228, 229–239
Movement Against Illegal Immigration 231
Movement for Liberty – European Georgia 207
multiculturalism 9
Murtazin, Marat 229
Mykolaiv (Ukraine) 21
myth-symbols complex 131

Nagorno-Karabakh 3, 18, 94, 129, 133; and Armenia 94, 139, 141; cultural icons 136, 138, *140*; ethnic homogeneity 138, 139; influence of pop culture 141; national identity 136, 138, 139–141; political icons 135, 136, 138, *140*
narratives, and nationalism 133
nation, definition of 132
national identity: and icons 128, 130–133; and Moscow neighborhood church projects 244; and regime identity 55; and religion 209; *see also* de facto states, building identities in
nationalism 15, 148; anti-Muslim 231; ethnonationalism 10, 56, 89–91, 139; and icons 130, 131, 133; and narratives 133
nationalization of political parties 202–203, 204, 209; and de-personalization of party votes 204; and ethnic diversity 204–205; influence of electoral systems on 204; non-structural factors 204–205; *see also* Republic of Georgia, territorialization of political parties
Navalny, Alexei 238, 243, 245
Nazarbayev, Nursultan 56, 90, 91
near abroad (Russia) 2, 7, 12, 27
neo-Eurasianism 67
new regionalism 47
New Silk Road *see* US Silk Road
Nezavissimaya Gazeta (Independent Newspaper) 14, 16–17

NIMBYism, and religious construction in Moscow 227–228; Babushkinsky 241–244, *242*, 243, *243*; Mitino 228, 236–239, *236*; Tekstilshchiki 229–235, *230*, *231*, *235*

normative publics, and protest movements 228

North Atlantic Treaty Organization (NATO) 1, 2, 7, 197n4; and Eurasian regional integration 84, 86, 87, 94; and Russian World 33; and Russia–Ukraine conflict 181–182, 185, 187, 188–189, 190, 191, 192, 193, 194, 195, 196

North Caucasus, violence in (2010–2016) 3, 152–154; action-reaction models 165, 166–169, *167–168*, *169*, *170*; arrests 162–163, *162*, 167, *168*, *170*; budgetary transfers from Russia 158; change in violence models 165–166, 170–174, *171*, **172**; Chechen–Russian conflict 154–155; civilian casualties 163–164, *163*; conflict events 160–162, *161*; decline in 156–160; distance to road measure 172; economic problems 158–159; elimination tactics of Russia 156–157, 162; forest cover 171–172; geography of decline in 164, *165*, *166*; linear model 172–173; military/police violence 167, *167*, *169*; multilevel model 173; participation by North Caucasians in Syria and Iraq 159–160; prophylactic list 157; returning fighters from Middle East 160, 174; Salafists 157; spatial lag model 173; subsidies 165–166, 170–174, *171*; tit-for-tat pattern 168–169; war of Russia with 154–160; *zachistki* operations 155, 162

North Caucasus Federal District (NCFD) 154, 155–156, 157, 161

Northern Distribution Network (NDN), and US Silk Road 71–72

not-in-my-backyard *see* NIMBYism

Novorossiya 14, 18, 92

Nye, Joseph 81, 88

Obama, Barack 69

Odesa (Ukraine) 21

oligarchs, and Ukrainian automotive industry 105, 106, 107, 108, 111–112, 113, 114, 117, 118

O'Loughlin, John 1, 2, 3, 6, 127, 152, 153, 161

Open Source Intelligence (OSINT) 197n3

Orange Revolution 117, 118

Organization for Security and Co-operation in Europe (OSCE), Special Monitoring Mission (SMM) 185

Ostrovskii, Alexander 8

Ostrovskii, Efim 9

Ó Tuathail, Gearóid 182

Pakistan, and US Silk Road 69, 71, 72

Parsons' model of social systems 205

participatory citizenship 237

Partnership and Cooperation Agreements (PCA) 104, 110–111

Pavlovskii, Gleb 9, 10

Pelkmans, Mathijs 209

Pentagon 71, 72

Peter the Great 144

Petrtylova, Katarina 159

Poland 10, 25, 111

Poliakov, Vasyl 111, 112

political icons, and national identity 129, *135*, 136, 138; Abkhazia 146, *147*; Nagorno-Karabakh 139, *140*; South Ossetia 142, *142*, 143; Transdniestria 143–144, *144*, 145

Pomfret, Richard 73

Poroshenko, Oleksiy 106

Poroshenko, Petro 106, 107, 108, 111, 117, 118

Prokhanov, Alexander 16

protectionism 55, 105, 108, 112, 117, 118, 119

protest movements, in Russia 227–229

public sources of information 183–184, 188, 189, 190

public spaces 228; *see also* Moscow, religious sites in neighborhoods of

Pugacheva, Alla 143

Pushkin, Alexander 136, 139, 141, 143

Pussy Riot 243–244, 245

Putin, Vladimir 7, 21–22, 51, 67, 90–91, 139, 143, 227, 238, 244; and annexation of Crimea 89; and North Caucasus 153, 154, 155, 158; and Russian World 10–11, 12, 13–14, 15, 17, 19, 20, 27; and Russia–Ukraine conflict 187; and Ukrainian crisis 33

Putnam, Robert 46

Pyatnitskii, Valeriy 119

Ratelle, Jean-François 160

regime identity: and Eurasian economic integration 55–57; and regime security 46–48

regime security 42, 44–48, 56–57; as brake on authoritarian cooperation 45–46, 52–55; definition of 45; as driver for authoritarian cooperation 44–45, 49–52; and regime identity 46–48

regime survivability 44

regional economic integration *see* authoritarian cooperation; regime security; Russia, approach to regional economic integration; Ukraine, automotive industry of

regional hegemon, Russia as 2, 80–81, 85, 92–95

regionalism 95–96, 130, 131

regional security hierarchy 81, 95–96

religiosity 223–224

religious construction projects 226, 227, 245; *see also* Moscow, religious sites in neighborhoods of

religious freedom 4
religious groups 225
religious organizations 225
remittances, from Russia 88
rent-seeking 104, 105, 112, 120, 121
Republic of Georgia: democracy promotion 207; economic development of 209, 211; free and fair elections in 203; party system 206; Rose Revolution 206, 209; and Russian World 31; and South Ossetia 25, 141, 143; war with Russia (2008) 207
Republic of Georgia, territorialization of political parties 4, 202–204, 216–217; center–periphery cleavage 209, 215–216, 217; cleavage dimension 212–215, *213*, *214*; district-level vote share 215–216; electoral turnout 216; ethnic, religious, and linguistic minorities 214–216, 217; global statistical model of 215–216, **215**; higher education holders 216; political history of UNM 206–208; political polarization 211–212; religious composition 209, 216, 217; research data and methods 208–209, **210**, 211; Samegrelo region 214; spatio-temporal patterns 211–212, **211**, *212*, *213*; urban-rural divide 209, 216
resource distribution, and Russian World 27
Rethink Institute 69
Roberts, Alasdair 197n4
Roberts, Sean 2, 40
Robertson, Graeme 227
Rogozin, Dmitri 9
Rokkan, Stein 25, 33, 205, 209
Romania 111, 121
Rose Revolution (Georgia) 206, 209
Russia 48; annexation of Crimea 1, 7, 18, 19, 24, 33, 54, 89, 181; and Belarus, meat war between 52; and China 50; decline in oil prices 49; direct support to Belarus 51; economy of 82, 87–88; and Eurasia 65, 67, 75–76; and Eurasian Economic Union 49, 50, 52, 84; fifth column in 51; financial crisis of 1998 110; geographies of protest movements in 227–229; and identity of de facto states 138, 139, 143, 144, 145; immigrants in 88; interests in shared/contested neighborhood 103–104; irredentism 89, 90, 92; mosque construction boom in 225–226; reliance of smaller EAEU states with 53–54; resource curse 82; September 1999 apartment bombings 155; use of force 54; and US Silk Road 72, 75–76; war with Georgia (2008) 207; war with North Caucasus 154–160; *see also* North Caucasus, violence in (2010–2016)
Russia, and Ukrainian automotive industry 2–3, 105, 109–110, 114–116, 118–121; credits, prohibition of 109; fuel and gas prices 109; import bans 115, 116; interdependencies 120, 121; military-industrial complex 120; protectionist measures 109, 115, 119; recycling tax on imports 119; trade flows, decrease in 109–110; trade policy 120; and Western producers 115–116; *see also* European Union (EU), and Ukrainian automotive industry
Russia, approach to regional economic integration 80–82, 104; and Central Asian states 83; coercive tactics 82, 85–86, 87, 90–91, 92, 97; Commonwealth of Independent States (CIS) 82–83; cultural influence 85, 86; declining ability to use instruments of soft power 87–89; economic incentives and soft power 85–87, 92, 95, 96; ethnonationalism 89–91; Eurasian Economic Union (EAEU) 84; and global financial crisis (2008) 84; guarantor of stability 92–95; hard power 81, 83–84, 89, 91, 92, 93–94, 95, 96, 97–98; and internal threats 92, 93, 94, 96; military peacekeeping 83–84; regional hegemon 2, 85, 92–95; regional security hierarchy 96; remittances 88; Russian World (*Russkii mir*) 85, 86; security threats 96; study of regions and regionalism 95–96; subsidization of supplies of energy/natural resources 83; and Ukraine 85–87; and Ukrainian crisis 81, 87–89, 90, 91, 92
Russian All-National Union 231
Russian Council of Muftis 225, 229, 230, 238
Russian language 9, 10, 11, 22–23, 85
Russian Orthodox Church (ROC) 245; and Russian World 12; "200 Churches" program 239–245, *241*, *242*, *243*
Russian World (*Russkii mir*) 2, 33–34, 85, 86, 90; and Abkhazia 18, *23*, 24, 29, *30*, 31, 32; attitudes toward Russia 28; biopolitical definition of 10–11; citizenship 26; civilizational meaning of 12, 16; cleavages 25; and Crimea *22*, 23–24, 28, 29, *29*, 30, 31, 32; culture 10, 11, 15, 20, 23, 85, 86; diasporas 9, 10–11; and end of Soviet Union 21–22, 23, 24, 26–27, 29–31; as geopolitical frame 7–14; interests 27; level of trust in the Russian president 28; linguistic definition of 10; media representations of 14–17; *Nezavissimaya Gazeta* (Independent Newspaper) 14, 16–17; optimists 29, 30; political party choices 33; Rossian/*rossiiskii* (civic form) 9–10; Russian/*Russkii* (ethnic form) 9; and southeastern Ukraine oblasts **18**, 20–23, *22*, 28, *28*, 29, 30, 32; and South Ossetia *23*, 25, 28, 31–32, *31*; *Sovetskaya Rossiya* (Soviet Russia) 14, 15; spiritual meaning of 11; survey data and definitions of 17–20, **18**; and Transnistria *24*, 25, 28, 30–31, 32,

32; Turkic-Muslim migrants 16; and Ukraine 12–13; values 26–27; willingness to fight 31; *Zavtra* (Tomorrow) 14, 16
Russian World Foundation 11, 12, 15, 16
Russia–Ukraine conflict, storyline evidence of 3, 181–182; cross border rocket attacks (July and August 2014) 190–192; cross-border transfer of tanks and military equipment 187–190; descent into war 184–185; Google maps 190; information warfare 182; multiple launch rocket systems (MLRS) 188; official sources 188, 189, 192, 194–195, 196; public sources 188, 189, 190, 192, 195, 196; Russian army units entering Ukraine 193–195, *194*; satellite imagery 188, 192; secrecy 195, 196; T-64 main battle tanks 188–190
Russkii mir see Russian World (*Russkii mir*)
Ryzhkov, Vladimir 17

Saakashvili, Mikheil 206, 207, 208, 209, 212
Saari, Sinukukka 13
Sahakyan, Bako 140
Salafism/Salafists 157, 163, 233
sanctions: European Union 51; and regime security 45, 51; Russian, on Ukraine 52, 120; and Russia–Ukraine conflict 187; Western, on Russia 87, 88
Sargsyan, Serzh 139
satellite imagery 183, 184, 188, 192
Saudi Arabia 233
Schevchuk, Evgeny 144
September 11 attacks 68, 183
Shanghai Cooperation Organization (SCO) 41, 42, 51, 83, 94
Shchedrovitskii, Georgii 9
Shchedrovitskii, Petr 9
Shevardnadze, Eduard 206, 214
Sichinava, David 4, 202
Silk Road 64–65; and Eurasia allegory of Russia 67; metaphor 73–74; New Silk Roads 70; non-US visions of 65–67; *see also* US Silk Road
Silk Road Economic Belt (China) 66, 72
Silk Road Fund 66
Silk Road Strategy Act of 1999 (US) 68
Simón, Pablo 203
Skoda 113, 114
Smirnov, Igor 143
Smith, Anthony 127–128, 130, 131, 132, 148
Sobyanin, Sergey 224, 227, 238–239, 240–241
social conservatism 245
social media 3, 185, 195
societal cleavages 204, 205, 209
societal verification 183
Socor, Vladimir 89
soft power, use by Russia 2, 12, 85–89, 95, 96
Sorok Sorokov movement 243, 244
Souleimanov, Emil Aslan 156, 159, 164, 174

South Caucasus 68
South Korea, and Silk Road 65
South Ossetia 3, 129, 133; cultural icons 136, *142*, 143; national identity 136, 138, 141–143, *142*; political icons 142, *142*, 143; and Russian World *23*, 25, 28, 31–32, *31*
SOVA Center for Information and Analysis 233, 238
sovereignty pooling 56–57
Sovetskaya Rossiya (Soviet Russia) 14, 15
spatial autocorrelation 208, 211, **211**, 212, 217
spatial lag 208, 211
Spiritual Directorate of Muslims 225
Stalin, Josef 139, 144, 146
Starr, S. Frederick 68, 69, 71, 72, 73, 74
Stoop, David Christopher 231
storylines: adducing facts and attributing hostile acts 182–184; definition of 182; gold standard 183; public sources of information 183; Russia–Ukraine conflict 187–196; *see also* Russia–Ukraine conflict
Subrahmanyam, Sanjay 65–66
subsidies: effect on violence 165–166, 170–174, *171*; Russian, energy/natural resources supplies 83
Suleymanov, Magomed 157
Sumar, Fatema 69
super-ethnos 14
Suslov, Dmitri 91
Suvorov, Alexander 139, 144
Svyatash, Dmytro 111, 112
symbolism *see* de facto states, building identities in; icons
Syria 81; participation by North Caucasians in 159–160; Russia's' intervention in 93

Tajikistan: and Eurasian Economic Union 84; remittance from Russia to 88; threats to regime stability 92–93
Tansey, Oisn 43, 46
Tapiador, Francisco J. 208
Tatar Cultural National Autonomy (RTNKA) 234
Tavits, Margit 203
Tavria 108, 109
Tekstilshchiki (Moscow), protests over mosque construction in 228, 229–235, *230*, *231*, *235*; arguments of mosque's supporters 234; environmental activism 231–232; Islamophobia 233, 234; neighborhood rights 231; overcrowding 233; privatized ownership 232–233; property rights 233; Saudi Arabia 233; sense of ownership 232
Terlouw, Kees 130
territorialization, of political parties 202, 204–206; effects of institutional design on 204; and electoral cleavages 203; and ethnic/religious diversity 205; influence of

decentralization on 204; local context 205–206; role of societal cleavages in 205; and single-member constituencies 204; *see also* Republic of Georgia, territorialization of political parties
thick geopolitics 1–2
Tiemann, Guido 205
Timoshenko, Yulia 117, 118
Toal, Gerard 1, 2, 6
Todd, Meagan 4, 223
Tolokonnikova, Nadezhda 244
Tolstoy, Leo 136, 139, 143
Torfyanka Park *see* Babushkinsky (Moscow), protests over church construction in
trade liberalization 105, 110, 114, 117, 121
Transatlantic Trade and Investment Partnership agreement (T-IPP) 49
Transdniestrian Moldovan Republic (TMR) *see* Transnistria/Transdniestria
Transnistria/Transdniestria 3; cultural icons 136, 144–145, *144*, 146; and Moldovans 145, 146; multi-ethnic character of 144; national identity 136, 138, 143–146, *144*; political icons 143–144, *144*, 145; and Russian World *24, 25*, 28, 30–31, 32, *32*
Transport Corridor Europe-Caucasus-Asia (TRACECA) project 68, 76n5
Truman, Harry S. 7
Truman Doctrine 8
Turkmenistan 94
Turkmenistan–Afghanistan–Pakistan–India Pipeline (TAPI) 68
Twenty-First Century Maritime Silk Road 66
"200 Churches" program 228, 239–245, *241, 242, 243*

Ukraine 9; and Customs Union (CU) 85; Deep and Comprehensive Free Trade Agreement (DCFTA) 85; energy dependence on Russia 86; and European Union 85, 86, 87, 89; and Gogol 131; Minsk I (2014) 187; Minsk II (2015) 7; political figures 140; pro-Russian separatists in 7, 89; rent-seeking elites in 104, 105; and Russia 2, 83, 85–87, 88–89, 90, 97; and Russian World 12–13, 14–15, 16, 33; Russia–Ukraine conflict *see* Russia–Ukraine conflict, storyline evidence of; southeastern, and Russian World **18**, 20–23, *22*, 28, *28*, 29, 30, 32; trade restrictions of Russia on 85–86; Western integration 81–82, 97
Ukraine, automotive industry of 2–3, 104–105, 108; collapse (1991–1999) 106–111; complete-knock-down (CKD) assembling 114, 115, 116, 117; Daewoo 108, 109, 110–111; DCFTA negotiations 117–118, 119, 121–122; domestic strategies 106–109, 111–114, 116–118; European Union impact 110–111, 114–116, 118–120; expansion of sector, by

type *107*; *Gosstandards* 110; import of used cars 107, 108–109, 112; import tariffs 107, 112, 114, 115, 116, 117–118; investment climate 116; liberalization *vs.* protectionism 112; main destinations of exports *115*; multi-vectoral foreign policy 114; ownership structures, change in 111–112; purchasing power of consumers, increase in 112–113; recovery without restructuring (2000–2008) 111–116; rent-seeking elites 112, 120, 121; Russian impact 109–110, 114–116, 118–120; second collapse (since 2008) 116–120; semi-knock-down (SKD) assembling 112, 113–114, 115, 116; tax privileges 107, 112, 114; trade liberalization 105; Ukrainian financial crisis (2008) 116; WTO accession 116–117; *see also* European Union (EU), and Ukrainian automotive industry; Russia, and Ukrainian automotive industry
Ukrainian crisis (2014) 15, 33, 134; and Belarus 90, 91; and Eurasian Economic Union 48, 52; and Kazakhstan 90, 91; and Russia's approach to regional integration 81, 87–89, 90, 91, 92
UkrAVTO 106–107, 111, 113, 114, 116, 118
Ukrprominvest 106
Umarov, Dokku 155, 157
Union State Treaty (Belarus-Russia) 51, 53, 90
United Civic Party (Belarus) 54
United Islamic Congress of Russia 236
United National Movement (UNM), Georgia 203, 216–217; cleavage dimension of vote territorialization 212–215, *213, 214*; global statistical model of vote territorialization 215–216, **215**; political history of 206–208; spatio-temporal patterns of vote territorialization 211–212, **211**, *212, 213*
United States 92, 98n3, 183; Bureau of European and Eurasian Affairs (State Department) 68; Bureau of South and Central Asian Affairs (State Department) 68; and Donbas conflict 181; and Russia–Ukraine conflict 182, 185, 187, 188–190, 191, 192, 193, 195; Silk Road *see* US Silk Road; and Ukrainian crisis 33
US Silk Road 2, 66; and Asia 72; and China 72; and Eurasia 65, 67, 75–76; and Europe 72–73; geopoliticizing US involvement 74–75; and international trade 73; and Iran 72; and maritime trade 73; *vs.* New Silk Roads 70; and Northern Distribution Network (NDN) 71–72; origins and framing of 67–70; policy criticisms 70; presentism 73; and Russia 72; and think tanks 68, 69; *see also* Silk Road
Uzbekistan 98n3; and Eurasian economic integration 92, 94; remittance from Russia to 88; threats to regime stability 92, 94; and US Silk Road 71, 73

values, and Russian World 26–27
Van Herpen, Marcel 12
Vasadze, Tariel 107, 108, 111, 116–117, 118
VAZ 114, 115
Verdery, Katherine 128
Vidal de la Blache, Paul 130
Vieru, Grigore 145
Volkswagen 113
von Soest, Christian 44–45
voting *see* territorialization, of political parties

Wahhabism 233
Wahman, Michael 204
Walker, Shaun 159
Warsaw Pact 1
Wawrzonek, Michał 12
Weber, Max 14
Weiss, Michael 191
Whitefield, Stephen 217
Williams, Daniel 157
Winter Olympics (2014), and violence in North Caucasus 153, 155, 163
Witmer, Frank D. W. 3, 152, 153
world affairs 7–8

World Bank 70
World Trade Organization (WTO) 110–111, 114, 116
Wright, Robin 160

xenophobia 16, 233–234, 237, 245
Xi Jinping 66, 94

Yambayev, Gayz 234
Yanukovych, Viktor 7, 86, 87, 89, 118, 186
Yarovaya anti-extremism laws (2016) 225
Yatsenyuk, Arseniy 118
Yeltsin, Boris 9
Youngman, Mark 159
Yushchenko, Viktor 117, 118

zachistki (sweep operation) 155, 162
Zaporizhia automobile Building Plant (ZaZ), Ukraine 106
Zaporizhzhia (Ukraine) 21
Zarycki, Tomasz 25, 26
Zavtra (Tomorrow) 14, 16
Zhirinovsky, Vladimir 90
Zlotnikov, L. 56
Zyazikov, Murat 155